WONDERFUL WORLDS

AN EXPLORATION OF BEFORE TIME BEGAN,
INTO LOGICAL CREATION AND DEVELOMENT
OF OUR COSMOS, WORLD, BODY, AND BEING,

AND WHAT MAKES US HUMAN

BY
ROBERT GREENOUGH

Order this book online at www.trafford.com
or email orders@trafford.com

Most Trafford titles are also available at major online book retailers.

© Copyright 2012, 2013 Robert Greenough.
All rights reserved. No part of this publication may be reproduced, stored in a retrieval system, or transmitted, in any form or by any means, electronic, mechanical, photocopying, recording, or otherwise, without the written prior permission of the author.

Printed in the United States of America.

ISBN: 978-1-4669-3244-9 (sc)
ISBN: 978-1-4669-3243-2 (e)

Library of Congress Control Number: 2012907371

Trafford rev. 12/20/2013

 www.trafford.com

North America & international
toll-free: 1 888 232 4444 (USA & Canada)
fax: 812 355 4082

TABLE OF CONTENTS

WONDERFUL WORLDS iii

COPYRIGHT iv

TABLE OF CONTENTS v

CAUTION xi

DEDICATION xiii

THE FRONT COVER xv

SOURCES xvii

ABOUT THE AUTHOR xix

INTRODUCTION TO WONDERFUL WORLDS xxi

PART I— COSMOS, CREATION, ANCIENT EARTH, AND SPECIES

OVERVIEW OF OUR "WONDERFUL WORLDS" 3

SCOPE OF WONDERFUL WORLDS 6
Soul; Significant Findngs

TIME BEFORE TIME 21
Variable Space; Energy Instead Of Matter In The Big Bang; Cosmic Radiation Background; Time Before Time; Star Growth; Dark Energy and Dark Matter; Planets and Moons; Limits of the Universe; Stars Are The "Machines" Of The Cosmos; Further

Research; Basic Hydrogen and Helium; Energy to Dark Energy and Dark Matter; Possible Answers

THE BIG BANG AND THE UNIVERSE 34
Cosmic Energy; Words of the Gods; Expansion; Stars; Beginnings and Various Accounts of Origin; Expansion of Space; There is Evidence of the Big Bang; Different Big Bangs; The Possibility of 'Wormholes'; Creation of Space; Additional Dimensions in Space/Time; A Form of God

A SCIENCE VIEW OF CREATION 49
Here is a Summary View of Some Currently Theorized Events Following the Big Bang; Conventional Theory of the Big Bang; Pre-History to Modern Man; Early Cultures in the Americas; Soul; Soul as a Religious Value; Agrarian and Social Cultures

STAR FORMATION 64
Our Sun

FORMATION OF OUR SOLAR SYSTEM 72
Elements of the Solar System; Could It Be?

THIRTEEN BILLION YEARS OF HADEAN EARTH 78
Time Divisions

HOW TO MAKE A PLANET 86
Here are Other Observations Made Concerning Our Planet and Life; Hawaiin Archipelago; Where Are We In The Comos?; Inorganic to Organic; Vertebrates and Hominids; Geology and Archaeology in Egypt, Mesopotamia, and Greece; How to Devise a Human; Intelligent Evolution

GLACIERS, CLIMATE, AND ENERGY 104
Glaciers; Climate; Energy

FROM MINERAL TO LIFE... 113

PART II— WHAT IT IS TO BE HUMAN

EARLY SPECIES OF MAN AND PRE-MAN 119

LIFE I—THE STORY 128
First Life; Stars and Galaxies; From Hadean Heat to Snowball Earth; Ice Ages; First Life; Narrowed Earth Population; Inorganic to Organic in Chemistry; First Life; Hominins and Primates; Lake Turkana—Mankind's Origins; Homo Erectus; Time Progression of Cosmos, Earth, Primates, Hominids, and Hominoids; Enter The Hominoid; Races; First Hominoids; To An Agrarian Culture; Male/Female Subspecies?; Development Of Homo Erectus; The Presence Of Jesus; Hominoids And Dinosaurs; Homo Neanderthals And Homo Sapiens; Out Of Africa?; The Caucasian Race; The Hobbits Of Indonesia; Rise Of Agrarian Culture; Homo Erectus As A Base Species; Creation And Development Of Soul; Soul And Reincarnation; Sex In Anthropology; Soul And The Monarch Butterflys; An Industrial Microcosm Of Man's Evolution; Intelligent Design Differs From An Evolution Process; Erata; What It Is To Be Human

LIFE II—TIMELINE 168

PART III— BRAIN, MIND, SOUL, GENOME, ENZYMES, AND CONSCIOUSNESS

ADVENT OF THE SOUL 207
Incarnation; Prayer; Possessors Of Soul; Soul; Qualities Of The Soul; Functions Necessary For Life; Basic Functions Of The Body And Brain; Qualities And Emotions Purely Of The Mind; The Sequence Of Species On Earth; Events In Man's Progression

SOUL EXISTS 224
Evidences Of Soul Existence

SOUL AND MAN.. 238
Spiritual Return, Reincarnation, And Vested Abilities; A Proof Of Soul

WHAT IS SOUL?... 257
Description Of Soul; In A Nutshell; Intelligent Design; Two Philosophies At Time Of Christ; Conclusion; Physical Considerations; Why Soul?

WHEN AND HOW DO SOULS AND GENES INTERACT?..... 273
Prokaryotes And Eukaryotes; Eukaryote Cells; Sexual Reproduction; A Random Thought

CONSCIOUSNESS, BRAIN, MIND, AND SOUL 280

SPIRIT, SOUL, MIND, BRAIN, AND GENOME—
HOW THEY RELATE... 286
Spirit; Soul; The Mind; The Brain; Genome; Unity; Space/Time Dimensions; Miracles; Multi-Verses And Added Dimensions

DINOSAURS AND SOULS... 296

QUANTUM MECHANICS IN ANCIENT TIMES 300

SEX IN EVOLUTION.. 304

ADDITIONAL DIMENSIONS IN SPACE/TIME....................... 309

SCIENCE VERSUS RELIGIONS 312
Domain Of The Mystical (God); Domain Of The Natural And Explainable; Additional Dimensions in Space/Time

PART IV— AN AGRARIAN CULTURE

THE GREAT LEAP FORWARD— INTELLECTUAL DEVELOPMENT.............319
What Makes Humans Different?; Did Hunting Make Us Human?; Do Human Races Exist?; Emotions In Early Man; Modern, Human Emotions; Were All Worldly Hominids The Same?; Mathematical Support For Intelligent Design And Planned Evolution.; Intelligent Evolution; The Gods; Succession Of Incarnations

TO AN AGRARIAN CULTURE 344
Ancient Lake Turkana (Kenya) Prehistory

FROM AN AGRARIAN CULTURE Part I 355
From Neolithic To A Culture Of Kingdoms; Timeline Of History

FROM AN AGRARIAN CULTURE PART II 376
1500 Years B.C. To The Time Of Christ; Another View—600 B.C. To Time Of Christ; The Anthropologic Void In South America; Various Origins Of Races In The World; The Author's Outlook

PART V— MYTHS, QUESTIONS, REFLECTIONS

THE MYTHS OF GENESIS..............403

DID MOSES PROPERLY SERVE HIS GOD?...............413

A CONTROVERSIAL IDEA: WAS CHRIST BRAIN DEAD AT THE CRUCIFIXION? 421
Gnostics; Other Miracles; Unanswered Questions; Is Christianity The Apex Of Religious Thought?

RADICAL SCIENCE 432
A Scheme Of Evolution

WONDEFUL WORLDS IN REFLECTION 445
Explored And Found; Author's Review Of The Homo Erectus Diaspora; Origin Of Races; Global Warming; The Past 2,000 Years; To The Future; Colonization In The Cosmos; Cern; Biologic Changes; Homo Roboticus

AFTERLIFE—AN INTRODUCTION 462

CAUTION

DO NOT READ THIS BOOK

There are thoughts here
with which you may not agree.

—The author

DEDICATION

This book is dedicated to my wife of 60 years, Fay, whose soul will continue to exist beyond our moments of secular death. Whatever successes I have had in life, it was her beside me.

Robert Greenough

THE FRONT COVER

The author proposes that more than 13.7 billion years ago, with a Black Hole accumulation of cosmic energy, singularity, annihilation of opposite electrical charges in elementary particles, expansion into the universe of cosmic energy, billions of years in change from an energy cloud to a Hadean, molten, Earth, then cooling with an ice cover, like a cosmic snowball, a first single continent called Rodina was formed, about 225 million years ago, of accumulated minerals and rock. Rodina divided into Pangaea, and then into Laurasia in the northern, and Gondwana in the southern hemisphere about 200 mya. Both were joined near Gibraltar, Spain, and other locations that would geologically open into major Earth features of the future.

The present continent of North America separated from Europe and Asia (Laurasia) about 135 million years ago. Century-by-century and inch-by-inch the continent moved in an expansive ocean toward its present location, perhaps caused by a deep ocean crevasse accompanying movement of geologic plates. South America also split from Africa, as the former Gondwana, in the southern hemisphere.

About 65 million years ago North and South Americas were still separated, but were near joining, as depicted, being absent Central America.

Effects of a cosmic collision with Earth by a cosmic traveler existed with a dark blue water depression off the coast of Yucatan. Many features, including the Great Lakes and future major river systems are clearly visible.

This book provides comprehensive analysis of the development, over nearly 14, or more, billion years, of the cosmos, our universe and Solar system, our physical Earth, first life, primates, pre-man, modern man, culture, and man's spiritual development, up to the Central Millennium and time of Christ. Myth is subordinated by reason, logic and fact.

The Author

SOURCES

The material for this book has been assembled by the author from extensive research and reading of numerous written articles, with some interpretations held free of questionable dogma. The views are based on reason and logic directed to happenings, from before the "Big Bang" to the time of Christ. The author acknowledges the many sources for the wealth of information given here.

The expansive scope explores Creation of our Cosmos, the stars, our solar neighbors, our pre-life Earth, first life, growth of species, origins of Modern Man, and man's physical and spiritual development through Agrarian and Kingdom cultures, through Mesopotamian, Greek, and Roman events, to the time of Christ.

Opinions and interpretations are those of the author, based on reason, logic, observations, and human insight. It is believed that man is biologically the same, and the physical Earth is essentially the same, for at least the past 10,000 years. Man's spiritual and emotional views have experienced momentous changes during this time and before.

Sources of information include the following:

1. The Great Courses, Chantilly, VA (Several courses and professors).
2. Google—Wikipedia, The Free Encyclopedia (Primarily for dates).
3. Western Civilizations (College history textbook).
4. King James Bible.
5. Harper Study Bible.
6. Book of Mormon.
7. The Glorious Qur'an.
8. Archaeology magazine.
9. National Geographic magazine.
10. Scientific American magazine.

11. Discover magazine.
12. Nature magazine.
13. Encyclopedia Britannica.
14. Physics Annual Review, The University of Michigan.
15. Goode's School Atlas.
16. The Last Two Million Years, Reader's Digest Association.
17. Prehistoric Life, DK Publishers, 2009.
18. A wide variety of general reading.

ABOUT THE AUTHOR

I have pondered the subject of what makes us human for at least the past 20 years. My initial thoughts were published in Year 2001, in a book titled, "Do We Live In Two Worlds? This led to further exploration and expansion in the scope of reason and writing.

Science and the cosmos have been long-time interest, with thoughts as an 8 year old that we should be able to control gravity like mixing hot and cold water at the bathroom sink.

Alas, my science pursuits gave way to a college major in Business Administration, and a later degree of Master of Business Administration (MBA) from The University of Michigan. My working career centered in Personnel Administration, with an avocation in science, philosophy, theology, the cosmos, geology, anthropology, chemistry, physics, and development of mankind, physically and spiritually. I have assembled over 2,000 books on a wide variety of subjects in my personal library for research.

Many aspects in both science and religion were questioned. I set out to explore what really happened when logic, reason, observation, and findings in readings of what other knowledgeable people have advanced from their research.

This book portrays an extensive scope, from actions actually before the "Big Bang", and following through, 13.7 years later, to the time of Christ. Hopefully, a third related book will evolve in a trilogy, on the subject of "Afterlife". In the meantime, I request all to "keep an open mind" to accept alternative thoughts of our cosmology world and personal evolution.

INTRODUCTION TO WONDERFUL WORLDS

I am sorry if, in writing the book, I have "stepped on the toes" of a few ancient "Greats" in various fields of knowledge. My formal and practical education was twofold in finally receiving a Master of Business Administration degree from The University of Michigan, and a parallel, ceaseless personal pursuit of inquiry by reading in many subjects that piqued my interest and questioned my mind over the several years leading to my professional retirement, and after. I wish I could only better remember all the knowledge in that assembled personal library of over 2,000 books on various subjects, and countless science and culture-related articles, in my house basement library. Much of this is reflected here for my enjoyment, and your basis for inquiry.

In this book of cosmic, universe, solar, and earth history, written for the common man, I have summarily laid out, for sake of simplicity and desire to keep the book as simple in presentation as possible without excessive detail, the happenings of over 13.7 billion years, and possibly even more, which I shall explain.

Perhaps from my training in business and organization, I visualize in the Heaven the corporation president who originates or coordinates ideas, and looks to the function directors and managers to work out and implement the plan and details. Such would be the organization in Heaven. Is this parallel to God, Saints, souls, and persons on Earth?

The critic will notice in presenting my ideas that there is sometimes little detail to support the argument. To this I plead guilty, with the reason that I wish to proceed directly to the goal without undue space, time, or technical detail applied to the explanation. It is only necessary, for instance, to conclude that Christopher Columbus arrived at the new land, not to discuss the many details of the voyage. I did this for simplicity. Scientific researchers may complete the details in the future if they feel the end idea has merit and reasonableness.

These alternative proposals are presented here for one reason, and that is to make the world better by diminishing myth and fabricated so-called truths that many of us thought we understood. We offer fresh perspectives and during the course of the book we will discover several anomalies in history. As an old saying goes, "there is more than one way to skin an animal". Likewise, there are alternatives to interpret historic findings, oral history and misconceptions, especially when it is undetermined whether the event was fact or myth. The more we know about history, the more we know about ourselves.

I have suggested a time eon even before the Big Bang, and existence in the Big Bang of, not presence of celestial matter, but instead cosmic, or vacuum, energy, drawn down to a point of singularity into a relatively compact initial universe, the cosmic energy was projected outward into the universe, into to a seemingly endless space, space perhaps to be better defined later.

It is suggested that more than one Big Bang took place, each creating its own galaxy, at locus of supplementary Big Bangs, complete with its own resulting stars, space, and its components. Space between galaxies, measured in light years, was caused by gravitation of stars into star clusters, with agglomeration into galaxies and galaxy clusters. Astronomers are continually discovering "new" stars. It may well be that some of these are actually existing stars drawn into our Milky Way Galaxy, or other locations, by gravitational attraction.

Stars were formed from hydrogen and helium that had rapidly expanded into the cosmos, where the gases met cosmic localities of low density, were drawn into the low density areas, further agitated to generate tremendous pressure, heat, and eventual (after 200 to 400 million years) brightness. The extreme pressure, heat, and gravitational attraction in the newly formed stars caused supernovae explosions, followed by gravitational return of the then developed matter, with even higher pressure and temperature resulting in chemical changes that created even more complex atoms, now described in a Periodic Table of Elements.

The Hubbell telescope has viewed a Great Wall of Galaxies, lying at the extremities of the known universe, that would amaze worldly observers as perhaps possessing gravitational pull, and that may explain Edwin Hubbell's findings of a universe that is both constantly expanding, and at an increasing rate. And what is beyond that Great Wall of Galaxies that draws our universe to it?

The 20th century revealed to science existence of antimatter, with additional dimensions to time and space beyond our known four dimensions of height, width, depth, and time, into a projected 11, or more, dimensions that allow existence far beyond our known world.

A presence in additional dimensions includes that factor we know as God, and the unseen, but a real universe that surrounds that concept. I have identified later in this book several qualities in our known dimensions that reflect the existence in additional dimensions and properties of soul.

I have attacked myth and magic as existing in the absence of actual knowledge. Much myth has been engendered by many sources, ranging from ancient romantic tales, writings about ancient wars and kingdoms, geologic events, and Biblical history, not least of which were reports of ancient human storyteller-authors of the Old Testament, reports supposedly communicated by the word of God.

Books beyond that of Genesis seem to be valid reports of ancient history and theology. I have presented rational alternatives to reported myths that stand the test of rationality, logic, and with physical, archaeological and geologic evidence. A driving realization prevailed that the physical earth was, basically, as real 2,000 and 10,000 years ago as it exists today. Some counter of myths delve into Biblical history, questioning some events held as true for 2,000 and 3,000 years, and alternatively now supported by actual events and archaeology findings.

Now in the 21st century after Christ, and untold centuries after humans first imagined and told their versions in oral history of how the earth, man, women, emotions, free-choice, devotion to gods and God, animal and human sacrifice, and catastrophic events of then-known history were derived in folklore and myth, even though encased within the credulity of Biblical terms and were written as Holy epistles, rendered in supposition personally by God.

Records of science presented here as alternative proposals with credible versions are of how physical creation of Earth and mankind may have developed. The myths of over 25 centuries were displaced by science findings as new theories and proposals were presented for scientific review and endorsement. Biblical versions of creation were challenged by science, although still held as viable by a Christian-minded world.

Development of periodic different findings in geography of world life and resulting species lay out for historic study a continuous panorama of successive, and progressive, life forms, each adding to logic in physical, mental, and emotional complexities leading to today's humans.

Existence of soul is presented with reasonable arguments that validate surreal existence. I have attempted to explain this enigma of soul and to relate soul to the mind of each individual on earth. Science, through our 21st century technical knowledge, has proven the existence of antimatter, and existence of additional dimensions in space-time, with the science aura in which they exists. Mankind is slowly closing the gap between the known world existing in matter, and the unknown world surrounding antimatter and additional dimensions in quantum science.

It is amazing how our world and cosmos started in virtually nothing of matter, but only energy, in the vortex of a gigantic Black Hole 13.7 billion years ago, to a present civilization having the ability to now leave the surface of this Earth to explore, either in person or by nearly unbelievably complex instrumentation, our Sun, planets, stars, galaxies, and beyond. We explore not only physical matter of the cosmos, but are on the verge of gaining insight to a philosophic world of antimatter and dimensions of space-time undreamed of only a century ago.

Mind, being supplementary to the brain and soul, is a force yet to be explained in "psyche-mechanism" that determines emotions and conscience.

I have proposed that "Heaven" actually exists here on earth, but in another dimension of space-time, unavailable currently to us. I also propose that an essence of God exists within our respective minds, allowing our minds to absorb direction by God, without a miraculous "central control".

I suggest that each person read this presentation "with an open mind", even to the point of questioning almost everything that you ever knew about the subject, and testing thoughts alternative to present dogma. The idea and goal is to strengthen knowledge of actual facts that guide our concepts of soul and spirit of the Hereafter into future life after death. This knowledge will give a renewed appreciation of an existence beyond our secular understanding today, and strengthen our belief in an unseen element that exists in our worldly presence.

In our journey here through time and space we have taken another view of several events in the history of the cosmos, our Solar System, and Earth itself. We explored and suggested scenarios that stand to reason and we think have good probability of happening, events that would replace some time-honored misunderstandings in reported history. These topics are only mentioned here and are discussed more in the chapters following.

1. Absence of matter in the original Black Hole and Big Bang. (It was all vacuum energy).
2. The story of creation. (Biblical versus science).
3. Stars. (Their physical creation).
4. Myths versus dimensions. (A visit to alternative universes).
5. One Big Bang? (Numerous Big Bangs create multiple galaxies).
6. Belief in one continuous time. (Proof of time passage in rocks, geologic formations, continental drift, and decay of bodies).
7. One world versus many. (Multiple universes in additional dimensions).
8. Limit of our universe. (Great Wall of Galaxies: Gravitational attraction, another existence?, expansion as needed).

9. Dimensions of original universe. (Comprehensive, but compact. Expansion followed).
10. Animal to man. (Species development).
11. Creation of soul. (Evolutionary development, directed Qualities).
12. Reincarnation. (Perpetual life or immortality, of souls).
13. Soul affects actions. (Soul, mind, DNA, and proteins, to body organs and brain).
14. Heaven in the sky. ("Heaven" is on Earth but in another dimension).
15. Absence of physical bodies in Afterlife. (There are only non-physical souls).
16. Selection in Afterlife. (Tests of "goodness" and elegant choices).
17. Christ's death at crucifixion. (Possibly not brain-dead, with mind and soul continue living. Others revived from dead).
18. Moses's fallacy. (Egyptian Book of the Dead re-told).
19. Noah epic not original. (Previous epic of King Gilgamesh and a great flood).
20. Prayers answered within self. (Mind and Guardian Angel formulate actions).
21. Dead arising at Revelation. (Souls exist, but not bodies).
22. Hell. (Defined as not admitted to "Heaven").
23. Dark Energy and Dark Matter. Possibly excess of cosmic energy, at star distances, awaiting conversion to matter).
24. Periodicity. (Significant periods of advancement: Bacteria, to life, to man: hominids, hominoids, incarnation, intelligence of mind in art, architecture, technology, spirituality, religions, agrarian culture, empires, trade, periods of humanism and commonness).
25. Union. (Science and spirituality are really complimentary).

In the following chapters our searching minds will explore several time-honored precepts delving in to sort out myths and magic to substitute reason and logic. It is believed the physical world was essentially the same 2, 5, and 10,000 years ago as exists today, differing mostly in accumulated knowledge, culture, beliefs, technology, and spiritual ideals. Ideas, new and old, would be spread into the newly settled parts of the world, through various cultures, and through men, kings, and empires.

WONDERFUL WORLDS

PART I

COSMOS, CREATION, ANCIENT EARTH, AND SPECIES

OVERVIEW OF OUR "WONDERFUL WORLDS"

As this book goes to publication in early 2012, we learn of a sensational find of fossils from South Africa that sparks debate over how we came to be human. (Source: Scientific American, April 2012).

Recently discovered fossils from a site northeast of Johannesburg, South Africa represent a previously unknown species of human with both an amalgam of Australopithecine and Homo traits that suggests it could be a direct forerunner of Homo.

Sometime between three and two billion years ago our ancestor became recognizably human. For more than a million years their Australopithecine predecessors, "Lucy" and her kind, who walked upright like us, had thrived, but their world was changing. Shifting climate favored the spread of open grassland, and early Australopithecines gave rise to a new lineage.

One of these offshoots evolved longer legs, hands able to make tools, and a larger brain. The new species was assigned the name of Australopithecus sediba. This find may cause paleoanthropologists to rethink where, when, and how Homo started. The fossils were unique insights into the order in which key homo traits appeared, and would be ancestors to Homo erectus.

Some paleoanthropologists however remain unconvinced. Some anthropologists contrast the new findings with conventional wisdom arising from other, possibly conflicting archeologist interpretations, in features of fossils from the period.

Geologic evidence indicates that Homo sediba fossils formed about the same time Earth was undergoing a geomagnetic reversal in which our planet's polarity flipped, and magnetic north became magnetic south. The coincidence in timing may reveal whether, and how, change in mankind's development was caused by a geomagnetic reversal some 1.9 billion years ago.

In this book you will be exposed to numerous original ideas, some you may believe are unfounded, yet logical. Some ideas run contrary to conventional wisdom of learned men of science, philosophy, evolution, and men of various religions. Our only defense against the unwavering holdings of conventional wisdom is in recognizing the ability of modern humans to reason, to be logical, and to displace myth by fact, told, for example, in Homer's tales of "Iliad" and "Odyssey" as mere stories, filled with man's ability and desire to exaggerate to an unbelievable extent.

Much the same applies to certain stories reported in early manuscripts, questionably reported as the word of God, but more likely dreamed or created by humans of the period.

The world was physically much the same 10,000 years ago as it is today. This cannot be said of man's beliefs. Ancient Egyptians placed the physical core of man's living of the beating heart, with the brain material actually picked out by tools and discarded or placed in canopic jars.

The ancient world was believed controlled by the panoply of gods, held as the cause of nearly all actions and reversals happening on Earth.

During the agrarian period man transitioned from the Pleistocene cave dwellers to dwellers in first constructed houses, and in well-peopled communities, with an intricate society developing.

Even as late as 2,000 years ago forerunners of Greeks, Romans, and others believed in an extensive panoply of spiritual beings before the advent of Hebrew, Christianity, and later Islam who believed in one God, whether called Jehu, God, Allah, or some other manifestation of a Superior Being. Today the scientific, anthropologic, and philosophic worlds still have not fully agreed on man's development. It is advocated here that the soul in man, or sometimes referred to as the "spirit" in man, originated 600,000 years ago, within the species Homo erectus, with an original set of elementary emotions. Magnitude of emotions expanded extensively, probably with each generation, over the 600,000 years to the present state of modern man.

The author attempts to show a comprehensive and inter-related analysis of all the factors present over the eons of time that went to make us as modern humans. These factors include the original energy gases of the cosmos, the Big Bang, expansion into the universe, creation of stars and galaxies, formation of our Sun, then Earth and the planets, cooling, first life, physical development, first of fish, then plants, animals, and amphibians, and first mammals. Primates evolved to man-like species. Homo species developed from Homo erectus, after many "dead-end" species, having a first spiritual soul, mind, and mental ability to search and question, through sub-species of Neanderthal and Homo sapiens, to Modern Man.

With this development, a culture of agriculture and animal husbandry grew, followed by first kingdoms in Egypt and the Mideast of Mesopotamia, migration spreading eastward through Asia, China, and the Far East, and leading to the time of Christ.

This is a momentous presentation, an effort of more than 10 years, that should not be missed by any person who seeks answer to the question of, What Makes Us Human?

Persons with some expertise in these fields may determine this expedition to be a comprehensive exploration of cosmic, solar, and Earth existence, as well as the spiritual existence and development of world species leading to Modern Man. All facets are found to be interrelated.

SCOPE OF WONDERFUL WORLDS

I was sometimes asked while assembling the material for this book whether I will publish it. My often response was that I wasn't sure, there are so many subjects that are controversial that I might be laughed right off the face of the Earth. And who was I to question the Greats who wrote of ancient times in Earth's development, the time-honored reporting by authorities in many fields of study in science, philosophy, the origin of the cosmos, and the Biblical Book of Genesis.

But I felt that I could modify those theories by application of reason, logic, and what I would propose as normal sense. I would try to "boil out" the myth and magic that seemed to creep into the writings and replace that with reasonableness from a 21st century standpoint of basic science and a "look back" of what may have, and I think probably did happen, at least as an alternative view.

The scope of this book is extensive. To comprehend and understand the wide scope we must explore, to varying degrees, relationships in many fields of knowledge including physics, chemistry, biology, botany, genetics, metaphysics, astronomy, geology, the cosmos, cosmic rays, the universe, climate and atmosphere, thermodynamics, space/time dimensions, theology, society, cultures, philosophy, archaeology, anthropology, nature, soul, mind, computers, evolution, and more—and how they all intermesh over time.

I believe you, as the reader, will find this book quite comprehensive. I have drawn from many sources to make it as all-encompassing as I can, in a continuum of development, and of a scope in a single writing that I had not previously found. The book is comprehensive and develops from the cosmos, to first life on Earth, and to subsequent animals, reptiles, fishes, amphibians, and plants, and to first species following as primates, hominids, hominoids, grasses, grains, bushes, and trees, and to developing intelligence of man-like species, the forerunners of today's modern man, and explaining what makes us human.

I consider myself a "freelance" viewer and not confined by traditional limits of accepted dogma in science, philosophy, theology, multiverse, cosmos formation, or evolution and its processes. I started writing separate concepts for my own enjoyment and my better understanding, to formulate my ideas, and to be more precise in my understanding. As result, I write for public consumption for the common man (and women), if only for a reason to express my thoughts.

This is not a scientific textbook. The book is oriented to two groups. 1) The common man or woman who looks to the end result of actions in history, without getting into the complexities presented by science over the past 500 years, and 2) a challenge to those theories, both proven and questioned, advanced by scientists using complex formulas, which in themselves may be subject to question. The proposals are presented with the end result being paramount, rather than through scientific procedures.

One of the great joys of intellect is using it. Arguments are presented to seek truth, which is likely a Middle Ground in philosophy between known science, fact, theology, and faith. I am not a degreed scholar in philosophy, but try to be logical and reasonable. My purpose has been that of digging into the works of the Greats.

My formal and practical education was twofold in 1) finally receiving a Master of Business Administration degree from The University of Michigan, and 2) a parallel, ceaseless personal pursuit of inquiry by reading in many subjects that piqued my interest and questioned my mind over the several years leading to my professional retirement, and after. I wish I could only better remember all the knowledge in that assembled personal library of over 2,000 books on various subjects, and countless science and culture-related articles, in my house basement library. Much of this is reflected here for my enjoyment, and your basis for inquiry.

In levity, let me state my "Ph. D. is a self-endowed "Personal Human Destiny", on which I shall elaborate in this book, an exploration and presentation of ideas that are not necessarily supported by present dogma or experimentation, but by reasoned possibilities from logical thought. I am not a professional anthropologists, geologist, climatologists, cosmologist, archeologist, biologist, microbiologist, or theologian. I have assembled here the findings of many professionals in their respective fields. I have taken the liberty of asking questions relating to relative events, and preparing possible answers, some in opposition to accepted dogma of events, but instead substituted thoughts based on logic, reason, and present state of knowledge.

The reader will find that this book is a little bit different—in fact it is a LOT different—in that some new and possibly radical ideas are presented here, based on logic and reason, to substitute for myths and lack of specific knowledge of 2,000 and 4,000 years ago.

This book contains some thoughts that may seem radical and "fly in the face" of accepted knowledge. However, the author presents those thoughts after considerable reading, comparison, and analysis, blended with sensibility and rationality in today's world, and with 21st Century knowledge to displace Christ-era understanding of physical actions and advances in medicine and diagnosis.

We will visit, in writing, our cosmos, Earth, Sun, planets, and stars, developments of reptiles, mammals, fish, and plants leading to the status of today, and I have ventured extensively into the soul of man, its origins, existence, evidence, purpose, and actions. Development of man and our environment focus on the question, "What makes us human?"

In this book you will find that I disagree with accepted knowledge on certain events. Really I disagree with many events, based on reason and logic. A further premise is that physical man and the physical Earth was basically the same 10,000 years ago as is today. Most changes in man have been in man's culture, his outlook on life, his community, his place in that community, perhaps as a leader, but more usual as an ordinary person, and his spiritual beliefs, with changes in his "technology" of language, vocabulary, counting, mathematics, and later in science and philosophy development. Religious views and beliefs varied widely, handed down as oral history from elders over the centuries, by word of mouth, and eventually reported in writing, the occurrence of supposedly actual events in history.

Some scientists and philosophers may not agree with certain ideas and concepts presented in this book. However, the author believes the thoughts are realistic, logical, and reasonable, and will withstand examining experiments when it becomes possible to further explore activities of soul, antimatter, added dimensions, and other views of particle science activities.

This book is written, not for the professional scientists with complex technical explanations and reference notes, but for the interested analyst in self-study of creation, life, evolution, spirit, soul, mind, brain development, culture, religious beliefs, anti-mythology, relationship of alternative dimensions, our physical world, afterlife, the soul within ourselves, and the return of soul to Earth after death in reincarnation, all encased in the history of culture development in the era before Christ.

The general scope of this book is without limit. Our thoughts will range from one eon before the Big Bang, to the Big Bang itself, with its core elements of hydrogen and helium that expanded into the universe and created stars. We explore first signs of life with increase in complexity, to four-finned fishes and their environment, to four-legged reptiles and mammals. We explore first primates and their successors of hominids, that developed into man-resembling hominoids, and successively more complexity in physical, mental, and spiritual existence. Many species and sub-species branched out like some bush, with Homo erectus and its sub-species forming to ancient man, but still to acquire the qualities that made us human. Development of an agrarian culture affected man immensely, with advances in technology, including warfare against groups competing for land, power, and philosophical empire that has developed further even to the scope known at time of Christ.

Theory is the beginning of solution. Anything is possible. Roger Williams, migrating to the state of Rhode Island, proclaimed nearly 400 years ago, "to advance the Kingdom of God", and exists in this modern day to provide merger of natural science with the nature of God and its spiritualism.

I challenge ideas from a point of reason and logic, with smattering of knowledge in the various technical fields. I frankly believe there is a more realistic and true panorama of world creation supported by facts in geology, archaeology, biology, the genome, brain development, mind, soul, and philosophical thinking that developed into spirituality.

Myths evolved from unknowing men of 2,500 and more years ago, and have come down by word-of-mouth over the many centuries, endeared by storied events and methods of various philosophies that may not be documented nor supported. A chapter enumerates many events that are questioned, and alternatives are suggested that would result in far different results.

The book presents a comprehensive timeline of major events, spanning from energy acquisition of the relatively small cosmos, BEFORE our own original Black Hole, Big Bang, and Annihilation of electrically charged elementary particles, and with subsequent galaxies formed nearby (relatively) in a similar and progressive pattern—thus building an archipelago of galaxies.

What I have done here in this book is to question and test with reason and logic some of the precepts presented against past misrepresentation, myth and magic recorded in the ancient history that affects our actions today. The book is comprehensive in broaching topics that covers nearly every topic that I could readily think of in the birth, and even before the birth, of the galaxies, earth, stars, the Solar System, and the species and cultures that followed to our modern day.

Socrates (469 to 399 B.C.) stated, "The unexamined life is not worth living". It is what it means to live in the modern world, to develop ideas and ask questions. People imagine Socrates as the rather lofty, graybeard, parading in a toga. However, he lived a very vigorous and quite gritty life, challenging the status quo. Nearly 2,000 years later Benjamin Franklin is commonly credited with the adage, "Believe none of what you hear, and half of what you see".

Inspired by this early wisdom, the book examines various aspects of our existence, disassembling some accepted views in life, in our living, our cosmos, our history, thought, philosophy, and even religious concepts and dogma. Written here are alternative concepts of how events might reasonably have happened. Some presentations may withstand technical examinations and some may not, but the author believes the concepts are logical and reasonable. If only 1 or 2 out of, say, 10 thoughts do in fact change records, I will feel successful in presenting this book.

Members of Academia may be disappointed that I have not been specific enough on certain dates or that I have not thoroughly referenced my comments. Again my reply is that I have written this book for enjoyment of the common people, whom I anticipate do not really require those specifics. At certain writings I have indicated the source to add authority for the comment.

The author respects the holdings of experts in science, theology, philosophy, anthropology, and other fields of study. However, I also respectfully disagree with some facets, which are detailed in this book. Any disagreement is based on logic, reason and freedom to propose alternative thoughts.

The academically-minded reader may see certain parallels with a former 840 page book compiled by Alexander Hamilton in 1820, six years before he died at age 81. Jefferson pored over six chapters of the New Testament, in Greek, Latin, French, and King James English. He had a classic education at the College of William and Mary so he could compare the different translations.

Your author, unbeknownst of Thomas Jefferson's editorial efforts, has followed parallel procedures in viewing the reporting of the course of Earth, culture, species, and spiritual history detailed in this book, with some events omitted and others presented to comprise an understandable and logical continuum leading to "What makes us Human?" The first time the former President undertook to create his own version of scriptures had been in 1804, "His intentions", he wrote, "was the result of a life of inquiry and reflection". (Smithsonian.com, January, 2012)

This has been the intention of our work: That a stumbling block to faith might be removed, that our appreciation and understanding of God's changing a spiritual creation might be enhanced, and that God might be further glorified. (Source: "When Faith and Science Collide", by G. R. Davidson).

SOUL

The existence and purpose of soul and mind, and the electro-chemical interworking with the genome, enzymes, brain and mind, result in the change of bodily organs, and our thought processes that are different from the stimulus-and-result effects of the brain. Explaining the process is an important part in defining the process of the soul and mind. Soul is presented, not necessarily in a religious sense, but that soul is an integral part of our being. Soul is a part of our human entity, as much as the nose on your face (except that you cannot touch soul). Soul is not matter. It is like our conscience and consciousness. We know it is part of us, but is actually of another existence—in a world of antimatter, and with our mind which is a part of our being as a human.

The existence and purpose of our soul is explained in devoted chapters of this book, with several evidences of soul enumerated. Soul continues in our existence even beyond worldly death. Soul has been a factor in development of our culture and a driving force into what makes us human. Basic evidence of the existence of soul is present in our emotions and it emerged starting with the genus Homo erectus species 600,000 years ago, and became more refined and characterized as our species developed over centuries.

Recognition of the presence of soul in man, and the essence of the Spirit of God in man allows explanation of many spiritual events presently charged to myth or to unbelievable events in religion. With brain, soul, and mind working with the genome, this provides for communication to the essence of God, and action (see Prayer) to fulfill a remedial need.

In this book of cosmic, universe, solar, and earth history, written for the common man, I have summarily laid out, for sake of simplicity and desire to keep the book as simple in presentation as possible without excessive detail, discourse, footnotes, and citations, the happenings of over 13.7 billion years, and possibly even more, which is explained.

Perhaps from my training in business and organization, I visualize a Heavenly organization that parallels that of a large corporation with the corporation president who originates or coordinates ideas, and looks to the function directors and managers to work out and implement the plans and details. I seem to associate our understanding of the hierarchal "Heaven" as parallel with God (the Chairman of the Board), Archangels (members of the corporate Board of Directors or Division Presidents, with Angels and Guardian

Angels (Managers or supervisors over specific operations controlling Earth individual's conscience, consciousness, and mind with the common man (workers).

The critic will notice in presenting my ideas that there is sometimes little detail to support the argument. To this I plead guilty, with the reason that I wish to proceed directly to the goal without undue space, time, or technical detail applied to the explanation. It is only necessary, for instance, to conclude that Christopher Columbus arrived at the new land, not to extensively discuss the many details of the voyage. I did this for simplicity. Scientific researchers may complete the details in the future if they feel the end idea has merit and reasonableness.

In this book I frequently refer to "man". It should be understood that the reference is to both female and male, as well as to children.

These alternative proposals are presented here to make the world better by diminishing myth and fabricated so-called truths that many of us thought we understood. We offer fresh perspectives and during the course of the book we will discover several anomalies in history. As an old saying goes, "there is more than one way to skin an animal". Likewise, there are alternatives to interpret historic findings, oral history and misconceptions, especially when it is undetermined whether the event was fact or myth. The more we know about history, the more we know about ourselves.

I have suggested a time eon even before the Big Bang, and existence in the Big Bang of, not presence of celestial matter, but instead cosmic or vacuum energy, drawn down to a point of Singularity, exploding into a relatively compact initial universe, the cosmic energy was projected outward into the universe, into a seemingly endless and expanding space.

It is suggested that more than one Big Bang took place, each creating its own galaxy, at locus of supplementary Big Bangs, complete with its own resulting stars, space, and its components. Space between galaxies, measured in light years, was caused by gravitation into star clusters, with agglomeration into galaxies and galaxy clusters. Astronomers are continually discovering "new" stars. It may well be that some of these are actually existing stars drawn into our Milky Way Galaxy, or other locations, by gravitational attraction.

Stars were formed from cosmic energy containing hydrogen and helium that had rapidly expanded into the universe, where the gases met cosmic localities of low density, were drawn into the low density areas, further agitated to generate tremendous pressure, heat, and eventually (after 200 to 400 million years) brightness. The extreme pressure, heat, and gravitational attraction in the newly formed stars caused supernovae explosions, followed by gravitational return of the then developed matter, with even greater pressure and temperature resulting in chemical changes that created even more complex atoms, now described in a Periodic Table of Elements.

The Hubbell telescope has viewed a Great Wall of Galaxies, lying at the extremities of the known universe that would amaze worldly observers as perhaps possessing gravitational pull, and that may explain Edwin Hubbell's findings of a universe that is both constantly expanding, and at an increasing rate. And what is beyond that Great Wall of Galaxies that draws our universe to it?

The 20th century after Christ revealed to science existence of antimatter, with additional dimensions to time and space beyond our known four dimensions of height, width, depth, and time, into a projected 11, or more, dimensions that allow existence far beyond our known world of matter.

A presence in additional dimensions includes that factor we know as God, and the unseen, but also a real universe that surrounds that concept. I have identified later in this book several qualities in our known dimensions that reflect the existence of additional dimensions and properties of soul.

I have attacked myth and magic as existing in the absence of actual knowledge. Much myth has been engendered by many sources, ranging from ancient romantic tales, writings about ancient wars and kingdoms, geologic events, and Genesis history, not least of which were reports of ancient human storyteller-authors of the Old Testament, reports supposedly communicated by the word of God.

Books written later than that of Genesis seem to be valid reports of ancient history and theology. I have presented rational alternatives to reported myths in genesis that stand the test of rationality, logic, and with physical, archaeological and geologic evidence. A driving realization prevailed that the physical Earth was, basically, as real 2,000 and 10,000 years ago as it exists today. Some counter to myths delve into Biblical history, questioning some events held as true for 2,000 and 3,000 years, and alternative comments now supported by actual events and archaeology findings.

Now in the 21st century after Christ, and untold centuries after humans first imagined and told their versions in oral history of how the earth, man, women, emotions, free-choice, devotion to gods and God, animal and human sacrifice, and catastrophic events of then-known history were derived in folklore and myth, even though encased within the credulity of Biblical terms and were written as Holy epistles, reported supposedly personally, by God.

Records of science presented here as alternative proposals with credible versions are of how physical creation of Earth and mankind may have developed. The myths of over 25 centuries were partially displaced by science findings as new theories and proposals were presented for scientific review and endorsement. Biblical versions of creation were challenged by science, although still held as viable by a Christian-minded world.

Development of periodic different findings in anthropology of world life and resulting species lay out for historic study a continuous panorama of progressive life forms, each adding in its own way to logic in physical, mental, and emotional complexities leading to today's humans.

Existence of soul is presented with reasonable arguments that validate surreal existence. I have attempted to explain this enigma of soul and to relate soul to the mind of each individual on earth. Science, through our 21st century technical knowledge, has proven the existence of antimatter, and existence of additional dimensions in space-time, with the science aura in which they exists. Mankind is slowly closing the gap between the known world existing in matter, and the unknown world surrounding antimatter, and additional dimensions in quantum science.

It is amazing how our world and cosmos started in virtually nothing of matter, but only energy, in the vortex of a gigantic Black Hole 13.7 billion years ago, to a present civilization having the ability to now leave the surface of this Earth to explore, either in person or by nearly unbelievably complex instrumentation, our Sun, planets, stars, galaxies, and space beyond. We explore not only physical matter of the cosmos, but are on the verge of gaining insight to a philosophic world of antimatter and dimensions of space-time undreamed of only a century ago.

The complex mind, being supplementary to the brain and soul, is a force yet to be explained in "psyche-mechanism" that determines emotions and conscience. I compare the mind in projecting knowledge, judgment, and moral correctness of actions to software in a computer.

I have proposed that "Heaven" actually exists here on earth, but in another dimension of space-time, unavailable currently to us. I also propose that an essence of God exists within our respective minds, allowing our minds to absorb direction by God, without a miraculous "Central Control".

I suggest that each person read this presentation "with an open mind", even to the point of questioning almost everything that you ever knew about the subject, and testing thoughts alternative to present dogma. The idea and goal is to strengthen knowledge of actual facts that guide our concepts of soul and spirit of the Hereafter into future existence after death. This knowledge will give a renewed appreciation of soul and mind and an existence beyond our secular understanding today, and strengthen our belief in an unseen element (God) that exists in our worldly presence.

For years man has been confounded by the ability of the brain in its operations. I believe this is a matter of "delegation in functions". The brain can be visualized as little more than a processing center, much as the first status of brain in early vertebrate animal life. It might be compared to operation of the hardware and memory in a computer. Its operation can be systemized, routinized, and trained, as even in our pets.

The real "miracle" of the brain emanates from the soul and mind. From here emanates emotion, judgment, wisdom, empathy, love, the sense of summum bonum (belief in a higher spiritual being) and a myriad of other qualities considered wonderful and unexplainable. These are what make man human and different from animals of the forest.

SIGNIFICANT FINDNGS

In our journey here through time and space we have taken another view of several events in the history of the cosmos, our Solar System, and Earth itself. We explored and suggested scenarios that stand to reason and we think have good probability of happening, events that would replace some time-honored misunderstandings in reported history. These topics are only mentioned here and are discussed more in the chapters following.

1. Absence of matter in the original Black Hole and Big Bang. (It was all vacuum energy).
2. The story of creation. (Biblical versus science).
3. Stars. (Their physical creation).
4. Myths versus dimensions. (A visit to alternative universes).
5. One Big Bang? (Numerous Big Bangs created multiple galaxies).
6. Belief in one continuous time. (Proof of time passage is seen in rocks, geologic formations, continental drift, and decay of bodies).
7. One world versus many. (Multiple universes in additional dimensions).
8. Limit of our universe. (Great Wall of Galaxies: Gravitational attraction, another existence (?), expansion as needed).
9. Dimensions of original universe. (Comprehensive, but compact. Expansion followed).
10. Animal to man. (Species development).
11. Creation of soul. (Evolutionary development, directed qualities).
12. Reincarnation. (Perpetual life or immortality, of souls).
13. Soul affects actions. (Soul, mind, DNA, enzymes, and proteins, to body organs and brain).

14. Heaven in the sky. ("Heaven" is on Earth, but in another dimension).
15. Absence of physical bodies in Afterlife. (There are only non-physical souls).
16. Selection in Afterlife. (Tests of "goodness" and elegant choices).
17. Christ's death at crucifixion. (Possibly not brain-dead, with mind and soul continue living. Others revived from "dead").
18. Moses's fallacy. (Egyptian Book of the Dead, re-told).
19. Noah epic not original. (Previous epic of King Gilgamesh and a great flood).
20. Prayers answered within self. (Mind and Guardian Angel formulate actions).
21. Dead arising at Revelation. (Souls exist, but not bodies).
22. Hell. (Defined as not admitted to "Heaven").
23. Dark Energy and Dark Matter. Possibly excess of cosmic energy, at star distances, awaiting conversion to matter).
24. Periodicity. (Significant periods of advancement: Bacteria, to life, to man: hominids, hominoids, incarnation, intelligence of mind in art, architecture, technology, spirituality, religions, agrarian culture, empires, trade, periods of humanism and commonness).
25. Union. (Science and spirituality are really complimentary).
26. Basic evidence of soul in our emotions, emerged starting with Homo erectus 600,000 years ago, and became more refined as our species developed over thousands of years.
27. Other thoughts will be explored in the following chapters.

In the following chapters our searching minds will explore several time-honored precepts delving in to sort out myths and magic, to substitute reason and logic. It is believed the physical world was essentially the same 2,000, 5000, and 10,000 years ago as it exists today, differing mostly in accumulated knowledge, culture, beliefs, technology, and spiritual ideals. Ideas, new and old, would be spread into the newly settled parts of the world, through various cultures, and through men, kings, and empires.

Our exploration will further venture through the cosmos, the making of the expanding universe, our Sun and the stars, galaxies, asteroids and space debris, comets, Earth and other planets. We will experience what made the various stratifications, layering, and curvature of rock formations and minerals. We will experience changes in composition of the atmosphere including sulfur, oxygen, hydrogen, helium, and other elements. We will understand the movement of geologic plates within Earth, one as far as from Antarctica to the Indian plateau, the slow and relentless collision raising the Himalayan Mountains to the highest in the world, and rising higher even today. We will see the progression in life from one-cell animals, virus, and microbes through four-footed vertebrates, pre-man animals, and species development leading to Modern Man.

It is a fascinating exploration leading to the Agrarian culture and explosion of man's ability in technology including ability in speech, making of alphabets, words, storytelling, for elementary counting and higher mathematics, and other advances in technology. At the time of Christ, man was physically much the same as he is today, but he will continue to expand his knowledge in cultures, medicine, philosophies of various religions, and technologies such as putting man and instrumentation on other planets, and on other star systems, with transmitting instrument-gained information back to Earth in space exploration.

Some philosophies expressed in this book may not be agreed to by everybody. Some persons are so deeply ingrained in their beliefs through a lifetime of teaching, analysis, and conservative faith that their thoughts will probably never change. Other readers may be more open to new ideas and concepts from a convincing philosophy, to be willing to explore those concepts and evaluate the merits based on the material expressed in the presentations. They, in effect, say, "Yes, I believe in an idea and make a judgmental evaluation, and it is my free will to make those judgments. I believe the concepts are logical, moral, and reasonable. They provide a needed bridge between science and religious philosophy. They may be science oriented, but are spiritual in their association."

The story is divided into five main parts:

1) From before the Big Bang to first life on Earth
2) What it is to be human
3) Brain, mind, soul, genone, enzymes, and consciousness
4) From first life to the agrarian culture, and
5) From the agrarian culture to the time of 2,000 years ago. History of the past 2,000 years is a study in itself and is not within the scope presented here.

Sources:

1. The Great Courses (Several, lecture courses and professors)
2. Google Wikipedia, The Free Encyclopedia
3. Western Civilizations (College History Textbook)
4. King James Bible
5. Harper Study Bible
6. The Book Of Mormon
7. The Glorious Qur'an
8. Archaeology magazine
9. National Geographic magazine
10. Scientific American magazine
11. Discover magazine
12. Nature magazine
13. Encyclopedia Britannica
14. Physics Annual Review, The University of Michigan
15. Goode's School Atlas
16. The Last Two Million Years, Reader's Digest Association
17. Prehistoric Life, DK Publishers, 2009
18. A wide variety of general reading.

TIME BEFORE TIME

I have never been comfortable with the conventional explanation of matter being drawn into a Black Hole at an Event Horizon, swirled around and downward by gravity as if in a giant funnel or centrifuge, and matter reduced to elementary particles and forces at the Singularity, or union of all elementary particles of forces. The basic question in my mind was, "where did that matter come from to enter the Black Hole? Supposedly, an original cosmos or universe was void of anything, and certainly void of matter. There had to be another explanation!

It is proposed that any Event Horizon of a Black Hole did not fill with matter because there was no matter in the cosmos to absorb. Instead, a Black Hole absorbed energy, called Vacuum Energy, by gravitation, reducing the forces of energy to a unification that, upon reaching its critical point of pressure and temperature, resulted in a sudden expansion, known today as the "Big Bang".

Common understanding of the Big Bang event 13.7 billion years ago seemed misleading. It follows the assumption that matter was collected into the Event Horizon, as result of gravity and "deconstructed", with a consequent Singularity exploding and expanding into the universe.

I can visualize another way that the Big Bang came about. In the before, or pre-Big Bang era, there were "islands" of energy in the cosmos, as opposed to the concept of general space of the cosmos that were in fact completely void of any forces or matter. What "nothingness" of space there was did consist, not of matter but of clumps of energy, specifically now known as "Vacuum Energy".

In the formation prior to the Big Bang, at a "time before time", there was not even space as known today, but pockets or islands of energy called Vacuum Energy in the mostly vacuum of space. Subsequent "Big Bangs" in other locations of the compact cosmos, also resulting from Black Holes of Vacuum Energy, formed various stars and galaxies, relatively near each other, later to expand in distance and create voids in space between galaxies and stars.

Two islands of energy, perhaps only clouds, could well have rubbed together, brushed the clouds of energy to agitate each other, forming a source of Vacuum Energy in the vortex of the Black hole, and that would become the Singularity.

Vacuum Energy are pockets of energy that can be stimulated to collectively produce momentous concentrations of energy, such as was present in the Big Bang at moment of the Singularity. That was largely energy that would later largely become matter in the stars and universe. Vacuum Energy is believed still a force in the cosmos, specifically as Dark Energy, and particularly in formation of stars, galaxies, and galaxy clusters.

VARIABLE SPACE

Perhaps Vacuum Energy possesses a unique quality of force, a boson, to produce or create space. Possibly Vacuum Energy, and a force able to create space, are closely intertwined. Space was created at the Big Bang, with creation of space continuing at the forefront of the expanding universe, and creation of space continuing to the present time, the result of Vacuum Energy and its twin, Dark Energy.

Vacuum Energy, at the forefront of past universe limits, and still at the forefront of the present universe limits, is present as Dark Energy, continually expanding space that galaxies, such as the Great Wall on the frontier of the universe, will expand into.

Given that 1) Vacuum Energy was present near time of the Big Bang, and 2) Vacuum Energy possess a force that produces space, and 3) that Vacuum Energy is still present in the form of Dark Energy, space is still being created to this day. The oldest space is at the distance or locale of the Big Bang, 13.7 billion years ago at an associated position in space.

Conversely, the youngest space is where we exist today, with new space existing where our Milky Way Galaxy and our Solar System are moving us to today.

The frontier of the universe is not necessarily moving, but we on Earth are moving into new space, newly created by Dark Energy and Vacuum Energy with an allusion of speeding away from the frontier of the universe. As an analogy, when we drive an auto into or away from a city, it appears that the city is moving, but of course the opposite is true—we are moving and not the city.

Creation of cosmic space is analogous to the creation of physical space when a balloon is expanded by a force of air from our body.

There may well be a force, a boson to be discovered in Vacuum Energy, that is believed to have been present at the Big Bang, producing space with energy and consequently matter, as vacuum energy expanded outward. In this Vacuum Energy was an elementary particle, a meson, not presently discovered, that possessed unique features, the greatest of which was creating and controlling the limits of "space zero" at the beginning, and to grow today to the limits of our universe. The interior of the space definition, the walls or limits, was of course, space itself. Space is not defined, but the limit of space, its walls or limits, are defined much as the limits of space are defined as we blow up a hand-held latex or rubber balloon.

The necessary qualities of such elementary particles of Vacuum Energy to expand the cosmos are:

1. Ability to confine space.
2. Ability to expand and continually define increasing space.
3. Uniformity in its expansion.
4. Ability to contain matter.
5. Elusive, so far, to resist identification.
6. Unlimited potential in size.
7. A boson, or meson, or force, rather than a fermion or element of matter.
8. Permanence and resistance to aging, so far lasting 13.7 billion years.
9. Defined in three, if not four or more, dimensions of space.
10. Possibly containing superstrings and additional dimensions.
11. Ability to convert its energy into matter in accordance with Albert Einstein's definition of relativity ($E=MC^2$)
12. Possessing qualities that it can be defined in common or alternative dimensions of space.

Again, space is mostly a vacuum but is actually defined by its limits. At moments following the Big Bang the limits of space were relatively near, but quickly expanding.

I believe it will be shown by researchers in the future that the origin of singularity was not matter, but intense energy. The formula developed by Albert Einstein of $E=MC^2$ proves his claim that energy and matter were indistinguishable in the Singularity. Presently, ordinary matter discovered in the cosmos accounts for about 4.5 percent of the total energy/matter density. There was no matter in the vortex of the Black Hole, but there was tremendous energy. Restated, there was no matter, but only energy, prior to the Big Bang.

ENERGY INSTEAD OF MATTER IN THE BIG BANG

It was energy and not matter that expanded into the universe. However, energy once ejected into the universe, cooled into matter particles of hydrogen and helium that formed stars in areas of high density or concentration of cosmic radiation background energy (COBE). Space is nothingness, yet we are all familiar with space and nothingness each time that we blow up a balloon. Natural space is not defined by its content, but by its limits, just as space within an expanding balloon is defined by the walls that describe the balloon. The size is variable from zero to whatever the walls of the expanded balloon will hold without bursting.

As a comparison, we could add an amount of dust, or various types of dust, to the empty balloon, in which particles might physically join together to form some shape of structure similar to variations of matter and objects. As an analogy, the various types of dust would be synonymous with various elements of matter.

Conventional wisdom holds that the universe was created at moment of the Big Bang. A mass of matter had been accumulated from gravity exercising its force on various matter of the universe, and near the Event Horizon plane, was attracted by gravity into a Black Hole at such a force of gravity that not even light could escape its gravity. This matter was successively pulled into a relatively minute mass of the Black Hole, called a Singularity, a unification of elements reduced to no larger than a modern-day basketball, or even the size of the period at end of this sentence, as described by various sources.

The mass of the Singularity, or uniformity, was unbelievably great from the gravitation and heat generated. Energy from space attracted into the Black Hole had been "disassembled" into many elementary particles, and eventually caused to explode in a Big Bang and expand into space of the empty cosmos, in time creating stars, galaxies, and galaxy clusters. From this process of gravitational attraction, followed by consequent explosion and expansion, stars formed, became more dense, heated, exploded into what we call today a star supernovae, and were further re-attracted back into the star by gravity, producing chemical and metallurgical changes, with successive elements, that are today defined in the Periodic Table of the Elements, such as hydrogen and helium, elements found, mined, and processed on earth, such as carbon, iron, copper, and over a hundred other elements.

COSMIC RADIATION BACKGROUND

The process of the Big Bang is evidenced today by the tell-tale remains of the ancient explosion in the radioactive gases remaining present in the cosmos, detected by probe missile flights into space to collect specimens and to scientifically analyze the radioactive material, providing proof of the Big Bang event 13.7 billion years ago.

Question is asked: From the original vacuum of space, void of matter, where did the matter come from to form a Black Hole? Space was a vacuum, the story goes, void of any matter.

However there was energy, in the form of Vacuum Energy, in the cosmos. Certain areas of the cosmos could be agitated in action of adjoining clumps of energy that would cause creation of greater energy. In other words, the original Black Hole did not consist of matter, but of tremendous energy. It was an explosion of energy, not elementary particles of matter, that erupted and expanded.

Albert Einstein, perhaps the greatest scientist of our past century, proved that energy will become matter. At least in the original phase, energy and matter are interchangeable, which happened upon the tremendous temperature, compaction of mass, and successive cycles of star formation, explosion into nebulae, re-attraction of elements to form new elements, new matter originated in the stars, agglomerated into planets, gases, and particles within the cosmos, producing planets in orbit of the star, our Sun.

TIME BEFORE TIME

Let us explore here that era of time before time. In the pre-Big Bang void of the mostly-vacuum original cosmos there were locations or centralities of not matter, but energy. The energy was a non-physical existence of power potential. Upon agitation, perhaps caused by two centralities of energy coming together, latent energy was put into action to become an active energy, and gases developed into the gravitation-induced transition from Event Horizon to the Singularity, to be diminished in size, and to become a relatively minute, but of tremendous mass and temperature, of energy converted to matter, mostly hydrogen gas, and a lesser proportion of helium.

Clumps of Vacuum Energy in the event horizon were drawn by gravity into the vortex of a Black Hole and further reduced by gravity to a Singularity of particles and forces. These particles may have been rudiments of elementary particles that would later become matter, or may have actually been particles of antimatter, derived from energy. Albert Einstein had proposed a theory that has withstood the challenges of time, that in the beginning energy and matter were interchangeable—that energy would become elements of matter. It conceivably was not matter particles that existed at the Singularity, but were actually forces of energy, to later become matter. This was the Singularity that exploded in the Big Bang and expanded into the cosmos.

The exploding gas (gaseous energy) from the Big Bang travelled outward in the cosmos at tremendous velocity. The expanding hydrogen and helium encountered, again by gravitation, certain locations of potential energy in centralizations, like that existing pre-Big Bang, that attracted the expanding gases. The attracted gases gained further heat, changing to matter (an Albert Einstein principle) and gaining mass. With increased mass and temperature the gases, now turned into matter, and being the forerunner of a star, ignited and exploded, like what we today call a star Nebulae or supernovae.

STAR GROWTH

The matter from the exploded star was attracted (pulled back) by the gravitation of the primitive star in nebulae state, to become an even greater substance of the star in mass and temperature. With this action of events of increased density, temperature, and gravitation return, new elements were created.

The cycle of gravitational attraction and subsequent explosion of a star nebulae continued, each time increasing mass and temperature and joining of elementary particles in a fusion of elementary particles, quite similar to today's formation of a hydrogen bomb,

After approximately 200 to 400 million years of these cycles, the stars gained illumination, to give off light and photons in the void of the cosmos, becoming full-fledged stars, each being of various volumes and magnitude of brilliance. Collections of such stars were grouped into what today we call galaxies, and galaxies grouped into galaxy clusters.

DARK ENERGY AND DARK MATTER

We come now to an added theory, that of Dark Matter and Dark Energy. It is proposed the excess energy of star formation, and elsewhere in the cosmos, converted from Dark Energy to Dark Matter, to agglomerate at certain Vacuum Energy locations, near the stars within the cosmos. The interaction of energy and matter in cosmic "neighborhoods" of "lumpiness", remaining unchanged still after the Big Bang expansion, except for formation of new stars, the author believes, goes far to explain Dark Energy and Dark Matter today. The early "particle soup" phase, with definition of matter particles, may have been indication of the early existence and growth of Dark Matter.

The excess energy/matter was tremendous, estimated at about eight times the mass of otherwise presently-known matter in the universe, even dwarfing the accumulation of matter that would form stars and loose matter within the cosmos, that would eventually become planets, comets, asteroids, and various orbiting space matter. The quantity of energy/matter, that did not form stars, yet cooled in the universe, may be the Dark Matter of today.

Dark Energy is the excess residue from the energy of the Big Bang that failed in its conversion to matter and to stars. In this concept it is Dark Matter, from Dark Energy, in the far reaches of the cosmos, that travelled furthest, and possibly fastest from the Big Bang, that accelerated expansion of the universe. This matter became the essence of various planets, which orbited certain stars, including our Sun.

PLANETS AND MOONS

Some planets also attracted certain matter and drew, by gravitation and centripetal force in orbiting the Sun, moons of various numbers, Earth having only one moon.

Our Moon, alternatively, may have been created from collision with a large cosmos traveller, gouging out from the early semi-viscous Earth planet, a mass that formed our present moon.

The composition of our moon has been found to be radically different from that of Earth. Instead of concentric masses as within Earth, of core, outer core, mantle, and surface, the moon is composed of mostly rubble, attracted materials that did not heat enough to form concentric layers. Moons of other planets in our Solar System appear to vary from one another in composition and nature and may have had different origins.

This activity of formation of stars, planets, space particles, orbiting comets, asteroids, and various rock-like space travellers, may well have been duplicated in stars, now many light years away from our Solar System. In fact, astronomers have sighted by space-based telescopes, and theorized, planets orbiting stars of cosmic distances from Earth.

The natural question arises: Do any of the planets of similar "solar systems" exist that harbor life in parallel to that of Earth? Time and continued scientific observation and research may reveal answers to this question.

LIMITS OF THE UNIVERSE

The extent and dimension of the cosmos or universe changes with the expansion and location of cosmic bodies such as galaxies. A Great Wall of galaxies, observed at the present limit of the cosmos, undoubtedly did not exist at time of the Big Bang, 13.7 billion years ago. It is believed that space itself is created, as needed, by some unknown element.

Again to illustrate, picture a child's balloon that is limp (like before the first moments of the Big Bang) and has only small dimensions. As it is inflated the dimensions change, increasing in dimensions further as it expands from becoming further inflated. Such is believed to be the character of space, increasing as some unknown element is provided to bring about expansion. Space, the cosmos, or the universe were of relatively small dimension 13.7 billion years ago.

Various stars are determined to be of different ages, as well as in size, density and brilliance. Various stars were created at various ages, and in fact are being created today, slowly as in cosmic time, largely from the presence of Dark Energy and Dark Matter at various locations within the cosmos.

STARS ARE THE "MACHINES" OF THE COSMOS

A "grand machine" for producing matter existed in the formation of stars, galaxies, and galaxy clusters. As described elsewhere, space was actually created at the forefront of the expanding "cloud" or wave of energy generating from the Big Bang. The heat induced by exploding and retreating stars would change or create new elements. It is tantamount, on a much lesser scale, to the chemical change in transforming iron to steel, and of changing to plutonium from uranium.

FURTHER RESEARCH

I believe it will be shown by researchers in the future that the Singularity was not matter, but intense energy. The formula developed by Albert Einstein of $E=MC^2$ proves that energy and matter were indistinguishable in the Singularity. Presently, ordinary matter discovered in the cosmos accounts for only about 4.5 percent of the total energy/matter density in the universe. Some kind of unseen Dark Matter accounts for 25 percent, and a full 70 percent is a smoothly distributed Dark Energy. There was no matter in the vortex of the Black Hole, but there was tremendous energy. Restated, there was no matter, but only energy, prior to the Big Bang.

It was energy and not matter that expanded into the universe. However, energy, once ejected into the universe, cooled into matter particles of hydrogen and helium that formed stars in areas of high concentration of cosmic radiation background energy (COBE).

Space is nothingness, yet we are all familiar with space and nothingness each time that we blow up a balloon. As stated above, natural space is not defined by its content, but by its limits, just as the space with an expanding balloon is defined by the walls that describe the balloon. The size is variable, originating from zero to whatever the walls of the balloon will contain without bursting.

As an analogy, we could add an amount of dust, and various types of dust, to the empty balloon, representing energy in space, which might somehow physically join together to form some shape of structure or ultra-high energy concentration, similar to variations in matter and objects, In analogy, the various types of dust would be synonymous with various elements of matter. Such is description of expanding space, with energy/matter added to it that forms the structures in space.

BASIC HYDROGEN AND HELIUM

Some hydrogen and helium atoms were activated in localized areas of the cosmos by high cosmic radiation background energy (COBE). This extreme atom activity produced intense density,

WONDERFUL WORLDS

pressure, and temperature in those localized areas through local atom activity, forming stars.

Then, through a process of star explosion and re-forming caused by gravity, the hydrogen and helium, and now other elements in greater concentrations, produced by re-forming at even higher temperatures and pressures in the star by the increased density and atom activity action of hydrogen, helium, and new elementary particles of matter, by fusion, and producing illumination by photons. The fusion process producing starlight was in much the same manner as the fusion process in today's hydrogen bomb.

In star formation, volumes of energy expanding in space are attracted to energy densities within space to create even greater energy, and perhaps by creation of additional stars and galaxies, more space. Energy is initially attracted to a density variation by gravitation. The gravitation increases the density further, with increased temperature and pressure. Increased temperature and pressure causes explosion of the star, which today we call a star supernovae.

But force of gravity re-calls elements of the exploded star back towards the core of the star, which now become additional, and new, elementary particles of matter. The re-called matter, re-called by gravity, surround the core of the star, now creating even higher temperatures within the star. The higher temperatures promoted formation of new elements. Hydrogen to helium, to carbon, to oxygen, and so on, to heavier elements, by successive star explosions, consequent re-calls by gravity, increased temperatures and atom descriptions for new elements that we know and understand today.

ENERGY TO DARK ENERGY AND DARK MATTER

Dark Energy, is at the forefront of past universe limits, and is still at the forefront of the present universe limits, is present in Dark Energy, continually expanding space that galaxies, such as at the Great Wall of galaxies on the frontier of the universe, will expand into.

Given that 1) Vacuum Energy was present during time of the Big Bang, and 2) Vacuum Energy pressures a force that produces space, and 3) Vacuum Energy is still present in the form of Dark Energy, space is still being created to this day.

As we know, new elements fitting into voids in the Periodic Table of Elements, were discovered and added over periods of scientific discovery, and are being further discovered and described even today. Earth, new stars, galaxies, and galaxy clusters have been formed, and are probably still being formed, in the eons since first formation. Stars are consequently the source of all elements of matter that researchers find.

Was Dark Energy, and subsequently Dark Matter (energy transforms into matter per Einstein), the source for energy of the Big Bang? Further, is Dark Energy the unused residue from formation of the stars and galaxies? The original elementary particles were hydrogen (mostly) and helium (to a lesser proportion). Dark Matter would result from Dark Energy in this residue that formed stars and galaxies.

Presuming that energy, not matter, was the source of the Big Bang, such energy (and resulting matter) could account for Dark Energy and Dark Matter observed and calculated today.

Increased temperature of stars generated photons of light. A dark period of approximately 200 to 400 million years after the Big Bang transpired from first union of energy to stars, initial formation by heat and pressure of first elementary cosmic particles, explosion or expansion, recalls by gravity to increased density of the star, until the star gained brilliance and emission of photons to be current (presently or extinct) "beacons in the sky". At about 200 to 400 million years after the Big Bang the stars began to light up from the fusion process.

In the pre-Big Bang era, there were "islands of energy" in the cosmos, as opposed to the general space of the cosmos that were in fact completely void of any forces, energy, or matter.

Any Event Horizon of a Black Hole did not fill with matter because there was no matter in the cosmos to absorb. Instead, a Black Hole absorbed cosmic energy (now known as Vacuum Energy), by gravitation, reducing the forces of energy to a unification (Singularity) that, upon reaching its critical point of pressure and temperature, resulted in sudden expansion known, as the Big Bang.

POSSIBLE ANSWERS

Often when I tried to explain the Big Bang to some of my Fundamentalist Religion friends, they would typically reply, "yes, but where did the matter come from that entered the Black Hole?" Until now, I had no explanation for them.

With the substitution of energy for matter in the Singularity, I now have an explanation, except they might still ask, "yes, but where did the energy come from?' Hopefully this explanation answers their questions.

Energy was the "mechanism" employed in specie development by affecting change in DNA that in turn gave intelligent direction (Intelligent Design) for the specie to adapt new qualities in its path toward becoming Modern Man.

In this concept, energy is the means to affect new and changed DNA, depending upon the new qualities designed to be added or changed for each specie or sub-specie.

The interaction of energy and matter in "lumpiness" in the total cosmos, remaining unchanged still after the Big Bang expansion, except for formation of new stars, may, the author believes, go far to explain Dark Energy and Dark Matter today. The early "Particle Soup phase", with definition of matter particles, may have been indication of the early existence and growth of Dark Matter.

The quantity of matter that did not form stars, yet cooled in the universe, may be the Dark Matter of today. Dark Energy is the excess residue from the energy of the Big Bang that failed in its conversion to matter and to stars.

In this concept, it is Dark Matter/Dark Energy, in the far reaches of the cosmos, that travelled furthest and possibly fastest, from the Big Bang, that accelerated expansion of the universe by gravitation.

THE BIG BANG AND THE UNIVERSE

It always seemed illogical to me in explanations of the Big Bang that all matter of the cosmos was contained in the Singularity, that was described as no larger than a basketball today, or much smaller, depending on the person writing. Even if the package was solid with particles, antiparticles, quarks, forces, and like substances, it seemed inconceivable that all items of our cosmos, including earth, other planets, sun, other stars, galaxies, other articles of the cosmos, and now Dark Matter and Dark Energy, could conceivably be contained in such a package.

Looking back on the cosmos and Big Bang, where on the cosmos map do the various vectors seem to originate and converge? Where in the current cosmos did the Big Bang occur? Where was the central point or locus? We note that as we look to a certain constellation in the sky (Sagittarius) we are looking towards the present center of our Milky Way Galaxy.

COSMIC ENERGY

I now believe that at the Big Bang and after, it was not matter that emanated from the Singularity, but that energy was the prime component.

To better comprehend this magnitude of Albert Einstein's calculation of energy in the formula $E=MC^2$, we can understand from the formula that the amount of energy existing would be something less from mass, but squaring C (speed of light) the amount of energy becomes almost inconceivably high.

This is the amount of energy created at moment of the Big Bang. At that time energy and matter were indistinguishable, and that is how the matter of the cosmos, galaxies, and other objects of the universe became existent to form stars far far away, and in tremendous quantity. Albert Einstein's formula of $E=MC^2$ was again proven correct.

Energy, not being a physical substance, would not be a factor in determination of physical dimensions. Einstein has stated that, at inception, matter and energy were indistinguishable. In fact, later matter was essentially energy from moment of the Big Bang. No later magnification of matter, such as Dark Matter that could become matter of space, was necessary. Physical matter would not need to be transported through space faster than the speed of light as was energy, even at the miniscule time interval to accomplish the Big Bang expansion or inflation. To restate, at the Big Bang and after, it was not matter that emanated from the Singularity, but Energy, more specifically Vacuum Energy. It was not matter that agglomerated into stars, sun, planets, asteroids, comets, and other matter in the cosmos, but Energy.

As Energy dissipated quickly at the Big Bang, there was reaction, like a shotgun pushes backward as the shell explosion exits forward,

Energy was dissipated in all directions to create a somewhat uniform cosmos, and Dark Energy is result in the cosmos.

It would follow that energy was the primary factor in Big Bang inflation. The energy was created when quarks appeared in negative electrical charge form, meeting to annihilate most positive charge particles. The remaining survivors formed protons (positive charge) and neutrons (negative charge), which formed the core of atoms. Electrons with negative charge annihilated protons with positive charges leaving one billionth or so of all that was created, and the rest expired in a flash of cosmic energy. Annihilation too results in radiation energy.

A few minutes after the Big Bang the universe would be larger than a galaxy, consisting of plasma, a mixture of energy, and charged sub-atomic particles, not unlike the center of the sun. The universe was hot and particles and energy were moving fast and interacting violently. Subatomic particles of energy and matter, after annihilation, were all charged with negative electrical charges. The whole universe would be sparkling as energy and matter separated, although separation was not complete.

At about 380,000 years after the Big Bang a crucial phase change took place as the universe cooled enough to allow electrons, having equal negative charge, to capture remaining protons having negative charge (positrons). They formed simple elementary particles of hydrogen and then helium. The particles were neutral because the two charges of negative and positive canceled each other out.

All the matter of the Universe suddenly turned into atoms, from energy to matter, and went electrically neutral. Energy could flow through the universe freely without getting entangled in the charges of the loose subatomic particles. At that point, Big Bang theorists hypothesized; there would be a flash of energy. This would be the forerunner of Cosmic Background Energy, later to be discovered by the COBE (Cosmic Background Energy Explorer) probe in 1960.

In a cyclical action, eventually the universe cooled and went into a form of "Dark Age" that lasted a few hundred million years of empty space, with a faint glow from energy.

How did the stars begin? It would seem the question is beyond evidence, a question that could be answered only in philosophy, or perhaps within religious traditions. (In the Biblical reporting of the Creation, there is no discussion regarding origination of the stars.) Now science has the evidence! As far as is known, all human societies have asked this question regarding beginning of stars. Most of the answers that were offered in earlier societies were based on very limited evidence. Explanations were widely varied, but we should not summarily dismiss those answers simply because they may seem naïve to us. Often imbedded in them was a huge amount of good, empirical, observational evidence. The ideas were rich, intelligent, and based often on as much evidence as an ancient society could muster. It is very important when we look at the scientific account of origins to also pay due to the many other alternative accounts that have developed in the past.

WORDS OF THE GODS

There are many alternative accounts. Traditional answers to the questions of the origin of everything often began with a god or gods, with the universe created by a panoply of deities.

1) A passage from a Mayan text portrays the bearers of the origin are in water, in a glittering light, enclosed in quetzal feathers, in blue-green, an erotic scene.
2) Many Creation stories posit a sort of initial chaos out of which something emerged. Rig-Veda in India is a basic Hindu scripture which was probably written about 3,200 years ago. Again, the universe emerged out of primeval chaos.
3) Many creation stories talk about a birth from eggs. Some accounts describe the universe as a sort of living organism. A story from the Karraru people of southern Australia relates that a Great Father spirit awoke a beautiful women and told her to stir the universe into life. As the "Sun Mother" opened her eyes, darkness disappeared as the rays spread over the land. She took a breath and the atmosphere changed.

There is a problem with deistic accounts of these origins of the universe. There is no evidence science knows of showing the existence of a creator god. We are still faced with the question of what was there before the beginning.

EXPANSION

At the beginning, the modern account of creation is really as arbitrary as any other. We just don't know! We can tell a well-grounded scientific story based on masses of carefully tested evidence. At the very beginning, matter and energy were indistinguishable, as in a phase change, or a whole series of phase changes, much like what happened in the early universe as it cooled. Particles, and the four fundamental forces of nature, appeared. (1) Gravity appeared with (2) the Strong Force (that holds the nuclei of atoms together), (3) Electromagnetic Force (that holds atoms together), and (4) the Weak Nuclear Force responsible for radioactive decay of sub-atomic particles.

In inflation, the universe expands at the speed light travels. Nothing is actually moving, but the fabric of space-time itself is expanding at the speed of light. The universe grew from the size of an atom, to the size of a galaxy, and within seconds and minutes.

Particles started to appear out of the murk. Quarks appeared in opposite form, eliminating or annihilating each other, virtually negative versus positive, and they formed simple atoms of hydrogen and helium, the atoms were neutral because of the electrically negating action.

Scientist know there is much "stuff" in the universe called "Dark Matter and Dark Energy" that is exerting a powerful gravitational pull, and may now account for up to 95% of matter in the universe, 70 percent consisting of undetected forms of energy, and 25 percent may consist of matter in forms science cannot detect because they don't emit radiation (such as with cold, dead stars or planets), or perhaps because they are too small (such as subatomic particles known as neutrinos).

With joining of elementary particles, all the matter of the universe suddenly turned into atoms and went electrically neutral. Energy could flow through the universe freely without getting entangled in the charges of the loose subatomic particles. At that point, Big Bang theorists hypothesized, there would be a flash of energy. This would be the forerunner of Cosmic Background Energy, later to be discovered by the COBE (Cosmic Background Energy Explorer) probe in 1960.

In a cyclical action, eventually the universe cooled and went into a form of "Dark Age", that lasted a few hundred million years, of empty space, with only a faint glow from energy, before stars illuminated to their full ability in photon output and brilliance.

STARS

How did the stars begin? It would seem the question was beyond evidence, a question that could be answered only philosophically or perhaps within religious traditions. Now science has the evidence! Energy entered agglomerations of gases within our galaxy, increasing the pressure and temperature of the gases and decreasing the volume by the pressure. Between 200 and 400 million years after the stars formation, gaining from temperature and mass, produced photons of light that illuminated each star.

BEGINNINGS AND VARIOUS ACCOUNTS OF ORIGIN

As far as is known, all human societies have asked this question of beginnings. Most of the answers that were offered in earlier societies were based on very limited evidence. Explanations are widely varied, but we should not summarily dismiss those answers simply because they may seem naïve to us. Often imbedded in them was a huge amount of good, empirical, observational evidence. Some ideas were rich, intelligent, and based often on as much evidence as an ancient society could muster. It is important when we look at the scientific account of origins to also pay due to the many other alternative accounts that have developed in the past.

EXPANSION OF SPACE

Were there factors in the Big Bang that, in addition to elementary particles, actually created space? Is space still being created to exist at the far outreaches or present limits of space? What new element of matter or antimatter may be accountable to create space? Is such elementary particle related to Dark Energy or Dark Matter? Is there any limit to how much space could be created? Was such an element present before the Big Bang from which Vacuum Energy developed? Is such an elementary particle existent in a universe of energy and antimatter, and a source of matter? Is a universe of energy and antimatter a true source of matter? Is there a separate universe, beyond our universe limits, that accounts, by gravitational attraction, for increase in the rate of our universe's attraction?

An element of "space" in a universe of antimatter would have quality to expand the dimensions of space, although nothing physical would exist within those dimensions. Space is not defined by what is contained within it, but by its dimensions, like space in a balloon is defined by the walls of the balloon. In the case of the balloon, space is created from expelled air from the body. In creation of a universe, space would be created from some new particle, yet to be discovered, within a universe of antimatter.

The definition of such an "elementary space particle" would be as elusive as locating the reasoned Higgs particle. As each particle of antimatter has its opposite in matter, the opposite of "space particle" would be full containment of some particle in matter, perhaps similar to the reasoned Singularity from the Big Bang. Therefore any new element that would have quality of creating space would really have a quality of expanding the limits of space, perhaps to a never-ending extent.

The cosmos, before the Black Hole and Unification, was not entirely smooth. Some locations in the compact space were more dense in energy (Vacuum Energy) than other locations, thereby allowing certain energy concentrations to form the Big Bang. Other energy locations (other Vacuum Energy locations) may very well have formed, either before or after our Big Bang, or other Big Bangs, at other locations of the cosmos, each originating what would become other galaxies, or even multiverses, that we don't even yet understand. Would another multiverse serve as the gravitational force and the cause for our universe increasing in expansion rate?

THERE IS EVIDENCE OF THE BIG BANG

1. The Red Shift revealed by spectrometry is like a Doppler effect. This shows that remote objects in space are moving away from Earth, and the further away they are, the faster they are travelling away.
2. The Cosmic Background Radiation exists as shown by the Cosmic Background Explorer (COBE) and spectrograph information from space.
3. Views through modern telescopes can presently view back nearly to 12 billion light years, which in effect is looking to status of the cosmos 12 billion years ago, near time of the Big Bang. The situation is indeed different from today's universe.
4. The age of the universe is scientifically estimated at 13.7 billion years, determined by modern telescopes, including the Hubble telescope in space. The new Webb telescope will allow viewing of objects at a distance back to nearly the moment of the Big Bang.

5. Big Bang cosmology states that, in the Big Bang, huge amounts of hydrogen and helium were made, but there was no ability to make any elements more complicated than that. It predicts that perhaps three-quarters of the early universe should consist of hydrogen, and the rest of helium. That, in fact, is what is seen when astronomers look at the universe.
6. Big Bang cosmology offers a scientific, well tested explanation of how the universe was created. Fluctuations in temperature at different locations of the cosmos were achieved by NASA's Cosmic Background Explorer (COBE) satellite in 1992. Since then additional, and more precise measurements, have been made by astrophysicists using balloons and ground-based radio telescopes. A comprehensive, precise map of the Cosmic Microwave Background (CMB) has been produced. Data provides a treasure trove of information regarding the CMB and locations in the cosmos of warmer and cooler areas.

The author believes Earth is currently entering a cosmic-caused warm condition, which will possibly be followed in several decades from now by increased glacial conditions. There is little that earth population can do presently to counteract this condition. We can only hope the diversion in temperature from "normal" is mild.

Perturbations have evolved from early times into visible galaxies. Consider an over-dense region of CMB in the early universe. Its tendency would be to collapse under its own gravity, however it won't, simply because the expansion rate of matter in the cosmos is too great.

If expansion of the universe were to slow down (which is unlikely) dense regions of CMB could begin to collapse, they would heat up and the pressure within them increases. This would produce a force in the opposite direction to cause the previously over-dense region to puff up and become less-dense. The decreasing force process is like that of a sound wave, and continues to oscillate back and forth, losing force with time passage.

Source: Great Courses, Big History, 2008

DIFFERENT BIG BANGS

It is commonly believed, and conventional wisdom holds, that the entire universe or cosmos was created by the one Big Bang. I believe this is an overstatement.

Our Milky Way galaxy contains a core of star brilliance that is believed to be location of the singularity, the space origin following the Black Hole, where universe cosmic expansion originated. Analysis of cosmic radiation also indicates the intensity source of singularity and expansion. Looking back on the cosmos map, where do the various vectors, emanating from the explosion, originally seem to originate? Where in the current cosmos did the Big Bang occur? Where was the central point? Origin is believed to be in the Constellation Sagittarius.

Other galaxies have been discovered to possess or have possessed Black Hole remains, identifiable by the brilliance at core of the galaxy.

It would seem to follow, then, with various galaxies and Black Holes, and probably other past Singularities, that there were many Big Bangs, each common to its own galaxy.

Our Milky Way Galaxy had our Big Bang, as other galaxies had their own Big Bang. Our Big Bang created our own Milky Way Galaxy and nothing more. Big Bangs of other galaxies created a universe around their own Big Bang space origins. When we read here of the Big Bang, this is to be interpreted as our Big Bang, the Big Bang of the Milky Way galaxy, and not from other galaxies.

Gamma ray bursts emanate from the birth of Black Holes. These form intense eruptions of high energy radiation from random spots in the sky and are now thought to be associated with the formation of Black Holes in distant galaxies. Their visibility from so far away means they are truly gigantic explosions.

This interpretation of separate Big Bangs in other galaxies, occurring separately, would seem supportable by reason. Within our Milky Way Galaxy there are untold millions of stars, or components of the galaxy. Creation of the Milky Way Galaxy alone would be of nearly unbelievable scope in cosmos development. To reason that other galaxies were also created by our Big Bang would seem illogical, if not impossible.

The outer space between galaxies would support logic that each Big Bang was "geographically" independent, with thousands of light years defining the space between galaxies. Analysis of interiors of various galaxies would indicate wide variance in star composition.

Each galaxy would possess certain similarities, but also many differences. In shape, the Milky Way Galaxy is spiral. Other galaxies might be spiral, round, oval, or possess seemingly disorganized star nebulae in process of reforming.

The cosmic history of our Milky Way Galaxy might well be duplicated in other galaxies, even to the extent of life on their planets orbiting their "suns".

Might life on planets, at other galaxies, be comparable in time and development to us, be eons behind our studied development, or be eons ahead of our life, culture, and technology state of Earth?

THE POSSIBILITY OF 'WORMHOLES'

Mathematically, worm holes seem to connect our universe with another universe or provide shortcuts within our universe. Wormholes are a postulated method within a general theory of relativity, of moving from one point in space to another without crossing the space between, to travel long distances in short periods of time. Several science fiction stories have expanded ideas of wormholes.

For a simple visualization of a wormhole, consider space-time as a two-dimensional surface (flat). If the surface is folded along a third dimension, it allows one to picture a wormhole "bridge", appearing much as two funnels connected at the necks, forming a tunnel between. A wormhole is, in theory, much like a tunnel with two ends, each in separate points in space-time. However, tremendous energy is concentrated at the inner point of Singularity where the two Black Holes would join, where a person would be vaporized. The wormhole itself would be destroyed by unification.

CREATION OF SPACE

Did the Singularity, the seed that expanded at the Big Bang, contain a highly expansive element, in addition to hydrogen, helium, quarks and other minute elementary particles? Was this unknown element the cause of expansion of the cosmos to accommodate the expanding gases from the Big Bang?

Limits of space is defined as the furthest distance from the Big Bang source of any matter. The universe was relativity compact at time of the Big Bang event. Except for the rapid expansion of a high energy gaseous cloud, there were no objects of stars, galaxies, matter, or anything else on the horizon of the universe. Unlimited space was non-existent, as it was before the Big Bang event. More extensive space became existent only as needed to house items of matter (e.g. future stars, star clusters, or galaxies) new to the location. This is when close or limited space at the Big Bang was enlarged, expanded, or created. Space was enlarged by some force yet to be discovered.

However, it might be argued that, like the sailing of Christopher Columbus, the unknown American continents were always existent. It was just that they were not populated by Europeans or people, as we understood people, although they were populated by other civilizations. The open seas (like space) and the New World continent (like the "Great Wall of Galaxies) were there and in existence all the time. Again, it may be determined in the future that a force existed to create space as required by presence of expanding galaxies.

Expansive elements could have accounted for the following:

1. The rapid expansion into the cosmos after the Big Bang.
2. The source material for the stars, galaxies, and galaxy clusters, and cosmic components (that are so hard to imagine as coming from a relatively miniscule Singularity, that various scientists describe as being no larger than a basketball or period at end of this sentence.
3. The expansive element of the Singularity would have the power and energy to create, such as future stars and galaxies by hydrogen and helium.

4. Expansive elements would of necessity need to grow exponentially.
5. At a beginning matter and energy are indistinguishable. The expansion element within the Singularity would be surplus energy, not matter.
6. Phase change would transform surplus energy to surplus matter.
7. What possible element of antimatter or matter has such a power of expansion, to expand space? Theoretical Physics may produce an answer to this conundrum in the future.

The author believes that Dark Matter is unused primal matter and Dark Energy is excess energy. The excess matter serves as the attraction by gravity that speeds up the cosmic expansion. Are dark energy and dark matter related to factors that create space?

ADDITIONAL DIMENSIONS IN SPACE/TIME

There are many theories regarding various stages of birth and development in cosmology. These include concepts of String Theory, extra dimensions in space/time, Vacuum Energy, limits of the universe, and more, not to mention the mysterious Dark Energy and Dark Matter. Each has potential to change our views on the birth, development, and the laws governing the universe. Any one of these concepts could change time-honored theories of the universe.

String Theory holds that energy in space is not located in precise points of space, but instead consists of chains or strings in the cosmos that affect qualities and actions of the cosmos. Under String Theory many additional dimensions besides the four we presently recognize of width, depth, height, and time must exist with a brane, or membrane-type structure, for each. It has been proposed by cosmologists that as many as eleven different dimensions exist and are necessary to validate String Theory. Others have ventured to state that the number of dimensions is unlimited.

Vacuum Energy is energy that existed in space, and continued to reside in the relative vacuum of pre-existence. This was an early concise space. It is not yet clearly defined, but Vacuum Energy is believed to be associated with Dark Energy and Dark Matter.

Vacuum Energy also could be source of the Big Bang, reflected in creation of not only the cosmos and our Milky Way Galaxy, but repeating also in creation of other galaxies in space.

The author ventures to propose that Vacuum Energy is the source for creation and of constantly expanding space in our universe, defining the changing limits of the universe, currently 13.7 billion years old. The phase of expanding space seems to be attracted to the Great Wall of Galaxies which has been observed at extremes of known space, which conceivably could account, by gravitation attraction, for the increasing velocity of space growth. This also leads one to wonder where and how this Great Wall of Galaxies originated.

What seemed relatively simple to Copernicus, Newton, and other giants in cosmos, study of astronomy is becoming increasingly more complex, it seems, with new findings and developments in astronomy and cosmology each year.

A FORM OF GOD

The author also proposes a definition of God as unlike any physical human, but is a non-physical entity, as perhaps a cloud or wisp of energy, or even as an actual portion, or spirit, of each man's soul and mind.

Biblical writings state that "Man was created in God's Image" (Genesis 1.6-27). Ancient believers in God falsely reasoned that god therefore was human. It is reasoned in this book, however, that God is reflected in the soul.

A cloud of energy might also define existence of one's soul in a state of Afterlife, having non-verbal, non-physical abilities in communication, without benefit of sight, hearing, smell, speaking, or touch, and in a space/time dimension different from matter experienced on Earth.

During lifetime on Earth, soul is a non-physical entity existing in the mind and brain process of man, to a large extent modifying the emotions and views of the individual.

In astrology, imagined thousands of years ago, It is suggested and supported by many testimonials over time, that reincarnation of soul to humans occurs, and that like characteristics of certain souls adhere to groups of similar souls in Afterlife. Each individual soul in the Afterlife group is assigned a similar period for subsequent birth and reincarnation. The time for birth is related to certain positions in the orbit around our Sun. With having similar personal and mind characteristics, their thoughts, fortunes, and activities of his or her soul, perhaps guided by a Guardian Angel, will be similar to other souls in the group. This theory relates and supports the ancient and time-honored field of Astrology.

Mankind did not develop without intelligent direction. Minute and individual changes did occur constantly from first life to the state of human progress today, and will continue to evolve in the future as environmental and other needs in development arise. Changes will be orderly in new mutations, but directed toward specific mankind goals, rather than in a senseless agglomeration of dead-end possibilities resembling chaos. Naturalists, or pure environmentalists, sometimes argue that unnecessary traits, developed in chaos, would wither from the genome over time.

It has been said that God is the greatest scientist of all, proven in directing the underlying path of evolution over eons, and formulated by specific variations in the genome that provide changes required by local needs, but in a directed path.

Perhaps developments in atmospheres, geology, botanical developments and evolution of the physical Earth, in early characteristics of Earth, were products in science, whereas creation and existence of soul, mind, and in birth and growth systems were planned by God, to be made physically actual through natural biologic means.

Fundamentalist religious individuals especially, may not agree with my views that challenge traditional religious Biblical dogma of long—standing, but hopefully individuals of these groups will at least understand there are arguments that alternative possibilities exist. Bringing realism, to replace folklore and myth, should actually strengthen beliefs in God through truths in the concepts of spirituality. After all, it was once believed, before evidence to the contrary, that the stars and sun revolve around the earth. The weight of evidence has changed with increase of accumulated knowledge at present, with all respect given to the necessary preliminary thinking of the past by giants in development of philosophic thought.

A SCIENCE VIEW OF CREATION

Let me take you back in time in your mind's eye, and even before time started—back before you were born, before your parents were born, and even before life existed on earth—back to 13.7 billion years ago and even before that, a supposed impossibility.

This is a science view of Creation—How the Sun, earth, planets, stars, galaxies, and unseen and unrealized space features came to exist. In the conventional wisdom science time started 13,7 billion years ago, at moment of the Big Bang.

The Big Bang's existence is accepted in science today, with physical and chemical evidence of cosmic background radiation (CMB) in space, as a remnant of the energy released by a Big Bang occurrence, and still present today, 13.7 billion years later. The radiation has been "sampled" by deep space probes and examined later by scientists specialized in this analysis.

As a radical thought, perhaps the Big bang actually was a massive Supernovae, a pre-history star that exploded with all similarity to an expanding Big Bang, with its capacity of expanding energy or even resulting matter, to form a cosmos, or universe, or Milky Way galaxy.

The universe, created like an exploded Supernovae, would expand and consequently withdraw into itself, perhaps shaping the present future of our own cosmos. The expansion would push out a wall of cosmic energy (the observed "Great Wall in the Universe" of galaxies) around the expanding cosmos. This wall of energy could conceivably create a source of attraction that would serve as a source of gravity and that would actually pull the expanding universe outward, speeding up the rate of expansion as currently detected. A Supernovae would contain the essence of Dark Energy and Dark Matter

The universe created like an exploded supernovae, would expand and consequently withdraw into itself, perhaps telling a different, unknown future of our own cosmos.

It is proposed that Dark Energy, the forerunner of Dark Matter emerging from Dark Energy, is actually energy in excess of that needed to form our present scope of stars, and is waiting to make formation of new stars and galaxies that would appear in our cosmos in the future.

This thought implies that a star, perhaps a very old star, existed prior to the event we call the Big Bang. If the reader thinks tracing history back 13.7 billion years to the Big Bang was a long time, try comprehending an existence even before that event, and exploring what the ramifications of that possible history could be.

A recent photograph as viewed through lens of the new Chandra Space Telescope, looking back billions of years ago, reveals a new possibility in cosmos science. (Smithsonian, February 2001, page 77).

In this book I propose the thought that prior to the Big Bang, and before the existence of any matter that would have been attracted into a Dark Hole and an eventual Singularity, there existed Space Energy, which actually expanded following Singularity and expansion to form our universe, matter, and possibly Dark Energy and Dark Matter, some still existing today.

In this scenario, It would be an impossibility that any matter would have been contained in that original Black Hole, simply because, supposedly, no matter existed in space prior to that moment. A Black Hole would have contained only Space Energy, but still could have erupted, with an event of Singularity, to form our universe.

These views may conflict with conservative or fundamental views of religion. The science views are well documented with physical evidence to support theorized events.

HERE IS A SUMMARY VIEW OF SOME CURRENTLY THEORIZED EVENTS FOLLOWING THE BIG BANG

13.7 billion years ago

The Big Bang forms gasses of hydrogen and helium, expanding at the speed of light, with energy in the cosmos to form what will later be numerous stars, galaxies, and galaxy clusters, and in later explosions of individual stars into nebulas, and re-forming of those stars by gravitation. The defined space changes with the location of distant cosmic objects. Gamma rays populate the cosmos following the energy expansion after the Big Bang, perhaps a major factor in changing inorganic minerals to organic first elements of life. We realize today that Energy can be created from a single particle, or from a star.

13.5 billion years ago

After a dark period of 200 to 400 million years, stars composed from gases chemically changed to elements produced by tremendous pressure and heat, have increased their energy, matter, temperature and mass, to illuminate, in a manner similar to modern-day hydrogen bombs, by fusion of elements. Gamma rays are of tremendous significance in the elementary particle world; Gamma rays are of the maximum possible potency with each unit of Gamma ray having energy 100,000 times more than a photon of visible light.

Creating such extreme elementary particles takes commensurately extreme processes: The collision of particles moving at nearly the speed of light; the annihilation of matter (negative-charge) and antimatter (positive-charge) occurs, which can convert their remaining mass entirely into matter, per Einstein's formula of $E=MC^2$; the movement of energy out of Black Holes; and the release of nuclear energy in radioactive decay of fusion actions.

As extreme as these processes may seem, we bask in their glow in the form of sunlight that started off from the Sun as Gamma radiation in the Sun's core, and degraded its visible light during its tortuous passage through overlying layers of atmosphere. (Source: Scientific American magazine, September 2009).

It is significant that in the Biblical account of Creation, no mention is made of the origin of stars. Was this beyond comprehension of the biblical writers who composed or reported in the Book of Genesis, going back to the life and writings of Moses?

6.1 billion years ago (7 1/2 billion years later, in a period of Solar and planet formation)

Our Sun ignites, millions of planetesimals form, that collide to help form a molten earth 4.56 Billion years ago. More than 200 minerals formed within earth from planetesimals melting, and shocks from collisions.

5.8 billion years ago (7.9 billion years after the Big Bang, in a period of star formation)

A Hadean-like earth is mostly black Basalt rock formed from molten magma and lava erupting from the interior of a molten Earth. The next 2 billion years see about 1,500 minerals formed by repeated partial melting, chemical joining, and solidification of rock concentrates, chemical reactions, and weathering on earth by early oceans and atmosphere. Minerals, formed by high pressure and temperatures, are brought to the surface by volcanic action and contained in later plate tectonics.

4.4 billion years ago

Our Moon was supposedly formed in this era from collision of Earth with a large cosmic traveller. But the question arises of, what accounts for the many other moons of planets elsewhere in the Solar System? It would seem improbable that all were caused by some similar chance collision. Perhaps moons of other planets were originally sub-planets or space debris that were captured and went into orbit of the larger planets.

2 billion years ago (2.4 billion years later)

In a period of red earth (high ferrous oxide content) earth atmosphere consisted of high sulfur content. Later photosynthesis processes gave earth's atmosphere a small percent of oxygen, dramatically altering its chemical action. Ferrous (iron) minerals in Basalt are oxidized to rust-red Ferric (Fe_3) and other compounds.

This great oxidation event paved the way for new minerals. Microorganisms laid down sheets of stromata made of minerals. Geologic plates began to move in response to the void left from the moon substance removal.

CONVENTIONAL THEORY OF THE BIG BANG

At this point I would like to express my personal view of the Big Bang actuality. Conventional theory supports an aggregation of space material drawn into the vortex, and by gravity within a Black Hole producing Singularity, decomposed into the component elementary particles and forces, to be finally expelled as a Singularity, a greatly reduced-in-size particle in a uniformity of those physically minute, separate, elementary particles and forces. At moment of the Big Bang all the matter is expelled and expands into the cosmos, at that time being primarily hydrogen and helium. These two simple elements travelled outwards into local low pressure areas of the cosmos, producing elements of stars at momentous high temperatures, later being violently expelled from the core gasses of resulting stars as nebulae and being pulled back into the star core by gravity, this cycle transpiring many times in formed nebulas and Supernovae.

This is believable in its explanation. However I have often felt uncomfortable with the explanation. I wondered, "wait a minute, where did the matter originally come from that was sucked into the vortex of a Black Hole, and expelled as a unification of elementary particles and force (a unification, or Singularity)?".

I theorize that there was no matter to arrive into a Black Hole at an original stage. Albert Einstein theorized that energy becomes matter—they are inseparable at the beginning. The Black Hole must have consisted of nothing but tremendous energy. Energy in vacuum of space is recognized today in science as "Vacuum Energy".

From a high school Physics class demonstration I recall that an energy force can be created in a vacuum container. This principle would be exercised by agitation of cosmic-size forces in an area of vacuum in space (space vacuum), thereby creating energy of tremendous magnitude.

What would cause cosmic-size agitation of an area of space vacuum is still unclear. Perhaps a non-vacuum area, perhaps a rogue area of energy in space, agitated another energy module, thereby creating the energy of tremendous magnitude. Matter from a Black Hole would not have been required, thus solving my question of "where did the Black Hole matter come from? There was no matter in the Black Hole!

The original Black Hole was not filled with matter but with energy. The Big Bang was an expansion of energy, which was the predecessor of matter. Energy changed to matter during the chemical formation of stars.

Negative and some positive electrical charges would constitute the Uniformity of the Big Bang, and the energy source, that later turned to matter in the cosmos and stars, eventually included our own Sun and planets of our solar system.

Science recognizes an Annihilation process in which negative forms of the uniformity annihilate positive forms almost completely. This raises the question of "where did the positive-charged elements go?" I theorize that, like negative-charged elements served as source of matter, positive-charge forces and elements became the basis of antimatter,

Like matter became a central force in elements that eventually formed bacteria, plants, fish, and animals of earth, positive-charged elements became a central force in antimatter. The antimatter became a central force in formation of what today we call soul, perhaps the antimatter equivalent of what would become Man, to eventually become, again, united with matter in the form of hominids and hominoids, and much later to become Homo sapiens. Soul would evolve in a universe of antimatter, as pre-humans would evolve on earth in forms of matter. Meaningful exploration of antimatter (and consequently soul) has only recently begun with a tremendous CERN laboratory and process center located at a mountainous border location of Switzerland and France.

But now back to describing the path of pre-life development in pre-history, revealed in the formation of rocks, minerals, geology, chemical analysis, and intelligent suppositions of earth study. A first single large land mass on Earth was Pangea, which later segmented to form the land masses of Gondwanna and Laurasia.

800 million years ago

The land mass on earth, Pangaea, broke up into various geographic tectonic plates that would move about the global earth, dramatically changing the topography, movement, and location of land masses, thus causing uprisings and creation of various mountain ranges.

For the next 200 million years earth cycled between extremes of a hot and cold planet, perhaps two to four or more times, due largely to earth's position as it travelled in the cosmos and within the solar system. Distribution of many surface minerals changed drastically with each new glacial cycle. The snowball/hothouse cycles had profound consequences for life. The ice ages shut down nearly all ecosystems, whereas the warming periods saw abrupt increase in temperature and biologic productivity.

The Coelacanth, a different species of fish living 400 to 135 million years ago, is possibly a common ancestor of all earth vertebrates. The Coelacanths are lobe-finned fish with the pectorals and anal fins on fleshy stalks supported by bones, and the tail or caudal fin divided into three lobes, the middle lobe includes a continuation of the notochord tail. Some scientists believe that the pre-evolutionary state of this creature once dwelled on the land surface of earth. A Coelacanth specimen was discovered alive in 1938 in the ocean off the east coast of South Africa and one again discovered in 1952.

At the end of a glaciation, atmospheric oxygen rose sharply from no more than a few percent to about 15 percent, produced in part by coastal algae blooms. Such high levels of oxygen may have been essential to the evolution of large animals (including dinosaurs at about 350 million years ago).

The earliest known multicellular organisms (Eukaryote cells) appear in the fossil records just 5 million years after one of the great global glaciations. Carbonate skeletons of sea animals led to deposition of large limestone reefs in countless cliffs, canyons, and ocean shores on earth.

700 million years ago

This was a period of earth climate change that covered the entire earth with one compound, ice, for centuries. Eventually carbon dioxide from volcanoes would trigger run-away global warming, and surface cooling would result from cosmos temperatures, as our planet cycled between hothouse and snowball phases. In a hothouse phase, weathering adds large quantity of fine-grained clay and silicon particles, including Kaolinite to the landscape, with thin layers of this element found in geologic strata of uplifted mountains to evidence the changes.

First basic and elementary nematode worms, having basic organs for food intake, digestive, and excrement systems for continuation of life produced reduction of soil. Sea life of basic plants displayed movement for fishes to feed upon the plants and on other fish or sea life.

There is current theory that in first life, organic microbes emanated from fissures in the floors of the seas, affecting chemical change of elements, combining, binding, and oxidizing inorganic matter to organic matter (e.g. carbon from earth and mantle combine with oxygen from water, earth, or atmosphere, to form carbon dioxide, CO_2 a compound required in growth of sea and land plants.

400 million years ago (after 300 million years of evolution)

Earth turned green by chlorophyll from multicellular plant organisms that had emerged, and plants had colonized the land, followed by many species of oxygen-breathing animals in a symbiotic relationship. Biochemical breakdown by plants and fungi increased the weathering of rock and production of clays by mixture of hydrated minerals. Earth's surface took on its modern appearance for the first time.

Chemical interactions directly produced additional minerals including widespread calcite, a compound required in growth of sea and land plants, and rare minerals layed down by microbes which had entered the life scene on earth.

First elementary four-footed mammals of lemur, eomaia, and other mouse-like creatures with backbones (vertebrate), fur, and unified interior organ systems, are classified as Vertebrates. Basic reptiles formed from amphibians and from previous fishes. (Source: Scientific American magazine, March 2010, "Evolution of Minerals")

350 to 65 million years ago

Dinosaurs, created perhaps from genetic mutations caused by burst of gamma rays, had been formed from both reptile and mammal origins, with wide complexity from many environments. Dinosaurs were extinguished at about 65 million years ago, believed by collision of a cosmic body with Earth, a cloud of Earth debris causing loss of plant life and food supply for land-living creatures.

65 to 6 million years ago (after nearly 60 million years)

Numeric expansion of mammals in many forms occurred upon demise of dinosaur competition and movement from sea to land environments, with amphibians evidencing transition to air-breathing mammals, four-footed, and fur-covered and scale-covered animals of complex design and species. Bi-pedal primates developed from tree-loving animals. Various forms of Great Ape lived as Gorilla, Chimpanzee, Monkeys, and Bonobos.

Culturally there were varying degree of intelligence, cranium size, and primate animal cultures, living methods, care and teaching of offspring, and providing for existence of animal family and conflict of species or attempt to gain leadership within social groups.

6 million years ago and first Homo species

First forms of man-like predecessors in Australopithicus africanus and later Austalopithicus afarensis (Lucy), and others at 3.9 to 2.9 million years ago, followed by "improved" variations of primate-like, Paleolithic man of Homo erectus (1.8 million years ago), after Homo habilis (2.3 million years ago), and Homo ergaster (1.8 million years ago).

Homo erectus, about 1.6 million years ago, may have been one of our direct ancestors, followed by Homo Neanderthal and Homo sapiens, which might be considered sub-species of Homo erectus.

PRE-HISTORY TO MODERN MAN

Prehistory, starting about 2.5 million years ago, was probably a clash of cultures. It began with human-like origins more than 2.5 million years ago with appearance of first tool-making hominids, and development of controlled fire about 2.0 million years ago. Sometime, about 1.8 million years ago, (the date is still uncertain), this dispersion resulted in later archaic Homo erectus in Asia and Neanderthals in Europe, to all parts of the Old World except the offshore Pacific islands, that occurred about 15,000 years ago.

We, as modern Man, are different because we keep adapting in new ways as result of our capacity for symbolic language and collective learning. Fossil evidence for most of the last million years is dominated by Hominine species of Homo erectus and Homo Neanderthal.

H. ergaster evolved 1.6 million years ago. Some migrated to Indonesia and China. They probably used fire, and certainly used "Acheulian" stone tools, which were better made than the "Oldowan" tools of Homo habilis.

This is evidence of Homo ergaster's use of technological creativity, but not of exceptional creativity. Other species (including apes such as orangutans) had migrated from Africa to Asia. Evidence of H. ergaster's control of fire remains limited, and their stone tools barely changed over 1 million years.

Homo erectus and Homo neanderthals seem even closer to us. Neanderthals lived in ice age Europe and Russia. They were as tall as us and had brains as large as ours (perhaps even larger). They also manufactured more delicate and precisely made stone tools described by paleontologists as "Mousterian". They probably used fire and hunted large ice age mammals such as mammoth and woolly bison, which was no small feat. Yet technologies of H. neanderthal show limited variation over 200,000 to 300,000 years, and there is no proof that they had symbolic language. Indeed, studies of Neanderthal skulls suggest that their larynx would not have allowed them to speak like we do. Recent analyses of DNA, removed from Neanderthal skeletons, suggest that Neanderthal and H. erectus lines split more than 500,000 years ago. Though tantalizingly close to us, neither ergaster nor Neanderthals display the technological creativity that is the birthmark of our species. Both species "disappeared" about 30,000 to 20,000 years ago, probably under pressure from Homo sapiens. A sub-species developed from Homo erectus, and possibly incorporating qualities present in another sub-species of Homo erectus, Homo sapiens.

Modern genetic dating techniques show that humans are closely related to Homo erectus and probably evolved within the last 250,000 years. The fact that the greatest variation appears within Africa suggests that is where pre-humans have lived longest, although it is believed Homo erectus may have emanated from Russian Georgia, between the Black and Caspian seas and near the Caucasus Mountains. Finally, the earliest fossil evidence of anatomically modern humans comes from Africa, and the oldest remains of modern humans are about 160,000 years old.

Archaeological evidence seems to show an acceleration in technological change in Europe and Russia about 50,000 years ago. Improved stone tools appeared, as did new materials including bone and skins. Cave paintings and carved objects provide evidence of symbolic thought that appeared in modern-day France and Germany. Some specialists argue that even if Homo sapiens evolved earlier, modern human behaviors appeared only 50,000 years ago, perhaps as result of tiny changes in the "wiring" of the brain. If this is correct, then the critical threshold may have been crossed, and human history would have begun, just 50,000 years ago.

Evidence from Africa suggests that the technologies that appear in the Upper Paleolithic era had already evolved in Africa. From almost 300,000 years ago, new technologies, and even hints of symbolic beliefs (such as the use of red ochre with burial of the deceased) appear in association with a new hominine species, Homo neanderthal.

The details of how our species evolved remains unclear, but currently the belief is of a rapid appearance of human-like creatures (Homo neanderthal) about 300,000 to 200,000 years ago, somewhere in eastern or southern Africa.

Contrary to the black, negroid skin of Homo neanderthal, Homo sapiens had skin pigments in providing lighter color (or white) skin of Caucasian race. There was evidence of man's belief in a superior force, or being, within or affecting his physical body and environment. First evidence of soul may have been by symbolic illustration on cave paintings and art forms.

Food production first turned to Agrarian culture about 12,000 to 8,000 years ago. By the time of Christ, subsistence farming was practiced everywhere in the Old World, and an aristocratic class ruled by virtue of their proclaimed spiritual relationship with their ancestors. Rank and ancestor's spirits were to be closely associated.

There were pre-empire civilizations, of proclaimed divine rulers, kings with strong centralized government, teeming cities, writing, and metallurgy, but also social inequality, poverty, warfare, despotism, and slavery. Earliest historic records of warring kingdoms go back 10,000 to 8,000 years in Egypt and Mesopotamia.

It is quite well documented that Abraham left the ancient city of Ur, near the delta of the Tigris and Euphrates rivers, in the latter part of the 2nd millennium B.C., travelling the Fertile Crescent to the area known as Canaan, being one of the first to recognize and worship Yahweh, to be later known as God, or by the Islamic name of Allah.

EARLY CULTURES IN THE AMERICAS

Much archaeology remains yet to be performed in the Americas, but several marks of South and Central America culture have been identified. Study of these cultures may lead to earlier cultures dating back to early fossils of man found in the Americas.

At the Americas, in Central Mexico, Central America, and along the west coast of South America, separate civilizations came to exist in the 2nd millennium B.C., and even before to about 20,000 B.C. Archeological records show the Olmec culture of the 2nd millennium

B.C., the state of Monte Alban in Central Mexico, and the great city of Teotihuacan, which dominated the Valley of Mexico near the present site of Mexico City. This was followed by the later Toltec culture of the 9th century A.D. and the Aztecs coming in the 14th century A.D., There are undocumented cultures between 20,000 and 4,000 B.C. to be yet discovered.

In the northwest coastal region, and south in the high elevations of the mountain areas of South America, an Andean civilization took root. This was the Andean equivalent to the Mesoamerican Olmec culture. The Andean civilization later incorporated the Maya of today. Extensive trade and communication took place between these cultures. Regarding soul, there was early evidence in the Americas of man's belief in a superior force or being, through gods, within or affecting his physical body and environment. Details of the cultures prior to the early Olmec culture remain for archaeology study.

SOUL

I have defined 50 or so qualities that describe elements and exhibitions of soul within man. These are different than those acts necessary only to sustain physical life, such as energy intake, digestion, excrement, motivation, limb movement, sensory acts, and the like. These qualities are demonstrations of emotions, determination, goal setting, acts, thoughts, viewpoints, beliefs, social actions, and recognition of a supreme power. These qualities are presented elsewhere in this book.

The question arises of when these elements of soul became a part of the mind in man. Conventional Biblical writings are not specific as to time, period, or era for creation. The Irish Anglican Archbishop of Armaghwhich, Reverend James Ussher, ventured to calculate, based on generations and genealogy cited in the Bible, deduced and proclaimed in Year 1640 that the Biblical Creation and Garden of Eden occurred on October 23. 4004 B.C.

First evidence of the presence of qualities of soul may well be depictions on walls of caves inhabited by Homo Neanderthal man or Homo sapiens man at 75.000 to 30,000 B.C. in France and elsewhere. New species (or sub-species of Homo erectus?) of man came upon the earth at about 120,000 years ago as Homo sapiens, and about 250,000 years ago for Homo Neanderthal, preceded by Homo erectus man at about 1.8 million to 600,000 years ago. From this it would appear that a new species of man, incorporating at least some advanced qualities of soul, was present on earth starting perhaps between 250,000 and 120,000 years ago, and perhaps 600,000 years ago when considering the Homo erectus species, a radical contrast from the Biblically-derived 4004 B.C.

SOUL AS A RELIGIOUS VALUE

No doubt changes to foster soul in man were not instantaneous, but evolved in presence and degree over tens and thousands of years of sub-specie development, determined by presence of genetic variations over many successive changes. Acts of reincarnation may have first taken place soon after.

It is the author's belief that soul is not an entity sent by God to necessarily be a religious article. Soul is a part of man's existence. The presence of soul precedes, by far, the origin of any religions. It might be said that religions more or less "adopted' the existence of soul and Afterlife as their own, to be part of their religious dogma. The soul creation in man is related to functions of RNA and DNA in the genome. Soul is an integral part of man, to be as much a part of the earthly make-up of man as our DNA, RNA, facial features, or the heart of our body. It is a sub-set of our brain, our mind, and has non-physical and unusual powers. In a religious sense, the existing soul is to be recognized, respected, and nourished.

The soul has a connection with the physical brain of man and the entities of mind of that person for an Afterlife—a continuation after life, and secular death on earth. This variation in the genome has been discovered and referred to as the "God Gene", which can go far in explaining the "mechanics" of soul incarnation. Some believe a God gene was created, as shown in the recorded growth of the genome, just as DNA changed, specie by specie, in the mutation history of the genome.

This presence of soul may even have been the difference that accounted for an accepted competitive superiority of Homo sapiens man over Homo Neanderthal man, climaxing at about 20,000 B.C. when Homo Neanderthal man became extinct and Modern Man (Homo sapiens), alone on Earth, later blossomed in culture out of the Stone, Iron, and Copper Ages from Paleolithic and Neolithic man, to an agrarian culture between 10,000 and 8,000 B.C.

Agrarian culture developed generally about the same approximate time in Egypt, Mesopotamia, Indus Valley, Southeast Asia, China, and Europe, although at various periods. Native American Indians are held to have arrived in North America from Asia with withdrawal of North American glaciers, at about 15,000 B.C.

Olmec and Maya civilizations of Central and South America date to the 2nd millennium B.C., or possibly to 30,000 or 20,000 years ago, originating perhaps from migrating civilizations of Southeast Asia or the Orient.

AGRARIAN AND SOCIAL CULTURES

Complexities of language, technology, and various cultures developed from 8,000 to 4,000 B.C. following a time of change from a Neolithic hunter-gatherer culture to an Agrarian culture leading to expansion of farming, animal culture, city-dwelling, warfare, and society in all aspects of communal living.

Ancient reportable history will start in Mesopotamia with Akkad culture in Babylonia, and in the Egyptian upper Nile River valley at time of the Old Kingdom. A multitude of kingdoms followed, to rise and fall for a variety of reasons, in Mesopotamia, Egypt, the Grecian peninsula, islands of the eastern Mediterranean Sea, India, China, Southeast Asia, Africa, Europe, and other countries as civilization spread, many with their own particular cultures and religions.

With world glacial periods largely ended in the temperate latitudes of earth, growth of civilization, population, communication techniques, and technology in life improvement would experience great advancement in the expanding world. This occurred without hindrance of extreme temperatures of excessive cold or excessive heat on earth's surface. Many of civilization's advances occurred first in temperate climates.

STAR FORMATION

In the early to mid 1940's of the previous century I was introduced to the fascinating subjects of Physics and Chemistry in high school classes. At that time we understood that the atom was the very smallest of elementary particles existing in matter. Later, in the 1960's, there was revision in scientific knowledge and the Quark was discovered, to be smaller and part of the atom. Quarks constituted part of the nucleus of atoms.

Quarks are elementary particles and a fundamental constituent of matter. Quarks combine to form composite articles (called Hadrons), the most stable of which contained protons and neutrons, the components of atomic nuclei. Due to a phenomenon known as color confinement, quarks have never been found in isolation. They can be found only with other hadrons.

There are six types of quarks, known as flavors: Up, Down, Charm, Strange, Top, and Bottom. Up and Down quarks have the lowest masses of all quarks. Through a process of particle decay there was transformation over time from higher mass to a lower mass state. Up and Down quarks are generally stable and most common in the universe, whereas Charm, Strange, Top, and Bottom quarks can only be produced in high energy collisions, such as those involving cosmic rays, and in particle accelerators.

The quark model was proposed in 1964. Further discussion of quarks becomes quite technical, leading to topics of electric charge, color change, spin, mass, interactions of fundamental forces, electromagnetism, gravitation, strong interactions, and weak interactions.

For every quark flavor there is a corresponding type of antiparticle, known as an antiquark. Complexities of quarks also related to Pauli's exclusion principle, bosons, leptons, hadrons, baryons, valence quarks, gluons, and antiquarks, the building blocks of the atomic nucleus, all quark properties which are more technical to discuss than is intended in this outline. Further discussion can be found at Google or from modern textbooks of Physics.

As the extreme temperature existing in the Big Bang cooled in the cosmos, quarks formed atoms in huge clouds of hydrogen and helium. These two elements were components of the energy that radiated outward into the cosmos.

Some of these concentrations of hydrogen, and to lesser extent helium, became attracted to locations in the cosmos of lower density. Also, clouds of hydrogen and helium atoms were herded together, squashing each cloud until its center began to heat up. In their gravitational attraction, hydrogen and helium atoms of increasingly higher temperature formed the first stars.

At about 10 million degrees Celsius, hydrogen atoms fused together, similar as they do in a hydrogen bomb, releasing huge amounts of energy, which stopped the clouds from collapsing any further. Attraction and temperature build-up continued, the process chemically changing atomic composition of hydrogen and helium to create more complex elements of carbon, iron (ferrous), gold, sulfur, oxygen, and forces producing gamma rays, all still the properties of stars.

As a star forms, it attracts a swirling disk of gas and dust. Most of that material collapses into the star. Some remains in orbit, where it clumps to form planets. And some shoots like a jet from the disc's center at velocities up to 30 times the speed of sound, triggering supersonic shock waves.

Stars gathered in huge societies called Galaxies, which gathered into even larger structures that are the largest things in the universe—Galaxy clusters. Stars provide the energy that keeps us living and the chemicals from which we are made. Gravity magnified these differences, splitting vast clouds of matter into billions of smaller clouds. As each cloud contracted, the pressure and temperature at its center rose, and atoms collided with increasing frequency and violence. Eventually, in a sudden phase change, the violence of these collisions overcame the electric charges among protons. Hydrogen nuclei fused to form helium nuclei, and the first stars were formed.

From then on, the star's stability would depend on a constant interchange between the heat at its center, which prevented further contraction, and the force of gravity that pressed it together. Fusion at the center explains why stars emit energy. Fusion can continue within a star until it has used up its store of hydrogen, which may take millions or billions of years.

Gravity herded galaxies into clusters, and herded clusters into even larger structures of super clusters. However, at great distances between millions of galaxies, gravity is too weak to overcome the expansion of the universe. Observational astronomy can identify regions of our own Milky Way Galaxy where huge clouds of matter appear to be collapsing and forming new stars.

But there was no emission of light from the beginning of star formation for a "Dark Period" of 200 to 400 million years after the Big Bang. Increasing gravitation and temperature eventually caused explosion of the star mass into a supernovae out of the star substance. But gravitation returned the star's mass to the star core, again increasing mass, temperature, and subsequent explosion and attraction back to the original star. At a time of 200 to 400 million years, hot stars emitted photons, producing first a glow, and then a brilliant illumination.

Spectroscopes can identify the relative amounts of different chemical elements within a star by its "absorption lines" shown in the star's spectrum. Spectroscopic studies show that stars consist almost entirely of hydrogen and helium, with a star's brightness, considering its distance from earth, tells of its mass. Temperature at the surface of stars can be estimated by its color. The key determinant of a star's life is the size of the initial cloud of matter (derived from energy) from which it is formed.

Why are stars so important in the modern creation story? First, stars create the preconditions for new forms of complexity by pumping energy out into the cold cosmic space, creating powerful energy flows. Most of the chemical elements from which we are personally composed were formed in stars, exploded, and transmitted to Earth as minerals and elementary particles. Stars laid the foundation for the now chemical level of complexity. Stars also represent the first large, complex objects created by our universe.

Most elements were not formed in the Big Bang. By the time protons and neutrons had been created, within seconds of the Big Bang, temperatures were still too low to forge elements heavier than hydrogen and helium. Temperatures high enough to forge heavier elements would not be re-created until the appearance of stars. The death of a star occurs when it runs out of hydrogen.

All stars are in motion. Stars are not stationary in the cosmos. They appear stationary because stars, as we are in our sun orbit, are all in generally respective motion and in relative positions within the rotating galaxy, or within other space outside our Milky Way Galaxy. (Einstein's Relativity Principle).

The stars (including our Sun) travel in the cosmos alternately through areas of extreme cold (zero degrees centigrade) and warmer areas (even a few degrees, e.g. 5 to 10 degrees warmer than average temperature of the cosmos would affect average temperature of the star (Sun) and planets of the Solar System, including Earth).

The orbit around the created force within the Milky Way Galaxy is periodic, and repeating in a period of about 80,000 years, the approximate period of glacial cycles of Earth.

Although the Milky Way Galaxy is also spinning in the cosmos, its repeating spin period is great enough in an interglacial period, to evidence a period comparable to the cycle of glacial action of earth. The period of entire galaxy revolution is estimated at 15 to 50 million years, whereas repeating periods of the sun within its travel through the galaxy, is estimated at approximately 80,000 years.

Within the period of orbit of the sun, the solar system, and earth traveling within the cosmos through "warm" and "cool" positions in the cosmos, causes global warming, global freezing, and intermediate periods. Intermediate periods are connective time spans when most changes take place in hominid and hominoid species, and in technological development.

Contrary to modern belief, we indicate that cause of modern global warming over a long term period, is considered fallacious as reflected in data trends of average temperatures over the past several thousands of years, as shown largely by probes down into ancient ice, review of ancient rock formations, and by other investigative means.

Earth has transitioned from a recent coldest era of the Wisconsin Glacial period, about 20,000 B.C., decreased from temperature maximums of about 15,000 B.C., and has experienced since that time a constantly falling average Earth temperature, even through our present cultural era of 10,000 years. Projection of the trend indicates record low average earth temperatures will decrease about 20,000 A.D., bringing an interglacial period to last into a subsequent future warming period. There is little doubt that glaciers are melting, which will cause rising of present ocean shores around the world. But the cause presented to the world population is fallacious. Man-made causes blamed for Earth warming are relatively inconsequential.

World temperatures are dropping because of an eon-old cycle whereby Earth, as well as our sun and planets of our solar system migrate to locations in the cosmos of colder temperatures, approaching zero degrees Centigrade, consequently building glaciers. Earth, in the future, will also emerge in cosmic travel out of extreme cold location of the cosmos as it has in the past. This cycle will cause centuries of glacial building and glacial withdrawal periods. There seems to be no remedy in sight to change or temper this future for cyclical geothermal existence.

Each of many cycles of Earth temperature and change of mass from supernovae explosions had constructed new atomic formation of elements, with dispersal into the cosmos of these elements at each supernovae explosion.

OUR SUN

One of these stars, relatively minor in size as stars go, was our own Sun. Our sun, as did other stars, must have exploded in supernovae, perhaps several times, and matter drawn back to the original star (sun) by its own gravity. Components of the exploding sun would have formed the cores of orbiting planets, later to increase in size by their own gravitational attraction of space debris, developing different orbits, different orbital periods around our sun, and of molten temperatures. From this creation, planets cooled in the cosmos during a time billions of years later, allowing the first formation of life on Earth, and perhaps on other planets, since supposedly lifeless.

Our Sun was created about 4.55 billion years ago. The Sun's temperature is 1.8 million degrees Fahrenheit and is only mediocre in the "hotness" of stars. The Sun is only 8.3 minutes away in terms of time for its solar energy to reach the Earth. The Sun is a spinning ball of gas, large enough to obtain 1.3 million Earths. It converts 6.55 million tons of hydrogen into helium every second, at an internal temperature of 28 million degrees Fahrenheit. The fusion is similar to that occurring in an exploding hydrogen bomb, but creates energy that reaches us in the form of sunlight. The core and inner layers are so dense that it may take a million years for a tongue of gas to fight even two-thirds of the way to the surface.

Above the surface of the sun is a thin layer of solar gases thick enough to extend far into space. A continuous solar wind blows through the entire solar system.

The Sun rotates at different speeds, about once every 24 days at its equator and, as a gas, more slowly, about every 30 days at its poles. This difference in velocity shears the gas and tangles its electrical currents, fueling the Sun's magnetic fields.

The Sun's field is full of curves and kinks, and about every 11 years the magnetic field flips, with the North Pole becoming south, and 11 years later it again flips with the same result. Alpha Centauri is the brightest star in the southern constellation of Centaurus. Alpha Centauri is actually a star system of three stars.

The overall magnetic field of the Sun has a direction, just as Earth's north and south poles attract our compasses. It is a dynamic cycle, but it is at the heart of most efforts to understand how the Sun behaves. During these flips the Sun's magnetic field rises up and pokes through the visible surface to create sunspots. These are dark patches of cooler gas that act as barriers, preventing some of the Sun's energy from escaping into space.

When a coronal mass ejection of the Sun reaches earth, solar particles stream along magnetic field lines, energizing gases in the atmosphere and shine as Northern (or Southern) lights.

Oddly enough, the stars we presently see may not even be in existence at this moment. Stars are so far away from Earth that their light takes millions of years to reach us. When we look at the stars we are viewing into the past. We observe the way the star was millions of years ago, not the way it looks today. Our nearest star system is Alpha Centauri which is 4.4 light years from Earth.

Alpha Centauri is the brightest star in the southern constellation of Centaurus.

The Andromeda Galaxy is 2.2 million light years away and is the greatest distance of any object viewed by the naked eye. When galaxies collide there is rebirth of a new galaxy—a rebirth of another universe.

A quasar is the brightest, most distant, most powerful object in the universe. It is the origin of a Black Hole that tears apart a surrounding cosmos.

Within our sun, the sun releases energy around the sun spots. At the dark center (the umbra) the magnetic field is so strong that it blocks the solar gases that typically bubble to the surface, radiating solar energy into the solar system. Weaker magnetic forces pull plasma outward at up to 60,000 miles per hour, again radiating solar energy into the solar system. The sun is in effect, an internal nuclear reactor, fusing hydrogen into helium.

Exploding stars, or supernovas, can be generated by various events. A "thermonuclear" supernova is created when a white dwarf star reaches a certain mass and begins to fuse carbon. A "core-collapse" supernova occurs when a large star is overcome by its own gravity and collapses in on itself, and producing an explosion into the cosmos, releasing huge amounts of energy.

Our earthly civilization will "freeze out" within approximately the next 10,000 to 20,000 years, but to be "re-born" into some future form of man and culture thereafter as the solar system and earth again become "warm".

In this presentation we note that there is only little conflict with Biblical accounts for creation of stars, and the reasoned theories of science. In Genesis 1:16 the Bible states summarily: ". . . he made the stars also." and ". . . God set them in the firmament of the heaven to give light upon the earth . . .", and Deuteronomy 10:22 states, ". . . has made you as the stars of heaven for multitude". Science shows and details that man is composed from minerals and elements from the created stars.

Ancient viewers of the stars imagined lines between stars and galaxies that resembled various earth objects. Astrological symbols were imagined that proceeded through the sky, each prominent in its own season as earth orbited the sun. The subject of Astrology has ancient origins and attempts to tie together the birth time of persons to their astrological sign that was prominent at time of birth, in guiding the person's lives and fortunes.

There may be some correlation of birth date, astrological presence, date of reincarnation of soul to the body of the new-born child, and integrated characteristics in mind, character, and life passage of that person. Since ancient times, persons associated with astrology have had deep beliefs and convictions in this correlation and prediction. Fortune telling involves study of astrology with birth time, as also the outline of creases in one's hands as tangible indication of one's fortunes on earth. The validity is strongly correlated, but has yet to be proved popular beyond coincidence.

FORMATION OF OUR SOLAR SYSTEM

From a hellish, molten mass, an accumulation cast out from our Sun and added to by extreme amounts of matter existing and circulating in the Solar System and cosmos, Earth, other planets, and solar bodies gradually cooled, forming concentric metallic and basalt rock cores, high temperatures, sulfur-rich atmosphere, and a quasi-plastic viscosity. Earth and other planetoids continued to grow in dimensions as cosmic particles and bodies collided with Earth. It is believed our sun formed about 4.55 billion years ago.

After the Big Bang energy expanded into the cosmos to form stars, galaxies, and galaxy clusters, there occurred later a cosmic process today called a Novae or Supernovae. There could have been possible later process of successive like explosions, followed by return by gravitation of matter to the star, with following addition of not only the original hydrogen, helium, and also resulting new successive elements.

Over a period of 200 to 400 million years, before stars, galaxies, and galaxy clusters heated to light up the cosmos, there was an absence period of visible light rays, but there was prominence of radioactive energy. The "re-created star" would chemically form new minerals (recently identified by spectroscopic studies), with loose matter in space from stars to be attracted and integrated into planets, such as Earth.

The energy had formed matter in stars, with perhaps the excess of energy forming today's Dark Energy and Dark Matter of our cosmos. The energy, thought making up Dark Energy and Dark Matter, preceded the star formation process.

The Dark Matter could be the source of gravitation by the stars, galaxies, and galaxy clusters, that cause our cosmos to continue expansion, and at an increasing velocity, as galaxies race outward to a source of gravitation, such as the observed "great wall of galaxies" near the extreme limits of the cosmos. Dark Energy may continue to form Dark Matter for new stars and new galaxies in the cosmos.

First stars may have been radically different from what has been previously thought. It is proposed in scientific study that the first stars were "Dark Stars", giant objects perhaps powered by Dark Matter annihilation rather than by the standard fusion that takes place in the sun. After the Big Bang, the universe was a sea of primordial, smoothly distributed particles. Roughly 15% of the mass is ordinary atomic matter, consisting of quarks bound together as neutrons and protons, responsible for the matter we encounter in our daily life.

Eighty five percent of the mass is something different and exotic known as dark matter. As time went on, this dark matter produced a web of enormous filament structures, whose nodes are the locations of galaxies and clusters of galaxies. The large, over-dense, regions of dark matter are known as "halos" of dark matter. Inside large halos of dark matter the first stars appeared and looked very different from today's stars: they were not luminous, made only of hydrogen and helium, as the other, more complex elements were formed by later generations of stars. (Source: Michigan Physics Annual Review 2008)

The pure energy had formed matter in stars (Einstein's transference from energy to matter) with perhaps the excess of energy forming today's Dark Energy and Dark Matter of our cosmos.

ELEMENTS OF THE SOLAR SYSTEM

The Sun: The only star in our solar system

The rocky planets: Mercury, Venus, Earth, and Mars

Composed of rock, they are close to the sun and have few or no moons. Rocky planets formed near the sun where it was too hot for ices and gases to condense. Composed of igneous rock, they are close to the sun and have various moons, some moons originated by attraction of large bodies in space and induced into planetary orbits.

The giant planets:

Jupiter, Saturn, Uranus, and Neptune are orbiting far from the sun. These large planets have gaseous atmospheres, moons, and Saturn having planetesimals ringing the planet. Jupiter and the other giant planets accreted in the region beyond the "frost line", where gas, ice, and rock were not available. These planets have gaseous atmospheres, rings, and moons. Being outside of the near attraction of the sun, they retained their nature as a gaseous cloud in their atmosphere, although solidification of matter, rotation, and orbit of the sun attracted matter in space far from the sun, some with moons formed by gravitational attraction. It is believed a moon of the planet Saturn shattered, with its components reduced in an orbital path around the planet forming a ring of matter in orbit of the planet.

The asteroid belt:

This is a band of small, mostly rocky bodies between Mars and Jupiter. Asteroids populate a zone where planet formation was disrupted by Jupiter's gravitational field. An Asteroid Belt, orbiting the sun between the orbits of Mars and Jupiter, consists largely of loose matter in space, or perhaps the broken remains of an early unknown planet. Rocky planets formed near the Sun, where it was too hot for ices and gases to condense. Asteroids populate a zone where planet formation was disrupted by Jupiter's gravitational field, prohibiting formation of additional planets.

The Kuiper belt:

The region beyond Neptune, are now known to be the reservoir of the short-period comets containing mostly icy bodies (including the defamed Pluto). The icy objects in the Kuiper belt and Oort cloud coalesced to the super-cold, low density conditions beyond Neptune.

The comets:

The comets in orbit around the Sun, loose matter from an exploded star, emit gaseous matter that is directed outward by the solar winds of the sun, the long form of the gasses sometimes taken erroneously as a trail of the comet.

The Oort cloud:

These are a reservoir of the sun's long-period comets. They are located almost a quarter of the way to the nearest star (Alpha Centauri). A recently discovered object, Eris, larger than Pluto, is in orbit twice as far from the sun. The Oort cloud extends 500 times farther than Eris. The icy objects in the Kuiper Belt, and the Oort cloud, never coalesced in the super cold, low density conditions beyond Neptune.

Moons:

One collision of a smaller planet-size cosmic body is believed to have struck the plastic-consistency of Earth about 4 billion years ago at a glancing angle, forcing a 2,000 mile-wide diameter portion of earth into orbit of earth, which became our Moon. Computer simulations show that our moon probably formed when a Mars-size body smashed into the molten, forming earth. The Moon, in turn cooling, also attracted cosmic bodies by gravitation, forming craters and geologic features, some seen even today through telescopes. Earth's Moon is absent of weathering action on the moon and is covered by eons of cosmic dust.

The absence of this large portion of earth likely caused the latter start in break-up and movement of geologic plates at the mantel of earth. Until this time, a single land mass covered a major portion of earth, today known first as Gondwana, and later Pangaea. The later geologic plates of mantle and rock layers shifted on the core of earth, the collision and movement creating present-day geologic and mountain-range formations.

Moons are a rich world unto themselves. Among the 7 large and 160 small moons, our moon is unusual in that it is one-quarter of the size of its planet.

Craters are found throughout the Solar System. Craters on the Moon can be dated to reveal a period of bombardment following formation of the Moon. On Earth, one recently identified crater is thought to be the smoking gun in the demise of the dinosaurs, and another on an Arizona desert is the remnant of the collision by an ancient space traveller. Possibly another is located in a wooded area (except in its depression) near our vacation cabin in Northen

Michigan. The depression is about 400 feet across and about 50 feet deep, its history is unknown, even to long-time residents.

Weathering:
Weather on other planets can be markedly more severe than weather on Earth. Jupiter has a gigantic, centuries-old, storm that could swallow several Earths, and the air temperature on Venus is more than 460 degrees Centigrade. Remains of Craters could well have been weathered to extinction.

Magnetic Field of the Sun:
An explanation by astronomers is how a star like the Sun produces a magnetic field. A theory exists that energy, in the form of electrical voltage, was the essence of the Big Bang, and that energy travelled in the cosmos to random, localized areas of high and low density. Hydrogen and nitrogen, eventually caused illumination of the sun and other stars. Some electrical voltage from the cosmic energy, having polarity from positive and negative orientations, with increasing pressure within the star Sun from gravitation, heat increase, changing locus within the Sun, and its rotation producing a magnetic field.

(Source: The Teaching Company, "New Frontiers: Modern Perspective On Our Solar System")

COULD IT BE?

It is generally accepted by cosmologists that, in the era of formation of the earth, there were millions of planetesimals that crashed into the matter comprising the forming Earth, growing Earth bit by bit, to its present size and composition of its many minerals that we have and mine today. This begs the question of, where these planetesimals of the various compositions came from?

Could it be one or both of two possibilities? First, were the various planetisimals the remains of a previous, unformed planet circling the Sun, that had exploded or disintegrated for some reason, possibly for lack of adhesion or inadequate gravity in the formative period of that ancient planet? The existence of this source, of many minerals and matter that would agglomerate to form Earth, would be a strong possibility. This would also open a new phase in ancient cosmology for the early formation of Earth, Venus, or Mars, and that gaseous planets are lesser proportion of iron, but greater proportion of other elements.

A second possibility could be that our Sun, like other stars, had transitioned to a relatively small supernovae, earlier than its recognized age of 4.55 billion years. As supernova, our Sun would have heated immeasurably, with increased internal pressures, and exploded, throwing its component matter out into its nearby cosmos, but later re-forming, say, into Sun II, but also many bits of Sun I would remain in gravitational orbit as planetesimals, which formed Earth and possibly other planets.

As a continuum, it may be determined that the planet Mercury is of an inordinately high percentage of iron, of a greater proportion than that of the rock-based composition of Earth, Venus, or Mars, and that the gaseous planets are lesser proportions of iron, but greater proportion of other, chemicals, and matter.

Such information and analysis could provide strong reinforcement to this theory and argument. The question regarding origin of the planetesimals might therefore be answered. The theory may have implications for the origin and formation of asteroids in solar orbit, and also the Oort Cloud, its matter gathered in orbit outside our Solar System.

THIRTEEN BILLION YEARS OF HADEAN EARTH

During Earth's Hellish Period, it glowed like a hot ember in a fireplace. It would have been a flaming pyre except for its lack of oxygen. The gases were inert, without active chemical or other properties.

After billions of years Earth cooled to the point of hosting elementary life, the source of life still unknown, but possibly from chemical combinations oozing up from subterranean crevasses in Earth's undersea surfaces, joining with oxygen molecules from waters of the sea, or possibly from chemical processes, instigated by electrical atmospheric lightening that turned inorganic metal to organic and living bacteria and green algae, or possibly from "seeding" by objects from outer space carrying life forms, all theories advanced at one time or another by scientists.

There was a long interim period of development between the 13.7 billion years ago time of the Big Bang and the 6 million years ago time when first hominids roamed the earth in Africa.

Much of that interim of 7.7 billion years ago is difficult for scientists to explain, but logic, reason, and study of Earth's features will provide future scientists with some provable answers. Other events surrounding the period are still unanswered, but may be described in science, based on evidence secured in rocks of the earth, and in exploration by drilling for rock samples deep into Earth's concentric surface, mantle, and core.

Presence of land or initial rock surface at various parts of the globe, have been traced by earth scientists. Earth's oldest rocks have been found in Australia and Iceland, originally located in polar areas. By borings into the earth, weather, climate, composition and atmospheric conditions can be revealed, and suppositions proposed of conditions in earth's existence millions and billions of years ago.

Early upon the earth there were presence of two substances, rock and ocean. Rock originated largely from stars and from outer space as cosmic travellers. Oceans originated from a tremendous sink of water resulting from chemical joining of hydrogen and oxygen.

First land of one unified body was Pangaea at 4.6 billion years B.C., followed later by further break-up and earth re-formations to what has been named Columbia (or Nuna) at 2.0 to 1.8 billion years ago, Rodina formed at 1.1 billion years ago to 750 million years ago, Panagea at 600 to 540 million years ago, Laurentia, Baltica, and later Gondwanaland that contained much of what today we recognize as South America, Africa, Madagascar, Antarctica, and India at about 570 to 510 million years ago, before geologic plate movements provided momentous relocation generally to their present locations.

In time, as the underlayment of earth cracked, like dry desert mud does today, several geologic plates formed throughout the earth. Movement of each geologic plate affected the features upon the surface of earth above the plate, many times subsiding or overriding a neighbor geographic plate. Geographic plates can virtually carry a whole continent from one earth location to another, such as the land of India was "floated" from near present day Antarctica to its present-day location, and forming the Himalayan Mountains upon its impact with the continental geographic plate of Asia. Like events occurred numerous times upon earth, to re-form the earth, and force upheaval to form mountain chains and various geographic formations.

Like a jig-saw puzzle, land masses of the earth can be projected back in time, considering comparison of similar geography, geology, and then presence of ancient plant or animal types, with suppositions of the climate, temperature, and even elementary and basic life forms.

From shortly after expansion from the Big Bang, after 13.7 billion years ago, to a period 600 million years ago, the cosmos and earth were in a hellistic phase of chaos, with development and growth of cosmic bodies, collision of earth with those cosmic bodies, star formations, supernovae explosions, galaxy formation, and cosmos expansion. The surface of Earth changed drastically from an extremely hot, molten state to a frozen "snowball Earth" covered with thick ice, in some places miles deep.

The cosmos, though expansive, through vectors of energy and matter excreted in all directions by the Big Bang, was initially relatively compact, much more so than exists today. Many stars existed, some merging with others, and groups of stars attracted into galaxies, and galaxy clusters by both centrifugal (gravity) forces and centripetal (outward) forces.

Among the resulting actions and forms was formation of our own Sun, growing in size possibly from merger with other smaller stars and with Vacuum Energy. Gases and cosmic rays emanated from the Sun, and from elsewhere in the cosmos into localized areas which formed the nucleus of planets that orbited the Sun.

Immediately preceding the moment of the Big Bang, space essentially did not exist. All basic elements of the cosmos existed in extremely close proximity to each other. Time and space were inseparable, each to acquire their respective qualities in expansion of the cosmos. Energy, and its component of matter, expanded and time was defined in terms of distance between elements of matter. (Albert Einstein proved in his Theory of Relativity that as distance decreases time is also reduced, and that as distance increases time also increases), cite the Doppler Affect in sound of moving objects. Time interval from Sun to Earth, for example, is defined as distance apart (average of about 93 million miles in its elliptical orbit).

It is believed that space was not always unlimited, and that dimensions of space are created as distant space objects require. The author suspects that an unknown element, force, or substance is present, but not yet discovered, that affects and determines the dimensions of the cosmos. Space, at time of the Big Bang, was relatively compact, and grew exponentially in all dimensions as the expanding forces required. Recently the Hubbell telescope has detected a Great Wall of galaxy clusters located at the supposed extent of space and the cosmos.

Defining this space would be like describing an ant walking along the inside of a minimally inflated balloon. As the balloon expanded, at that moment the ant would understand that that was the limit of space. However, as more need is applied (e.g. more matter of air is forced into the balloon, symbolic of galaxies moving outward) the limits of the balloon increases from minimal to expansive. This cycle would repeat several times as needed, without limit. There would be an unknown factor, yet to be discovered, that causes expansion of the cosmos. Such is the way that the cosmos will increase as need requires (as galaxies travel in the cosmos), growing from minimum volume at time of the Big Bang to expansive later and at present.

We know from research in geology that Earth, as a globe, experienced many changes in temperature and, moisture content, including a period of "Snowball Earth" during which the entire earth was encrusted with ice and snow. Science knows that there were interglacial periods providing medium atmospheric temperatures, weather, and plant growing conditions. Temperature variations are believed attributable to Earth's position in the cosmos and Solar System, from cosmic cold to Hadean hot conditions, and recovery for much the same, but opposite orientation. In all likelihood, the Earth was "like a true Hell", a Hadean earth. There was minimum-to-no existence of sea life in this era.

We believe other planets of the Solar System also underwent similar conditions, culminating in the loss of atmosphere, by loss of adequate gravitation, such as on planet Mars, and development of gases in atmosphere on planet Venus, possibly changing from life-supporting to a present desolate and dry environment.

There are many questions that can be raised in science and our mortal minds can only offer conjecture of events and cause. Rock samples of a Hadean era are non-existent, except possibly in Iceland or Australia where oldest rocks on earth have been found. Regretfully, inquiry raises more questions than can be answered regarding the 13.7 billion year to 6 million year era. We are limited and can only "mind travel" through this era.

The earth globe advanced from a gaseous cloud, orbiting our Sun, to a hot, molten, and viscous grouping of matter, absorbing a bombardment of cosmic travellers of all shapes, size, and chemical composition. Gravity of Earth in its rotation caused not only attraction, but also chemical separation of elements, with the earth's core being composed basically of ferrous (iron) material, which still retains its molten state.

Elements and minerals of lesser density comprised concentric layers over the core, with additional concentric layers leading upward to a layer of Basalt rock, still miles below the present surface of earth, but observable in the form of lava rock from virtually millions of volcanic eruptions throughout the earth. Above the basalt rock earth comprised about 50 miles in depth of mud, sands, limestone, and detritus, today compressed and heated into various types of igneous, metamorphic, and sandstone rocks or limestone.

Rock was physically changed by weathering and chemical elements, to organic material to make up the sub-soil, mostly clay, covered with further organic material comprising a wide variety of soils of the 8,000 mile diameter of earth.

One must raise questions regarding the relationship of Black Holes, Big Bang, and the galaxies. It is believed that the core of the Milky Way Galaxy is contained in the center of our galaxy. (The Milky Way Galaxy is a Spiral Galaxy, the spiral shape with many "arms" requiring millions of years to form and rotate, beginning from an initial spheroidal galaxy.)

Observation through the Hubbell telescope indicates a brilliance of light intensity at the supposed galaxy center, locatable by looking in the direction of the star constellation Sagittarius. The center of the Milky Way Galaxy was supposedly the origin point of our Big Bang. Its location would seem that an analysis of vectors (lines of direction) leading from a Big Bang to various stars could reveal the (then) exact location of the Big Bang origin.

As stated, the Milky Way Galaxy shows evidence of a Black Hole and a Big Bang event following development of a Singularity of elemental particles. There are, in the cosmos, many galaxies, and it is believed that each separately contains at its center a Black Hole. This begs the question of "Were there more than one Big Bang events, one at each galaxy?" If so, "What would be the relationship among the various Big Bang events?". "Were they caused by perhaps a wave of intense energy, travelling through the cosmos that "ignited" the various Singularities, one after another, or even with some time delay, in a chain of events through the then infant cosmos 13.7 billion years ago. Or were the moments of the relative Big Bangs coincidental to all be at approximately the same time in the different galaxies?.

This also raises the question: After explosion and expansion from the Singularity, does a Black Hole continue to exist, or has all of its contents disseminated? Does a Black Hole "re-fill?

During the glacial thawing periods, the years leading to an interglacial period, the glaciers were melting. This is result of tremendous energy and warmth from the cosmos that is being released to the earth as our Solar System, including Earth, revolves around in the Milky Way Galaxy. As it revolves to a relatively warm location in the cosmos, temperatures on Earth rise, energy is released to the earth, and energy is absorbed by plants, animals, and man, possibly explaining why man at about 15,000 years ago, when glaciers were thawing, energy was being released from the large glacial energy "sink" to energize man in his culture.

An additional effect on earth, besides glacial melting, is reaction of the genome DNA to resulting energy bursts or gamma rays from our sun and reduction of the glacier ice sink. The energy bursts would not come all at one time or it would "fry" all of life, including future humanity. But instead, energy existed in the earth environment (glaciers) over an extended period causing change in passive qualities of our DNA, present but unreleased, to change qualities of man through mutations and to initiate soul that would become part of man through incarnation. Steps in development of man may, by analysis, be related in history to interglacial and glacial thaws

The author holds that soul existed prior to man's incarnation and the supply of energy to the earth promoted the union, a slow process radiating heat energy over an extended period. So too, energy (gamma rays) released to earth and man, resulted in eventual DNA change to accept the inert Soul into union with the animal body of man. This is believed to have happened at about 600,000 years ago, and may coincide with a time of glacial thaw.

TIME DIVISIONS

EONS
Hadean Eon	13.7 to 4.0 Billion years ago
Archeazoic Eon	4.0 to 2.5 " " "
Proterozoic Eon	2.5 to 1.0 " " "
Phanerozoic Eon	1.0 " " " to present

ERA
Paleozoic Era	600 to 250 million years ago
Mesozoic Era	250 to 50 " " "
Cenozoic Era	50 " " " to 250,000 years ago
Paleolithic Era	250,000 to 10,000 years ago
Agrarian Era	10,000 " " to present

EPOCH
Paleocene Epoch	65 to 55 Million years ago
Eocene Epoch	55 to 35 " " "
Oligocene Epoch	35 to 25 " " "
Miocene Epoch	25 to 5 " " "
Pliocene Epoch	5 to 2 " " "
Pleistocene Epoch	2 " " " to 20,000 years ago
Holocene Epoch	20,000 years ago to present

PERIODS
Pre-Cambrian Period	Hadean to 580 million years ago - 4 billion years long
Cambrian Period	580 to 490 million years ago - Major groups of animals;

					Cambrian explosion
Ordovician Period	490 to 430	"	"	"	- Life expands in diversity
Silurian Period	430 to 410	"	"	"	- Glacial melting; stabilization
Devonian Period	410 to 355	"	"	"	- Age of fish & sea plants
Carboniferous Period: (Mississippian & Pennsylvanian)	355 to 275	"	"	"	- Vegetation for coal
Permian Period	275 to 220	"	"	"	- Greatest mass extinction
Triassic Period	220 to 180	"	"	"	- Transistion—Pangae exists
Jurassic Period	180 to 135 million years ago				- Dinosaurs, fish
Cretaceous Period	135 to 70	"	"	"	- Dino extinction, mammals, birds
Early Tertiary: (Paleolithic, Eocene, Oligocene)	70 to 15	"	"	"	- Rocks of Bryce canyon
Late Tertiary: (Miocene, Pliocene)	15 to 1	"	"	"	- Major mountains uplifted, canyons
Pleistocene	1	"	"	"	to Present - Glaciations and thaws

We live today in the Phanerozoic Eon, Agrarian Era, Holocene Epoch, and Pleistocene Period.

HOW TO MAKE A PLANET

The first 500 million years after the planet Earth formed were still known as the Hadean Eon. The young Earth was not life-friendly. As it formed, largely by accretion and by gravity attraction of space matter, the earth was heated by constant, wrenching collisions, by radioactive material, developed from matter of various supernovas, and by internal pressure that increased as the earth grew by accretion. Eventually, the surface of Earth cooled, froze, and later, ice melted to allow a temperate climate.

Different elements and minerals sorted themselves within planet Earth. Other concentric layers formed around the central core up to the surface of Earth to differentiate into distinct layers. The heaviest elements, mostly iron and nickel, were drawn to the center, forming a metallic core, its rotation of a super-hot metal liquid inside the earth globe increasing its gravitation attraction force, and increasing the earth's magnetic field.

Lighter, rocky material formed semi-molten middle layers, and the subterranean mantle. The surface congealed to form a comparatively thin coating that we call the crust.

Gasses and water vapor, as well as magma, belched from innumerable volcanoes to form the first atmosphere, which was of high sulfur content.

Then, very slowly, the earth began to change. As the surface cooled, water vapor composed of hydrogen and oxygen (H_2O), rained down successively to form the first seas, a tremendous form of storage capacity, or Sink, of water on earth.

After the Pliocene Epoch, from about 4 billion years to 2.5 billion years ago, the number of collisions from asteroids and other space matter declined. Free oxygen began to appear in the atmosphere generated by living plants and organisms—a first hint of the transformation power of living organisms. Oxygen is an extremely reactive element that eagerly combines with other elements, an action we observe whenever we light a fire. The process to originate free oxygen was photosynthesis, produced by plant-like organisms.

The earliest eon of molten earth history is known as the Hadean. It lasted from about 13.7 to about 4.0 billion years ago. It was the first of four eons in the earth's history. (The others were the next earliest Archeazoic Eon, from 4.0 to 2.5 billion years ago, the Proterozoic Eon, of widespread glaciation and great mountains uplifted, from 2.5 billion to 550 million years ago, and the Phanerozoic, or eon of multi-cell organisms, from 550 million years ago.

The young earth heated, and surface melted, by different forces of violent collisions, by radioactive pressure, and by location within the Solar System and cosmos.

The earth also acquired a satellite, the Moon. The fact that the moon contains few metallic elements suggests that it was gouged out of the earth's upper layers by a violent collision with a Mars-size object, after the differentiation when most metals had sunk to the core. The point of collision is said to have been where the Atlantic or Pacific Oceans are now, followed by numerous geologic adjustments at earth's surface and mantle.

A most difficult technique is study of what is inside the earth. This is done today largely by seismology, common in the study of earthquakes. Using seismograph sensors placed at many different locations on earth, scientists can tell us much about what is inside the earth's interior, and activity movement, somewhat like spectroscopes tell us in the study of stars, whereas seismographs measure vibrations, but spectroscopes measure light.

The effects of geology, winds, water, and aging work hand-in-hand to make physical changes of earth's surfaces that interrelate, and how these different disciplines of study overlap each other.

Measurement of Cosmic Microwave Background Radiation (CMBR) will reveal what the cosmos looked like at 400,000 years after the Big Bang—after an interval of full darkness, except for a glow, until stars started to give off photons and light.

A useful timeline is presented (see another chapter) courtesy of The Teaching Company, Big History, 2008. These show the diorama of the Cosmos, Earth, multi-cellular organisms, mammalian evolution, human evolution, human history, and agrarian societies. Additional events not included in the timeline are presented here.

1. Some scientists have argued that:

    a) Originally life did not come to Earth from outer space, but the heavier elements necessary for life,(such as silicon, potassium, and iron) were previously sprayed into the cosmos by exploding stars and supernovae. Once here, the heavier elements joined with carbon and oxygen— and life on earth had originated, ultra basic, but here!
    b) Another theory is that, in early life on earth there were microbes and/or biomolecules. Their origin was from the sea floor, from hot vent ecosystems in the seafloor, from hydrothermal vent systems. Most organisms were microbes. Inorganic elements combined to form organic compounds.
    c) Radiation helps drive mutation. Mutations may be driven by Gamma Rays. Mass extinctions, from the increase and decrease within the Sun and the Solar System's plane, possibly cycling about every 62 million years (per an ancient Hindu religion), clears the way for evolutionary change. (This would require much change in DNA from mutations.) For the first several billion years of its 13.7 billion year history, earth was a non-inhabitable entity.
    d) A theoretical multiverse possibly fed our known cosmos, or an additional Singularity, and consequently additional multiverses were formed.
    e) With seeding of multiple Black Holes, and resulting Singularities, pocket universes and multiverses are a theoretical possibility. Any universe that produces more than one Singularity will create more "daughter universes", and its physics will be passed on to those daughter universes. Such daughter universes could influence formation of one or more multiverses. The Milky Way Galaxy is one of the largest in the cosmos.

f) The multiple Black Hole theory can be tested: Since universes that give multiple rise to the largest number of Singularities have the most offspring, our Milky Way universe should be optimal for making daughter universes.

2. Subsequent major changes in earth were:

a) Permian-Triassic period (the era of plants that would form future coal) had duration from 275 to 180 million years ago.
b) Cretaceous-Tertiary period (the era of tremendous geologic formations on earth), had duration from 135 to 20 million years ago.
c) The Cambrian Revolution of 600 million years ago, occurred when former creatures of the ocean level, and prior to rise of the Rocky Mountains in present-day British Columbia of Canada, were fossilized, buried, and rose with future uplifting of mountains, to expose sea life, providing a wealth of sea life fossils for study at what is known today as the Burgess Shale.
d) The Tethys Ocean stretched continuously from modern Gibraltar to what is now India at 5 million years ago until a sharp cooling of the climate at 3.4 mya.
e) Other momentous changes were ushered in by major changes in earth's surface movements.
f) It is generally accepted that at about 4 billion years after the Big Bang our Moon was created from a momentous collision of some large object in the cosmos colliding with earth, the impact causing a mass of earth's matter to exit and fly out, settling into an orbit around earth. (This is reasonable but it bothers my intellect to a certain extent. Other planets also have moons, and the composition of those moons differ in physical make-up, not necessarily similar to matter of their related planet. Were those moons created in like manner to Earth's moon? If so, why

do they not all mimic their source planet? Why is each moon different in chemical and geologic make-up?)

3   Petradoctryls, bird-like dinosaurs that could glide from high perches and are believed to be forerunners of today's birds, lived in the period 135 to 70 million years ago, during the Cretaceous period, and were forerunners of the extensive variety today of all species of birds emanating from changes and mutations in their genome.

4   The planet Jupiter is a failed star that never acquired the stage of ignition. It contains all elements of a sun except for ignition of nuclear reaction within its particles. If Jupiter had acquired this status and erupted as another sun, all of the Solar System, including Earth, would have been baked into eternity.

5   About 65 million years ago geology of rocks formations showed a layer of iridium found as evidence of an earth catastrophe, with extinction of large animals on earth at the time.

6   Measurement of Cosmic Microwave Background Radiation (CMBR) will reveal what our cosmos looked like at 400,000 years after the Big Bang, after an interval of full darkness until stars started to give off light in the form of protons.

7   Classification of Taxonomy Groups are described as follows:
    In 1758 the Swedish taxonomist, Carolus Linnaeus dubbed our species Homo sapiens Latin for "wise man".
    Super Kingdoms (Example: Prokaryotes and Eukaryotes
    Kingdoms (Example: Animal, Plant, Fungi)
    Phyla (Example: Chordates, Vertebrates from worms)
    Classes (Example: Mammals)
    Orders (Example: Primates)
    Species (Example: Human beings)

HERE ARE OTHER OBSERVATIONS MADE CONCERNING OUR PLANET AND LIFE

First, there was change in Earth's polarity. There was changing dynamics of an early spinning Earth with change in magnetic orientation evidenced in ancient rocks. Positive polarity has been found and determined to be where negative polarity would logically be found. This would indicate that, at some time in earth formation, the magnetic North Pole and South Pole were located in different positions on earth than is found today. Even in recorded history the exact location of Magnetic North varies, possibly due to shifting internal molten mass within Earth. It is possible that the change in polarity was caused by movement of the ultra-hot liquid Earth core, thereby shifting the location of Magnetic north and Magnetic South.

HAWAIIN ARCHIPELAGO

There is evidence that the Hawaiian Archipelago progressed roughly from west to east, at a first location at the present, easternmost, large island of Hawaii. The Hawaiian Archipelago has been forming over the past 40 million years as specific locations of in line weak spots in earth's crust passed over a subterranean "hot spot", allowing underwater volcanic action, resulting in formation of the chain of various islands of the archipelago.

As stated, the archipelago forms a line of islands that run generally west to east. But there is connection to the archipelago, running generally north to south from end of the Alaskan Aleutian Islands, where the Alaskan Trench meets the generally north to south Kuril Trench of Eastern Asia. This is a similar evidence of volcanic action, located in a chain, connecting to the Hawaiian Ridge. This is the Emperor Seamount, a chain of would-be islands, except that the volcanic mass generally failed to break the level of sea. The Emperor Seamount continues southward until it connects and mysteriously bends to the eastbound Hawaiian Archipelago. It is calculated that the change in direction occurred about 40 million years ago.

The North Magnetic Pole location at some time was in Central Canada, possibly near Hudson Bay or the American Great Lakes area, or possibly present-day Iceland. The South Pole would consequently be in the general area presently occupied by Australia, where interestingly, some of the oldest rocks on earth are found. Similar-age rocks have also been found in Iceland.

Initial formation of the Emperor Seamount is estimated to date from 80 million years or earlier. The ongoing subduction (or undercutting) of the Pacific Plate has plunged unknown numbers of earlier seamounts under the North American Plate, forever obscuring the true age of the Emperor Seamount/Hawaiian archipelago ridge.

The epoch of the apparent change in direction of the Emperor Seamount to that of the Hawaiian archipelago may somehow coincide with the estimated 4 billion years after Earth formation collision to form our moon, and this change in earth-magnetic change dynamics could also account for change in location of Magnetic North and magnetism of earth rocks.

This progression is also about the time when the Indus plate started moving from Antartica towards the Asian continent forming the Himalayan Mountains, perhaps at 4 billion years ago. The dynamics of the shifting mass would greatly alter the axis of rotation of the rotating Earth and could be a factor to affect the location of Magnetic North and Magnetic South.

Change in dynamics of the spinning earth would likely occur if substance of the moon formation was removed from earth, supposedly by collision of earth with a large cosmic body in a "glancing angle". (Oddly and unlikely, moons of other planets were possibly formed by similar cosmic collisions.)

WHERE ARE WE IN THE COSMOS?

A second observation begs the question, "Where are we?", and shows that the "Man In The Moon" travels faster through space than a Doctor Seuss "Cat In The Hat" on earth.

To explain, one of the strangest physical phenomena occurs to our view nightly. That is application of mathematics of the surface rotation of the 8,000 mile diameter of Earth and that of the 2,100 mile diameter of the Moon. We can calculate this relative velocity and will use the formulae: C = πd (Circumference = π of 3.1414 times the diameter of the circle). The greater orbit of the Moon, 240,000 miles from Earth (radius of a circle), being many times further than when measuring Earth's surface to its center locus, of 4,000 miles, (a much smaller radius of a circle).

Diameter of Earth at the Equator is 7,916 miles, but if measured through the poles the diameter is only 7,901 miles. Thus the earth is a tad wider (25 miles) than it is tall, giving it a slight bulge at the Equator. The diameter of the moon at 2,160 miles is about one-fourth the diameter of Earth, and 400 times smaller than diameter of the Sun. The Moon is 238,857 miles distant from Earth. (This would be the radius of the circle formed by the moon orbiting Earth).

This is further complicated by travel in space a) of the entire Solar System, b) rotation in space of one arm of the Milky Way Galaxy, c) rotation of the Milky Way Galaxy, and d) travel of the Milky Way Galaxy through space. We calculate that Dr. Seuss, pulling a cat from a hat on earth travels 1,047 miles in an hour, while an astronaut standing on the moon travels through space at 64,141 miles an hour.

If we were to view Earth from above the North Pole we would see that Earth rotates in a counter clockwise motion. (Sunrise occurs in New York while Los Angeles is still experiencing night.) This is, of course, due to the spinning of Earth while in its orbital rotation. This direction of spin in space was originally determined by chance and physical forces, at moment-one, when gases that would become Earth started to spin, creating gravity, as opposed to maintaining a static or unmoving orientation in the solar System.

If there were any atmosphere and any weather on the Moon, it would be near impossible to transport or walk upon the surface at such tremendous atmospheric velocity through space. Clouds, if formed, would blow by quickly within seconds or minutes. On Earth currently, weather in Chicago today is a clue to weather in Detroit tomorrow, or Minneapolis, Minnesota weather today is probable weather tomorrow for Traverse City, Michigan.

Some other planets of the Solar System rotate in an opposite direction where the sunrise occurs in the west.

Complicating the relationship of rotation is that, again viewing Earth from above the North Pole, the Moon rotates around Earth in a clockwise direction. The Moon rises in our east and sets in our west, like our sun rises in the east and sets in the west. While the Moon doesn't rotate at all (due to gravity exerting force on an unevenly distributed mass in the Moon) the same side always faces Earth, the combined set of Earth and Moon rotate counter-clockwise in space (looking at the Solar System from above).

The net effect is that the surface of the Moon travels through space at a velocity greater than does the surface of Earth. Otherwise the moon would retain one position above Earth as if it were to rotate at a comparable speed as Earth rotates. Instead, the moon rotates around Earth once every 27 days, 7 hours, and 43 minutes.

We must leave the specific calculation to the mathematicians and astronomers to determine our actual velocity through space as we stand on the surface of Earth or the Moon.

INORGANIC TO ORGANIC

A third observation is that it is possible to demonstrate that inorganic minerals and elements can be made to show a transition from lifeless forms of, say metal, to living, organic cells. This is done by subjecting the inorganic metal molecules to oxygen, or free oxygen, in an atmosphere where it is exposed to lifeless metal molecules. Oxygen is a very unstable chemical and will easily unite, in oxygenation, with other elements.

About 2.5 billion years ago free oxygen came to exist in the atmosphere from photosynthesis. For example, independent atoms of oxygen interacted with independent atoms of various mineral elements. Free oxygen originated from water. The result was formation of amino acids, nucleotides, and phospholipids, as well as rust and oxygenated metals, the chemicals needed to form cell membranes and the start of life—a scientific change from inorganic to organic molecules.

The place or location of this transition is thought to be, not on the hellish surface of Earth, but from deep sea volcanic vents. There much energy, as magma, poured out from the interior of earth. There was a rich mixture of organic chemicals, and there was also protection from ultraviolet radiation. It has been discovered that very rich colonies of chemical-eating bacteria existed around mid-ocean undersea vents. Oxygen could be available, via water, to sub-surface levels of an ocean, to mix with chemicals and minerals.

The formation of DNA chains or double-helix DNA clusters, or single chains of nucleotides, provided chemical evolution that would be continuous and reliable for reproduction in preserving precious variations, adaptive changes, or mutations.

Photosynthesis produces oxygen as a by-product. By 2.5 billion years ago free oxygen began building up in the atmosphere. For many first prokaryotes, oxygen was poisonous. As consequence, Eukaryote cells having cell walls thus appeared about 1 billion years ago through the merging with once—independent no-wall-cell species of prokaryotes.

Internal organelles, including mitochondria, which can extract energy from oxygen, and chloroplasts, which can extract energy from sunlight through photosynthesis, were encased to be within a eukaryote cell. The merging of these entities through "symbiosis" anticipates the later creation of multi-cellular organisms.

Most Eukaryotes are 10 to 1,000 times larger than prokaryotes. Some can be seen just with the naked eye. Unlike Prokaryote cells, Eukaryote cells flourished in an oxygen-rich atmosphere.

Eukaryotes have a fatty membrane, a surface that is semi-permeable. It allows chemicals to flow inward for nutrition, and it allows waste products to flow out as excretion.

In reproduction of eukaryote cells, two organisms trade some genetic material before reproduction and the offspring shares genetic material from both parents. The offspring will inherit a mixture of genetic material from the two parent individuals. This is important because it meant that the offspring of eukaryote cell parents were no longer simply clones of their parents (Example: Yeast). Sexual reproduction introduces greater variation between individuals. The offspring are never exactly the same as their parents. As natural selection selects from such variations, the result of sexual reproduction is to speed up the rate of evolution, and is logically why evolution seems to have accelerated in the past million years. Reason for why substantial increase in world population over the past 2,000 and 1,000 years is mystifying but is perhaps related to a geometric increase of bi-sexual genes.

VERTEBRATES AND HOMINIDS

Vertebrates developed for need of strong skeletons, tough skin, apparatus to breath oxygen for living out of water, and other needs.

About 65 million years ago, at end of the Cretaceous Period, a massive asteroid impacted Earth, dinosaurs became extinct, and species of mammals increased. As part of this "mammalian radiation" There appeared a new order of mammals, the primates. Primates were tree-dwelling mammals, with eyes facing forward and stereoscopic vision, hands with opposing thumbs were available to grasp, and they had larger brains. Primates included all monkeys, lemurs, tarsiers, gorillas, Chimpanzees, Apes, and Bonobos. Primates had larger brains and lived longer, with ability to learn, and develop technology in living. Great Apes are a family within the order primates.

They were large, relatively intelligent primates that probably included many presently extinct species. The Great Apes evolved in Africa about 18 million years ago and some migrated to other parts of Eurasia. They had a basic and familial life.

It is believed another species, Homo erectus, originated or migrated to an Asian area between the present Caspian and Black Seas, near the Caucuses Mountains, and perhaps were the base for geographic variations of the Caucasian race in Mesopotamia, Egypt, India, China, Indochina, Sumatra, and into the South Pacific, and the Americas.

The hominids are a sub-family of ape hominid primates that lived about 7 million years ago. The human line of hominoids split from the Gorilla line about 8 to 10 million years ago, and the Chimpanzee line split about 7 million years ago. The hominids diverged rapidly in a new "radiation" that included 20 to 30 distinct species.

There are three major groups of hominids. 1) Australopithicus were a large and varied group that lived about 4 million years ago, including "Lucy", bipedal and short. 2) Homo habilis (skillful man), appeared about 2.5 to 1.5 million years ago. They made stone tools that implied intellectual ability and improved diet that allowed increased brain ability. 3) Homo ergaster, and Homo erectus both at about the same time, evolved about 1.8 million years ago, as shown by fossils found in the Caucasus Mountains area. They made more sophisticated "Acheulian" stone tools such as stone axes. They were the first hominids to migrate out of Africa and entered the colder lands of southern Eurasia, reaching as far as modern Beijing, China. They may have learned to use fire.

Recently a fossil of a new species, Denisovans, has been discovered. These people populated Northern China, Russia, and Papua New Guinea. They may well have been ancestors of North American Indians and perhaps other cultures. Denisovans lived in Asia about 50,000 years ago. They interbred with Neanderthals living in Asia that had migrated east toward Asia while some Neanderthals migrated west. Modern humans in Papua New Guinea still carry genome with nearly 5 percent Denisovan DNA.

By 1 million years ago most of the features that define us as a species were already present, but there was little sign of the technological and ecological creativity that makes us so different. (Author's comment: Homo erectus may have been recipient of incarnated soul that was nature of Homo Neanderthal and Homo sapiens in gradual advancement by sub-species leading to Cro-Magnon man and Modern Man.) Today we are the only survivors of this diverse group of Great Ape family that possessed bipedalism.

Evidence came together in the past 50 years, partly as new dating techniques appeared. The picture of human evolution has gotten much more precise and accumulated at a tremendous rate by 1) fossil evidence, 2) studies of closely related species such as chimps and gorillas to tell how they lived by studying close relations, and 3) genetic or molecular dating, including DNA, to determine when different evolutionary lines split, and mitochondria organelles formed within cells.

GEOLOGY AND ARCHAEOLOGY IN EGYPT, MESOPOTAMIA, AND GREECE

A fourth observation concerns geology and archaeology. Ancient archaeological findings of early man, hominids and hominoids are found in the Rift Valley of Africa. What caused the Rift Valley culture of 3 and 4 million years ago to move north by about 15,000 to 8,000 B.C. into the present Ethiopian and early Egyptian area, and even to where fossils were found hundreds of miles away in West Africa at Lake Chad, is unknown. Egypt and northern Africa is the mother culture with location of previous grasslands at the present Sahara Desert and a lobe of the Mediterranean Sea in that area.

Principles of science and mathematics were revealed in construction of pyramids 8 and 10 thousand years ago. This was before present and primordial reasoning in religion, soul, afterlife, gods, symbolism, government, empire, kingship, and culture development.

The Sahara area was then, not the desert it is today, but a grassy green, pasture land, with a lobe of the Mediterranean Sea, the source today of salt deposits when that lobe of sea water evaporated.

In more recent history Mycenae and its modern day ruins were located just off the western shore of Phoenicia. And an extensive and progressive culture existed there as revealed by presently submerged ruins or buildings, palaces, citadel, courts, and residential buildings.

There was at one time a culture, possibly destroyed when geologic earth movement, such as an earthquake, allowed waters of the western Mediterranean Sea to inundate the Mycenaean location, quickly destroying its buildings and culture. Is it possible that the sunken location is site of the fabled "lost city of Atlantis"? Is this flooding related to the Biblical Flood, or perhaps the flood told previously about King Gilgamesh?

Meanwhile in Egypt, Greece, and Mesopotamia communities, kingdoms, and empires of combined kingdoms arose. In the upper reaches of the Nile River and present day Ethiopia there arose the Old Kingdom of ancient Egypt. Like today, the population of Upper Egypt was mostly of the Black race. Kingdoms also arose along the Tigris/Euphrates river system and at the island empire of Greece.

HOW TO DEVISE A HUMAN

A fifth observation is in how to construct a human. All the physical materials contained in the body and brain of a human exists on Earth. We are made largely of carbon, hydrogen, potassium, water, oxygen and a wide variety of other chemicals and minerals. Study of Biology shows how simple chemicals formed larger and more elaborate molecules that would serve a wide variety of various organs in the body and brain, like those found in living cells. Further, a most difficult challenge to constructing a human is to explain the original planning, evolution, and details of DNA that controls human reproduction.

What would be lacking is ability in how to assemble an animal having logical motions and actions of intelligence—the ability to assemble and cause operation to make an inorganic collection of elements into a functioning metabolism, and further to change a body and brain from a life-subsistence level to a thinking organism, having a mind that sets him different from animals of the forests. Above all, he would need an ability in adaptability. This is the realm of Intelligent Evolution.

INTELLIGENT EVOLUTION

As a sixth observation, we are discussing Mystical Intelligence and Evolution. In this, God has been described as "the greatest scientist of all". In studies of evolution there seems to be two schools of thought. One is that most changes in life over millions and billions of years, from original amoeba, microbes, bacteria, yeast, worms, early vertebrates, and consequently more complex animals, birds, and plants, progressing forward to age of today's Modern Man, is the result of nature-driven changes to accommodate biological needs and manifested by mutations in the genes of DNA as indicated in the Charles Darwin studies. It is held by some evolutionists that the progression of changes did not occur by any source of intelligence of direction, but solely by natural needs of the organism.

There are many gruesome manifestations in and on earth, especially in beings of the deep seas, where darkness and intense pressure prevails. There are also many sub-programs recognized in animals, such as coloration, striping, shape of claw, foot, hand, and in skeletal arrangements, protective hair, skin covering, fat layers, skeletal and anatomic figurations, and virtually millions of variations. Many variations serve practical purposes and have undoubtedly been derived from environmental needs. The needs were manifested by changes in RNA and DNA over eons of time.

However unlimited mutations would likely result in purposeless, irrational, and simplistic modifications in life structures. One could imagine a group of technology-minded monkeys, set loose in a modern day junk yard, having a goal of producing something that is workable and useful. Without an overall plan, the product would be a failure.

Not all pleasures of living are accounted for by mutations, or even by some mystical force. They are purely products of nature and science. A beautiful day is a product of science, with a proper proportion of oxygen and hydrogen in the air, that at one time contained high percentages of sulfur. Clouds in a blue sky are products of weather, explained in science, to be somewhat predictable, as are rain and snow. Views, from or to, high mountains, hills, or even flat lands of growing grain undulating in the wind, are products of science, man, and geologic events. Abundance of food also is a product of man, weather, and science. Even our emotions and relationships in society are products of our minds and personalities, which we can control. Our health is the product of biological happenings that we know today to be largely controllable by ourselves in diet, exercise, and good medical practices. Events concerning our daily futures are largely combinations of skill and luck of our worldly make-up.

Nevertheless, and to which many physicians and surgeons will attest, there are some events in patient medical recovery that are mystical and not explainable. There comes a body's decision point of whether health will be sustained, or whether eventualities of personal health, or misfortunes, will prevail, leading to death or continued life. The cause to life is sometimes attributable to prayer and its mystical abilities.

One of the greatest examples of a directed plan of life exists in the animal body, brain, and mind itself, starting with the process of reproduction, either as in mammals or by egg birth. The sequence of change from sexual mating and exchange of single initial cells, through progressive development of a living organism, is nearly beyond belief in its scope. The progression is determined, predictable, and complete in the fetus growth and birth, generally without meaningless mutations.

Evidence of the wonders of the human body can be found in any textbook of medicine, anatomy or brain. There are numerous, separate systems, each of which is ingenious in its design, composition, function, and continuality. To list only a few are the pulmonary system, eating and digestive system, heart, blood, lung and kidney systems, eyesight, hearing, touch systems, glands, neuron system, biological repair, brain functioning, age changes, and many others, none of which can possibly be conceived as having happened by accident or without purpose, undirected mutations, or other purely environment-caused occurrences.

A second school of thought is that there is Intelligent Evolution. These are in the field of mystical, and not science. Scientific method requires experimentation to be reproducible. Some factors, such as mind and soul, appear to be in higher dimensions of space and life, and are not always available in reproducible experiments. They are more based on probability of supportive reproduction in any test results. To fathom answers to questions in this area of higher dimensions is mind-boggling and unexplainable in our three and four dimensions universe.

Changes result from changing needs, but instead of having, say, a hundred possible changes, one or only a few changes in qualities of body, brain, or mind are selected to be developed in programs leading to a workable change, and those future variations would refine the function desired, all within an overall plan of intelligence.

The size and shape of a grown tree without leaves, standing alone in an open field, is both beautiful and wonderful. Its height is of various but definite amount depending on its specie. And its shape too is of some miraculous determination of round, egg-shaped, square, or rectangular.

The tree with green or colored leaves is wonderful, but I think its best glory is in the winter when its naked growth and development is seen in its wonder of branching, with main arteries above the trunk, smaller off-shoots and branches, to thumb-size and smaller outer terminations.

Why does it grow, yet limit its size and shape as if sculpted by some unseen god or herbiferous artist? Of course this is determined by its species, but we wonder at how the original shape, height, and design, unique to its type, was specified in its genome.

This same wonderment applies to people. Why don't people continue to grow in height and shape throughout life? Certain people, like giant Redwood trees, would continue growth all during their lives. But most, like trees in a field or around our neighborhoods, have a determined and limited physical future, again determined by their respective qualities, God-given in original design.

So it is with evolution that parallels man. The tree started with a seed and simple sapling, and guided by its inherited genome grew in a certain defined pattern during its growth phases, perhaps bending at some phases, to mature into its wonderful glory and beautiful, matured branch growth, leaf, and coloration pattern, into its evolved wonderment.

The act of changing animal to human is Intelligent Evolution. The part which is Chemical evolution is the chemical change that turns iron to rust or hydrogen to illuminated stars, or even to expand the growth of trees and grasses, and combination of hydrogen and oxygen to form water. These chemical and physical changes would take place on Earth, planets, or to a lesser extent on the moon, or without thought or plan to providing hair on the body to allow warmth, and to eliminate it when body hair is no longer needed. These are all events and changes that form the arrow of improvement in the specie.

The author believes that living organisms are simply too complex to have been the product of blind, random processes, such as those that drive natural selection.

To illustrate the arguments of natural selection used against Intelligent Evolution, imagine a junkyard that contains all the bits and pieces of a Boeing 747 airplane, dismembered and in disarray. A whirlwind happens to blow through the yard. What is the chance that after passage of the whirlwind, a fully assembled 747, ready to fly, comparable to how Modern Man now lives, will be found standing there? Without intelligence and a plan, life as we know it, would be highly improbable.

Intelligence, combined with natural-inspired changes in body forms through genetics, is Intelligent Evolution. In essence, this is what happened in evolution.

GLACIERS, CLIMATE, AND ENERGY

GLACIERS

In the formative stage of Earth and the Solar System our planet was extremely hot and viscous. Earth had mostly completed transition from an extremely hot, gaseous mass to a melting hot sphere of metals, pulled together by gravity, collecting aggregates of cosmic residue into its spinning self, growing in size in the process. Heavy metals of iron and nickel had been drawn to the center of earth forming its molten metallic core, with lesser density mass of matter drawn to concentric levels above the core, each level being lighter, a lesser mass from igneous rock to soil.

Over the past 540 million years geologic records show at least 18 major spikes in average Earth temperatures, followed by severe lowering of temperatures. Spikes in temperature occurred at 515 million years ago (Ma), 510 Ma, 450 Ma, 375 Ma, 360 Ma, 359 Ma, 359 Ma, 340 Ma, 305 Ma, 260 Ma, 250 Ma, 230 Ma, 205 Ma, 200 Ma, 180 Ma, 150 Ma, 70 Ma, and 40 Ma. Each was followed by lower temperatures near average Earth low temperatures. The warm weather spikes occurred at no predictable time but averaged every 22.8 million years, with the last spike 40 million years ago, or anticipating the last global warming cycle starting about 17 million ago, or start of the present global warming period.

Like the stars were formed at areas found of high and low density, and pressure attracting hydrogen and helium atoms, local neighborhoods of the cosmos also projected higher and lower temperatures.

As our Solar System, with earth, passed through the respective hotter and colder areas of the cosmos, average temperatures on earth rose or fell, causing glacial formation and interglacial periods. Those cycles through the cosmos would have existed far back into prehistory, and will continue to exist into the distant future.

The temperature of earth changed and came into a cooling stage, resulting from different cosmos conditions. In travel through the cosmos Earth encountered locations of "hot spots and cold spots" which had extreme effect on Earth, our Sun, and other planets of our solar system. As the extremely hot earth travelled to, and through, a cosmic cold spot, it undertook extreme change in average temperature, changing gradually to an ice-covered planet.

Physical change happened to the rocky and metallic sphere of Earth. The surface of earth was subjected to relatively fast cooling, while the core of earth continued to produce heat. Indeed, it would have been much like placing a substance, such as very hot soup, into the freezer section of your refrigerator, or subjecting warm plate glass to sudden freezing that would cause crackle or shattering upon encountering low temperatures.

The solar coldness lowered temperatures of the surface igneous rock, crackling the surface and lower rock into various geologic sections or plates, creating voids at margins of various plates in the otherwise solid layers. The voids would later become important as collection areas of chemicals and liquids within earth, as deposits of oil, gas, iron, nickel, copper, and a wide variety of other compounds would be tapped, billions of years later, by drilling and mining, generally starting about 1800 A.D at the Industrial Revolution, and continuing today.

During the billions of years since original cosmic cooling, Earth, other planets, our Sun, surrounding planetoids, comets, asteroids, Moon, and space trash, underwent periodic cooling, followed by generally warmer climate change as we progressed, in an arm of the rotating Milky Way Galaxy, through alternate cosmic warmer and cooler locations;

Change has been observed through space exploration, by increasing complex telescopes, and launches of space missiles placing planet-surface robots and planet-orbiting spacecraft that transmit pictures back to earth for scientific analysis (and even viewing on home televisions sets).

Observed changes from previously existing conditions have been detected on planets of Mars and Venus, which space scientists believe once possessed atmospheres, water, and possibly life forms, before they were decimated of these elements by loss in gravitational attraction of their atmosphere, and dryness caused by rising temperatures.

History of changes on earth have been detected recently, not only by man-made thermometers, but also by observing growth rings of trees, even back a thousand years ago, by noting growth rings of timbers in buildings found in archaeological digs and studies. The Sahara Desert was once green with plant life, with lakes and rivers that supported life in waters, and a lobe of the salty Mediterranean Sea left deposits of salt by evaporation in present desert locations. Borings into the earth and ice produce samples at various levels as evidence of ancient weather, atmosphere, rainfall, geologic happenings, and other conditions at past ages, producing a "time gauge", so to speak, of past occurrences.

CLIMATE

Earth has experienced numerous temperature changes, ranging from extreme hot, then gradual cooling, periods of low temperatures forming glacial coverings of all or portions of earth, followed by warming trend periods, through moderate temperatures that aided life formation, with a slow but sure change over thousands of years, and with gradual technical and cultural improvements occurring in various life species, emanating from one-cell animals, and possibly in plants, then multi-cell animals and plants, more complex fish, amphibians, four—legged, fur covered vertebrate animals of the forests, somewhat intellectual, or at least complex in thinking processes, of the ape family, and division to several different species, some leading to Modern Man.

Even before universal acceptance of the Glacial Theory, which spoke of several great ice ages, some investigators were coming to the conclusion that the ice had advanced and retreated, not just once, but several times in the recent geologic past.

In the United States each glacial period is named for a state where deposits of that particular period were first studied or where they were exposed; the Nebraska, Kansan, Illinoian, and Wisconsin Glaciers.

One early species was called Homo (meaning a series of man-like creatures) erectus, living about 600,000 years ago. This species is believed to possess a first presence of the elements of soul, and became progressively more complex and complete in the qualities of soul. These qualities were exhibited in succeeding species (or possibly sub-species of Homo erectus) of Homo neanderthal, Homo sapiens, and Modern Man, each surviving the previous species, Modern Man surviving alone at about 20,000 B.C.

It is believed Modern Man (Caucasian) progressed, in sub-species, to cultures from southwest Asia (the Russian area of Georgia), migrating into Mesopotamia, Delta Egypt, Europe to the west, and to the east, India, China, Siberia, southeast Asia, Indochina, Sumatra, Papua New Guinea, Australia, New Zealand, Fiji, Hawaii, and perhaps to the western coast of South America, and to Mesoamerica, about 8,000 B.C., or even before.

The latest ice glaciers covered much of the northern and southern hemispheres, generally north and south respectively of the Tropic of Cancer and Tropic of Capricorn. This area was last at low temperatures on earth starting about 125,000 B.C., then gradual warming and thawing of glacial ice cover at recent world high temperature starting about 15,000 B.C. Paralleling moderate temperature conditions, rapid changes were made in technical and cultural improvements on Earth.

The last glacier was called the "Wisconsin", which started about 125,000 years ago, and thawing beginning about 20,000 years ago. A previous extensive glaciers was the "Illinoian" from 250,000 to 150,000 years ago, and after the "Nebraskan Glacier", at about 325,000 to 275,000 B.C. and others before.

Between these glacial cold years were Interglacial Periods, transitioning from low to higher, (or later high to low) moderate temperature years on Earth. Following a peak high temperature period, lowering average temperatures progressed to eventually start a new interglacial, and subsequent glacier period, or the reverse in evolving out of the severest points of a glacial period, through an Interglacial period, to an average high temperature.

It is observed that little significant advances are made during extremely hot or extremely cold conditions.

Other glacial periods existed on Earth, going back even to "Snowball Earth" which was one of earth's initial periods of cooling, that followed its Hadean Period. The Hadean Period was a time of transitioning from gaseous to solid Earth, with extremes in earth surface, a sulfur-rich atmosphere, and formative extensive heat that followed the gaseous state of earth.

Location of Earth and Solar System in warm areas of the Milky Way Galaxy, and in warmer areas of the cosmos, has caused glaciers eventually warming earth, and promoting transition of man, animals, plants, and sea life, with consequent and substantial changes in technology ability and culture of man.

There was much of Earth's changes, including observations of geologic change, tectonic plate shifting, species change in plants, development of primates, hominids and hominoids. Much of the Homo Specie's technical ability (use of tools) ability, and cultural developments, took place during earth's interglacial periods, absent extremes of hot and cold climates.

The Earth is currently in an Interglacial Period of lowering average temperatures, progressing toward a time of glacial cooling on one hand, and glacial and earth warming on the other. In charting temperature changes, Earth could be forecasted to be at a temperature low perhaps about 20,000 A.D., but man-made effects are producing, oddly, temperature warming during events of our current living in the 21st century.

Will Earth travel out of its current cold spot in the cosmos, with occasional warming spikes before further onset of a new extreme glacial period, as has frequently happened in the past during temperature-lowering trends of earth cool, proceed to glacial lows? Only time will tell which trend will prevail.

It would seem that man-made earth warming is not all bad as creating a delaying period from glacial freezing, except for disastrous rising levels of oceans and inundation of coastal locations over the next hundreds of years, or even sooner.

As our solar system with Earth, passed through the hotter or colder areas, average temperatures on earth rose or fell, causing glacial formations or interglacial periods. This cycle existed back into pre-history, and will continue to exist in the far-away future. Stored energy in glacier ice, with later release of world energy world from thawing, could increase cultural development and technical changes.

In our modern world, weather is largely a study in linear progression. Weather to the west of Michigan, west of Lake Michigan for example, will likely become the weather in Michigan at a later time. Rain in Minneapolis on a Monday will almost certainly predict rain in northern Michigan on Tuesday. High winds in Chicago on a Tuesday will anticipate windy weather in the southern tier of counties in Michigan on the next day. Snow and moisture evaporation, logically increases humidity of air parcels, which will result in 1) further snow, 2) rainfall, or 3) fog, depending upon relative temperatures at various altitudes or at ground level.

We have intimated that global warming caused by our cultural actions is a fantasy. Trends are reflected in data of average temperatures over the past 100,000 years. Earth transitioned from its latest and coldest period about 20,000 B.C., gradually increasing in temperature to a high maximum about 15,000 B.C., and since that time has experienced constantly lowering average Earth temperatures. We are currently actually in a temperature-lowering (cooling) period. Projection of the trend indicates earth temperatures will be at its lowest about 20,000 A.D., bringing return of still another following glacial warming period.

There is little doubt that glaciers are currently melting, which will cause inundation of present ocean shores around the world. But the cause presented to the world population is fallacious. World temperatures reflect lowering of an eon-old cycle whereby Earth, as well as our Sun and planets of our solar system, migrate to locations in the cosmos of colder temperatures, approximat5ely zero degrees Centigrade.

In the distant future Earth, travelling through the cosmos, will emerge in cosmic travel out of extreme cold locations to warmer locations of the cosmos, into subsequent warming withdrawal periods, thereby ending centuries of glacier building. Countering factors to low temperatures from human activity are such as industry and industrial operation, greater world population, burning of fossil fuels, destruction of forests, making of crop lands from reduction of those forests, animal flatulence, and natural gases from both animal and human body decomposition, or even natural feces from after eating. Would worldwide mandated human body cremation result in lower, or only equal, environmental pollution?

There seems to be no remedy in sight to change or temper this future for cyclical, cosmos, geo-thermal existence. It seems the only question is the extent of time and severity of global cooling.

ENERGY

As when freezing water or food in your home freezer or refrigerator, it requires energy (electricity) to perform the cooling. Frozen water in a refrigerator represents energy accumulation. Conversely, when you leave the freezer door open, or there is loss of electricity, there is loss of stored energy. Forming ice of extensive and thick land-covering by glaciers requires tremendous energy on the surface of earth. In this transposition, the ice of glaciers form tremendous sinks (or accumulation and storage) of energy. Conversely, thawing of ice in glaciers releases large amounts of energy from these frozen sinks of energy. Like the stars were formed by high attractive concentrations and pressure of hydrogen and helium atoms, local neighborhoods of our cosmos also projected higher and lower temperatures. This release of energy within Earth's confines affects advance of culture and technology advancements. Everything is energy in one form to another.

This attraction of energy into freezing of glaciers has effect of drawing energy from other energy-consuming actions, such as creation of and development of new species in animals, plants and sea life. Seeding, of what millions of years later would be beds of coal and volumes of natural gas, are other examples of energy conversion, or sinks.

This results in high world energy absorption during glacial formation, and comparable world energy release as glacial ice melts in the period of interglacial periods.

There is a relationship between the low earth temperatures with build-up of glaciers, and the opposite higher earth temperatures with interglacial periods, to the notable advancements and changes in botany, body, brain, physical species changes, geographic, and geologic events in the history of Earth.

The accumulation and abstraction of energy available on earth has far-reaching effects on acts requiring energy, such as production and maintenance of new species, developments in technology of stone, iron, and mechanical instruments, growth requirements of plants, trees, animals, and general improvement in cultures. (Question: Did the tremendous growth of large dinosaurs result from release of energy from melting ice following the Kansan or Illinoian Glacier, and were Mammoths, Mastodons, and large Sabre-Tooth Tigers result from release of energy following the Wisconsin Glacier, or did large Coelacanth fish and whales result from world-energy release about 400 million years ago, after the Kansan or some previous glacier?)

As our solar system, with Earth, passed through the relatively hotter or colder areas of the cosmos through Milky Way rotation, average temperatures on earth would rise or fall, causing glacial formation or interglacial periods. This cycle would have existed back into pre-history, and will continue to exist in the far-away future.

Average temperatures can be plotted and compared. The last glacier, the Wisconsin Glacier, started from earth cooling about 125,000 B.C. During this period glaciers covered northern and southern lands extending to approximately the 40th or 35th parallel. (The 45th parallel is marked in the state of Michigan at the county of Leelanau and horizontal counties.) It has been determined that the depth of ancient glacial ice was often a mile or more thick, and left moraines, valleys, lakes, rivers, and other geologic feature as the ice, and rocks frozen into the ice acting as scrapers, moved and melted over thousands of years.

Prior to this Wisconsin Glacier at geologic time of 125,000 years ago, a previous glacier, the Nebraskan, existed. The low temperatures were at about 325,000 to 275,000 B.C., prior to start of the 250,000 year long Interglacial period.

Likewise, a Kansan Glacier, preceding the Nebraskan, produced a peak in high temperature about 410,000 B.C. Temperatures then declined to form the Kansan Glacier around 300,000 B.C.

The peaks in temperature recur at 400,000 B.C., 325,000 B.C., 240,000 B.C., 125,000 B.C., and 12,000 B.C. respectively, for an average period between temperature peaks of about 97,500 years. Distance between temporary peaks in the last two glaciers however average 115,000 years between peaks of glacial duration.

It is logical to assume that prior to, or within a glacial period, there existed temperature rises and lowering to form minor, unrecorded glaciers and interglacial periods. As stated, frequency analysis shows temperature peaks occurring at intervals of about 97,500 years on average, but 115,000 years considering only the last two glaciation periods, and 113,000 years of interglacial period leading to the last, the Wisconsin Glacier,

FROM MINERAL TO LIFE

In my previous published writing (Do We Live In Two Worlds? ISBN 1-928623-16-6) I presented a theory for the creation and history in the development of matter and antimatter through the process of annihilation, and consequently the concept controlling the effect of electrons and protons in the matter and antimatter.

In the initial Singularity, which grew from a Black Hole quantum prior to the Big Bang, there existed quarks, as minute subdivisions of atoms. Quarks are the basic components of all atoms.

Atoms formed into a variety of elements, the chemistry make-up described in the now-known "Periodic Table of Chemical Elements". The first element to form was hydrogen, which qualities combined and formed a second element, helium, and subsequently more complex atoms of different elements in a complicated system of element and molecule development.

From chemical interactions of elements forming molecules of hydrocarbons, carbohydrates, and proteins, these inorganic elements formed into organic molecules, foretelling the wide variety that existed after, and today.

A theory of early life forms is that a 3-dimensional, but thin, flat, "sheet" of generated organic material, curved around and formed into a tube-like worm, that in turn ate and ingested other molecules that produced energy through conversion by the worm. The spent substances were expelled from the tube as excretion. Subsequent formed worms developed alimentary canal systems to convert products ingested, into energy. From this grew circulatory blood systems, a neural system, and other complex systems generally found in earthworms, other nematodes, and later fishes, reptiles and mammals.

From this initial form, first pulsations took place, with a basic tube or elementary heart developed, that gave a form of animal life. A liquid would have been formed, first blood, to circulate distributing energy from food sources.

Perhaps next would be development of first kidney to better extract waste sources of energy, and a notochord channeling a nervous system, including an elementary brain to communicate sensations throughout the living organism. Appendages of fins or legs would follow to allow physical movement other than the undulating motion, and organs that would reveal a glow of brightness and sound, forerunners of first eyes for sight and hearing.

Over time, perhaps centuries or millions of years, the developed worm would evolve into first animals, four footed quadrupeds of mammal, and reptiles. Additional functioning of organs, fur to maintain body heat, appendages of fins in fish or feet in land mammals to provide movement, and a refined nervous system including first brain, allowed a first recognizable animal, independent within itself, to exist in life.

There are many unproven theories of the beginning of life. One is that the source of life came from erupting crevasses deep in the sea floor. The sea floor crevasse eruptions may have occurred off the eastern shore of Africa where specimens of living, but ancient, Coelacanth fish were discovered. Inorganic minerals, perhaps carbon, would erupt to chemically unite with oxygen in the waters to form oxides of various sorts, such as CO^2 and other molecules.

Molecules of oxides in seas would evolve in complexity parallel to that of on land. Cells under the sea waters would develop into minute "sea worms" that would evolve to a fish family of species, with ability to move about in the ancient waters. From this early and basic configuration grew a wide spectrum of organic plant life, fishes, and reptiles, some later forming amphibians and hominids as first man-resembling animals. Pre-history, egg laying mammals formed first mammalian animal life on earth, a branch in zoology later forming primate mammals as a forerunner to man.

The "design" of organic, living organisms involved development of an extremely wide variety of cells. (Perhaps the recently discovered "T" cell, a general cell that can move to any of many cells in the human body and adapt the necessary cell qualities, was a simplification of needs for various types of body cells.)

First cells, called prokaryote, were absent the presently-known cell walls. Each prokaryote cell was a complex array of neutrons, protons, and other substances. Other refinements were introduced in a new eukaryote cell, such as the addition of cell walls and design of the genome carrying RNA and DNA in a complex innovation and development of the future evolutionary system.

The living substances of earth are carbon-based life, or "life as we know it". Carbon, like other elements and minerals, were forged by nuclear fusion inside stars and supernovae explosions at temperatures above 100 million degrees. Supernovae disperse carbon and other elements around the galaxy where they will form not only new stars but also planets, life, and today's greenhouse gases.

Only a dozen minerals are known to have existed among the ingredients that formed the solar system 3.7 million years ago, but today earth has more than 4,400 mineral species.

Earth's diverse mineralogy developed over the eons, as new mineral-generating came into play. More than half of the mineral species on earth owe their existence to life which began transforming the planet's geology more than 2 billion years ago.

Source: Scientific American magazine, March 2010

From this early and basic configuration grew a wide spectrum of organic plant life, fishes, and reptiles, later forming amphibians and hominids as first man-resembling animals. Pre-history, egg laying mammals, formed first mammalian animal life on earth, a branch later forming primate mammals as a forerunner to man.

The "design" of organic, living organisms involved developing an extremely wide variety of cells. First cells, called prokaryote, were absent the presently-known cell walls. Each prokaryote cell was a complex array of neutrons, protons, and other substances. At some later eon, eukaryote cells developed to replace prokaryote cells, thereby introducing the existence of a cell wall and containment, to the make-up of most cells. Other refinements were introduced, such as design of the genome carrying RNA and DNA.

After the Eocene, Oligocene, and Miocene periods had passed, the physical world and geographic world were in place much as we see them today. Climate would vary with growth or thawing of glaciers. Base species of animals existed and variation or sub-species would take place as they do even today. The factors that were not constant lay within cultures of the people, in their shelter, manner of living, hunting versus agriculture oriented, spirituality, communication skills, and development in basic technology. Man wondered about his origins and advanced many theories, myths, and oral history stories, perhaps similar to what might be told around campfires at hunting or fishing camps today.

WONDERFUL WORLDS

PART II

WHAT IT IS TO BE HUMAN

EARLY SPECIES OF MAN AND PRE-MAN

In understanding our position in this world, we must recognize that there was an incessant parade of life species from the first amoeba, bacteria and microbes, and from first primates, hominids, and hominoids, to the present day. These various species, some "side branches" in development, were developmental, not entirely known, but climaxing to 600,000 years ago in the species Homo erectus, the "developed capsule" that holds the tale of man's development, step-by-step to our status today.

During the very early periods of humanoid development, from 7 million to 120,000 years ago, several species evolved.

1. The very early development of life produced little more than "animals of the forests", walking on four legs, or two legs with arms for climbing trees, swinging from among the branches, and living mostly on tree fruit or digging into the ground for roots. Later, placental mammals, known as eutherians, shared the earth with dinosaurs, as tiny, shrew-like creatures that roamed China 160 million years ago, as the oldest known ancestor of placental mammals. Early primates included Old World monkeys, Gorilla, chimpanzee, and bonobo of the Ape family. Fossils of primates of 15 million years ago have been found in Namibia of southwest Africa, in France, Spain, and Austria, evidence of widespread diversity of forms across Africa and the Mediterranean basin during the warm climate region of the Middle Miocene, 15 million years ago. The oldest of the Miocene hominids is from 9 million years ago, fossils discovered in coal beds of Italy. All evidence of a wide diversity of life forms through Africa and southern Europe, support the Regional Theory (original diversity) of Early Primates, at least in Africa and southern Europe and Asia.

2. Old world monkeys split from the Great Ape family about 20 million years ago. Gibbons became distinct from Great Apes between 18 and 12 million years ago. Orangutans became distinct from other Great Apes about 12 million years ago. Between 8 and 4 million years ago, first gorillas and then chimpanzees split off from the line leading to humans.
3. Other prior fossils have been found, animals supposedly eaten by hominoid in Kenya, of antelope, hyena, dog, pig, elephant, and horse bones, all of about 8 million years ago. The Equatorial Belt of present hot and dry climate lands changed after about 8 million years ago, leaving reduced areas of green plant growth found only in isolated oasis, in a formerly "wet" Sahara Desert.
4. Early species at 7.0 to 6.2 million years ago were recently found at Lake Chad, 1,500 miles from Olduvai Gorge and the archaeological findings of most fossils, begging the question of why that particular specimen, Sahelanthropus tchadensis, was located such a distance away. Were they different, or the same species, from the others of Olduvai Gorge?
5. Species Orrorin tugenensis lived 6.15 to 5.9 million years ago. This species is considered the 2nd oldest known hominine ancestor that is possibly related to modern humans, and is the only species classified in genus Orrorin, an early bipedal hominine. If genus Orrorin proves to be a direct human ancestor, then Australopithicines, such as Australopithicus afarensis ("Lucy") may be considered a side branch of the hominid family tree, earlier by almost 3 million years, and lived in dry evergreen forests environments, not in a Savannah environment. The date of split between hominines and the African Great Apes is at least 7 million years ago. Human DNA is approximately 98.4 percent identical to that of chimpanzees.
6. A progression of species evolved including Ardipithicus kadaba of 5.9 to 5.25 million years ago, nicknamed "Ardi", that possessed such man-like features as bi-pedal walking, with long arm extremities, and a species Ardipithicus ramidus of 5.5 to 4.25 million years ago.

7. Six main species developed in progression of the Australopithicus genus. These genus fossils were found in Tanganyika of South Africa. Australopithicus are considered to be intermediate between that of the ape family (Gorilla, chimpanzee, monkey, and bonobo) and the Homo, or human-like species:

   A. A. anamensis was 4.5 to 3.9 million years ago.
   B. A. afarensis of 4.0 to 3.1 million years ago was named after the town of Afar near where these fossils were found.
   C. A. bahrelghaz (A. garhi) was 3.5 to 3.0 " " ".
   D. A. africanus was 3.3 to 2.5 " " ". ("Lucy" of this species is believed to be a first to walk erect and a common ancestor of following species).
   E. A. garhi was 3.0 to 2.0 " " ",
   F. A. sedibo was 2.0 to 1.75 " " ".

8. Paranthropus was a family of species that existed but was not determined to be on progression towards modern man. Their species were:

   A. P. aethiopithicus (A. ramidus) of 2.75 to 2.5 million years ago.
   B. P. boisei of 2.65 to 1.5 " " "
   C. P. robustus of 2.5 to 1.25 " " " "

Despite the image commonly perceived, not all fossils of early man-like creatures were found at Olduvi Gorge. Fossils have also been found in such diverse locations as at Lake Chad, hundreds of miles west of Olduvi Gorge in Kenya, northern Ethiopia, Russian Georgia, and numerous other locations of Africa, as well as throughout the Old World. Homo erectus and their various sub-species had left fossils in Africa, Mesopotamia, India, China, Korea, Japan, Indochina, Papua New Guinea, and westward to Polynesian Fiji, the Hawaiian Islands, and to South America, perhaps as ancestors 12,500 years ago of the more recent Maya culture. Some anthropologists theorize that the Maya were descendants of people from Japan.

Were early species of primates (Paranthropus) inadequate in qualities of survival to compete with other, more advanced species? Was their changing environment the cause of their demise? Including the time of Paranthropus (2.75 million years ago), there seemed to be an intelligent purpose in development of a line, with body and brain leading to Homo erectus and modern man.

This begs another question, "why did that species come to exist and grow?" Paranthropus evolved and grew over a time of 1.5 million years, but became extinct about 1.25 million years ago. Was this an example of uncontrolled evolution, based wholly on environmental dictates, and not by intelligent direction as held by some environmentalists, and even by at least one court of law? The author believes Intelligent Design projected existence of successive man-like creatures, and evolution provided the biology of changes to develop into new species or sub-species.

9. Genus Homo was a family believed to include elements of future man-types, with larger cranium size, and ability to develop elementary tools to make life easier. Tools were first of Oldowan type from Ethiopia, and later of the Acheulean type. Homo indicated a specie line with unique abilities in thinking, analysis, planning, technology, emotion, and spiritual recognition, all abilities that could be defined thousands of years later as modern man.

Later species (i.e., Caucasian)were known by much loss of body hair, with social and familial arrangements involving the rearing and training of their young, and with division of labor between males and females. With Homo erectus the specie lived almost exclusively bi-pedal.

Anthropologist have been divided in their thinking as to whether current human populations evolved as one interconnected population (the multiregional model), or evolved only In Africa, to engage in speciation, and then migrated out of Africa, forming human populations in Eurasia and evolving worldwide. The Out of Africa model holds that the Homo family of species originated only in Africa about 1.8 million years ago, whereas the species Homo erectus, which had demonstrated human qualities, including soul and spiritual life, had migrated eventually worldwide, about 600,000 years ago.

Homo sapiens (or more properly, the author believes, sub-species of Homo erectus) began migrating from Africa and southwest Asia, between 1.8 million and 1.7 million years ago or before, and eventually replaced existing hominoid species in Europe and Asia. (Mitochondrial genetic diversity of ancient hominoids is highest among African populations, indicating several species of hominoids originated in east Africa, at or near the Olduvai Gorge location.)

Also, after analyzing genealogy trees and studying Mt DNA, researchers concluded that all species development of the known world was descended from an individual women in Africa, dubbed " Mitochondrial Eve".

The super eruption of Lake Toba in Sumatra, Indonesia about 70,000 years ago had global consequences, killing many humans then alive and creating a population bottleneck that affected the genetic inheritance of all humans today.

- A. H. habilis lived 2.5 to 1.5 million years ago in south and east Africa, in the late Pliocene Epoch. This Homo species diverged from Australopithicines. Homo habilis had smaller molar teeth and larger brain cavity than the Australopithicines, and made tools from stone and animal bone.
- B. H. ergaster lived 1.8 to 1.4 million years ago. H. ergaster is believed to not be in progression to modern man.
- C. H. Erectus lived 1.8 million to perhaps 70,000 years ago or later—believed to have possessed first hominoid features of later man, and to possess genome that allowed expression of emotion and spirituality. They may have used fire to cook their meat, for light, and warmth. Earliest H. erectus of 1.6 million years ago seemed more elementary as hominids than as the H. erectus of 600,000 years ago, who were more developed in qualities later to be unique in H. sapiens and modern man.
- D. Peking and Java man were sub-species of Homo erectus. H cepranensis refers to a single skull cap from Italy, estimated

to be about 800,000 years old. H. antecessor is known by fossils from Spain and England that are dated 1.2 million to 600,000 years ago. Homo erectus soloensis fossils have been found in both Indonesia and Australia where culture was unusually advanced. H soloensis persisted until the Toba volcano catastrophe of 70,000 B.C. in Sumatra, Indochina, which had global consequences, killing most humanoids then alive in the area and creating a population "bottleneck" that affected the genetic inheritance of all humans today, and some until 50,000 years ago in Java.

E. They may have built crude ocean-going vessels or rafts, and possessed speaking and language ability. Homo florensis species and culture continued in Papua New Guinea.

F. Many anthropologists use the term H. ergaster for the non-Asian forms, and reserve H. erectus only for fossils found in Asia. H. neanderthal and H. sapiens had co-existed in Europe for10,000 years, but competition and superior abilities of H. sapiens led to H. neanderthal extinction.

G. Homo heidelbergensis fossils of 800,000 to 300,000 years ago were found near Heidelberg, Germany and elsewhere in Europe. H. rhodensis (300,000 TO 125,000 years ago—Rhodesian man) is placed within the H heidelbergensis group. The Gawis cranium fossil of 500,000 to 125,000 years ago is thought to be an intermediate sub-species of H, heidelbergensis.

H. Debate continues on whether H. neanderthal and H. sapiens were separate species that shared a common ancestor (H. erectus) 660,000 years prior. Since no mitochondrial DNA indicates that H. neanderthal did contribute to modern humans, Neanderthals were not considered ancestors, but sequencing in year 2,010 indicated that H, neanderthal did indeed interbreed with H. sapiens about 75,000 B.C. The two species could have co-existed in Europe as long as 10,000 years ago.

I. H. neanderthal existed about 300,000 to 28,000 years ago.

J. Homo sapiens (120,000 years ago to present) culminates to Homo sapiens sapiens, which is Cro-Magnon or Modern

Man. These people developed agrarian culture and life about 10,000 B.C. in Mesopotamia and Egypt.

K. In year 2008 the Denisova hominoid fossil (Little Finger) was discovered at the Denisova Cave (Alti Krai) in Siberia and was carbon dated to 41,000 years ago. This may be a previously unknown species based on an analysis of its mitochondrial DNA (Mt DNA). The cave in which the portion of the little finger was found was inhabited also by Neanderthal and modern humans.

This group (called Denisovans) shares a common origin with Neanderthals and did also interbreed with the modern ancestors of the Melanesians of an island group in the west Pacific Ocean (Papua New Guinea). Modern humans (H. sapiens, H. neanderthal, and Denisovan hominoids) last shared a common ancestor, with possibility that Neanderthal, Modern Humans, and the Denisovan hominoids may have co-existed. Little is known of the precise anatomical features of the Denisova analysis.

This DNA work provides an entirely new way of looking at the still poorly understood evolution of humans in central and eastern Asia during the Pleistocene Epoch. This find raises two questions: 1) Does this find suggest that the Native American Indians, who supposedly originated in Siberian Russia, was of a species dating from the Denisovans, and, 2) were the Denisova group in Papua New Guinea on a migration route to the Pacific Ocean islands of Fiji, Hawaii, and to the Pacific coast of South America?

L. The "Great Leap Forward" started about 50,000 B.C. when modern human culture started to change at great speed. Modern humans began burying their dead, making clothes out of animal hides, developing sophisticated hunting techniques, making technological advances, and engaging in cave image paintings. They left artifacts of fish hooks, buttons, and bone needles, specialization tools, use of jewelry, organization of living space, rituals, exploration of new geography, barter trade, and other evidences of a more complex culture. (Source: Google Wikipedia, Human Evolution)

The period of H. erectus, from 1.8 million years ago to 600,000 B.C. would likely have been a time of strengthening its basic qualities. H. erectus possessed unique qualities of intelligence, technology improvements, stereoscopic front-facing eyes, modern physical appearance, a more keen, but yet invisible and not understood mind and soul, ability in analysis, association of a recognized spiritual presence, and spiritual expression. These qualities would become increasingly predominant through 600,000 B.C. and forward, to blossom as Homo sapiens after 20,000 B.C. The author believes this developed presence of Homo erectus was manifest in a "flow-through" of Homo erectus qualities to Homo neanderthal and to Homo sapiens. Homo erectus would be recognized as the root species.

The body, brain, mind, soul, and genome is a marvelous and miraculous assemblage of organs and systems that allow present life. It appears to be assembled from billions of individual cells, with knowledge, and intention of what is needed to allow the entire mammal body to be born, exist, mature, and in total to work for accomplishment and life progress.

Scientists have recently estimated the number of species living on our planet at 8.7 million, not counting bacteria. Nearly 6.5 million of these species live on land, versus 2.2 million in the ocean. There are 5,500 species of mammals, plus reptiles, amphibians, plants, trees, microbes, virus, bacteria, yeast and others. Previous estimates were as few as 3 million species to as many as 100 million. "Until now we have not had a good idea of how many there are". (Source: Dalhousie University in Nova Scotia)

Sources:

1. The Great Courses (Several lecture courses and professors)
2. Google-Wikipedia, The Free Encyclopedia
3. Western Civilizations (College History Textbook)
4. King James Bible

5. Harper Study Bible
6. Archaeology magazine
7. National Geographic magazine
8. Scientific American magazine
9. Discover magazine
10. Nature magazine
11. Encyclopedia Britannica
12. Physics Annual Review, The University of Michigan
13. Goode's School Atlas, Rand, McNally & Company
14. The Last Two Million Years, Reader's Digest Association
15. Prehistoric Life, DK Publishers, 2009
16. A wide variety of general reading

LIFE I—THE STORY

> NOTHING IN THE EVOLUTIONARY DEVELOPMENT OF BRAIN AND MIND MAKES SENSE UNLESS THE CONCEPT OF SOUL IS INCLUDED IN THE MIX—The Author

FIRST LIFE

First forms of elementary life followed cosmos, solar, planet, and earth development that started with the "Big Bang" (although descriptions of a possible event prior to the Big Bang is explained elsewhere in this book it is essentially a Black Hole collection of cosmic energy, not matter, that constituted the "Big Bang", with some conversion of energy to matter). There was energy acquisition from the relatively small cosmos existing before in our own original Black Hole agglomeration of cosmic energy, with a subsequent Big Bang.

Other "Big Bangs" occurred in the cosmos providing the elements for forming stars and galaxies, in a similar and progressive pattern in space—"an archipelago in the cosmos of stars and galaxies".

From the aggregation of energy that provided the Big Bang, there existed a Singularity, or a partial cancellation of minute, sub-atomic, elementary particles. About half contained a positive electrical charge and the balance possessed a negative charge.

There developed in the Big Bang a process of Annihilation in which positive electrical-charge particles were annihilated, cancelled, or changed, allowing predominance of negative-charge particles that formed matter of our cosmos.

The cosmos expanded at the speed of light after the Big Bang. Energy from the Big Bang expanded and found locations of particular density. Cosmic energy was attracted to tho areas through gravitation in the cosmos. With increased pressure from assimilation, there was tremendous pressure and heat build-up, and after a period of 200 to 400 million years, the stars illuminated sending out photons of light. Some stars later exploded in supernovae, but were re-attracted by gravity to even more intense mass and heat, forming elements additional to the original hydrogen and helium from the Big Bang. Other galaxies formed nearby in the cosmos, both before and after the Milky Way Galaxy formations.

STARS AND GALAXIES

Among the stars later created was our own Sun, also experiencing supernovae expansions, gravitationally re-forming each time with greater pressure, density, heat, and creation of new physical elements. During the Sun's one or more explosions through supernovae expansions, many cosmic particles and planetoids were thrown out into the cosmos, that would later become gaseous planets of the Sun, together with numerous asteroids, comets, and space particles of various size, mass and density. Planets formed from the gas and solid matter in the cosmos, the inner four planets (Mercury, Venus, Earth, and Mars) growing by gravitational attraction of solid matter to the gas, and developing enlarging orbits around the sun in a growing balance of centrifugal and centripetal forces. One developing planet was that of Earth, growing with numerous varieties of physical elements and minerals, in a cauldron of creation.

FROM HADEAN HEAT TO SNOWBALL EARTH

After the Hadean Eon, lasting billions of years, Earth gradually cooled, even to the extent of becoming a giant ice-covered "snowball" in orbit of the Sun, the positioning of the Solar System, and the Earth itself into relatively cold spots, and alternate hot spots. These remnants of the cosmic gas of the Big Bang within the cosmos, as the Milky Way Galaxy rotated and its spiral arms traveled within the universe. This period of extreme cold in certain locations of the cosmos was an effect great enough to cause "Snowball Earth" periods, and also the opposite of generating higher—temperature earth periods from the relatively warmer cosmos of another location in space, all from the rotating movement of our Milky Way Galaxy through the universe.

ICE AGES

Start and completion of an ice age, and of interglacial periods, is at random over time. Our Solar System and the Milky Way Galaxy travel through the cosmos, connecting with alternately cold and warmer areas of the cosmos, the latter warmer periods caused by cosmic radiation, would account for irregularity in cycles of earth temperatures.

In a course (Great Courses, taught in Cosmology by Professor Milankovitch) likewise identified causes of ice ages and warming periods, due to variations in Earth orbit around the Sun, changes in polar axis, and relative location within the cosmos, from having alternating cold and warmer areas. Random positioning is made with cooler and warmer areas as our Solar System, and arms of our galaxy, travel through cosmic radiation. For example, 120,000 years ago, a cool period of the last ice age started, and at 100,000 years ago a warming trend started as the last interglacial period commenced. Earth completed and exited another ice age (the Wisconsin Ice Age) at 15,000 to 12,000 years ago.

A deviation of 6 or 8 degrees from an average cosmic temperature, and for a lasting period, can make a difference between extreme cold or higher temperature spikes in cosmic and earth's temperatures.

Temperature studies are available through Earth borings back to 4 billion years ago, well into the age of earth (4.7 billion years). Machines have made borings into ice thousands of feet thick at Antarctica that reflected temperatures of successive times. Interglacial periods occur about every 100,000 years.

FIRST LIFE

Three billion years ago, after a long period of hot, viscous, molten earth, and following periods of frozen Earth and later warm periods of 750 to 600 million years ago, life had started in the form of algae, bacteria, microbes, or virus. First hominid (animal) life appeared on earth at about 6 million years ago in forms of the Great Ape family of gorilla, chimpanzee, monkey, and bonobo.

"Snowball Earth" phases, at least once sometime earlier than 650 million years ago, may have killed off much of any life existing, but was followed by a vast increase in the diversity of living things, including sea life plants, fish, amphibians, and the later occurrence of land animals that had their origins following these unusual climatic events called "Snowball Earth".

NARROWED EARTH POPULATION

In year 2010 a civilization was discovered in southwestern Africa which had lived, or better stated survived, from about 164,000 to 35,000 years ago, in caves near Mossel Bay near the tip of South Africa, at a time when Homo species were in danger of dying out. At some point between 195,000 and 120.000 years ago, the population size of Homo sapiens plummeted due to cold and dry climate conditions that left much of their ancestor's African homeland uninhabitable. Many of Negroid Race alive today are descended from that group of people, from a single region, who survived this catastrophe. The southern coast of Africa would have been one of the few spots where humans could survive during this climate crisis, because it harbors an abundance of shellfish and edible plants. These people may have been the ancestors of us all. (Source: Scientific American magazine, August 2010.)

Other planets, too, changed from their original characteristics in topography and atmosphere. Inner planets of Venus and Mars may have been able to support elementary plant or animal life before losing their atmosphere and changes in other environmental qualities.

INORGANIC TO ORGANIC IN CHEMISTRY

With the operation of photosynthesis and oxygenation of minerals, transition was made from metallic, inorganic elements, to organic forms of plant, fish, and animals, including algae, bacteria, microbes, and vertebrates. Self-moving creatures of worms, with elementary organs for food (energy) intake, an alimentary canal for digestion and conversion of food forms to energy, and with means to expel food residue as waste following digestion, were probable as first multi-cell organisms.

Later developments involved formation of blood fluid, a heart, and of a blood-circulating vascular system to pump the fluid blood, a nerve system controlled by a central, simplistic nerve and elementary brain, with later appendages that would become head, legs, arms, fingers, toes, ears, nose, and more. Ingenious systems developed to successively perform various body, digestive, and breathing processes, to hearing by transfer of pressure and electrical impulses from future ear location to brain, with a hearing system to recognize changes, and to interpret different sound vibrations. Predecessors to eyes by means of sensors to absorb light existed with uncanny ability in light recognition to meet changing needs, and later discrimination of objects for sight, visual acuity and organs for magnification and sensory transfer.

FIRST LIFE

It is still not definitely known how and where life started on earth. There are different theories regarding form, origin, and location on earth in which first cells of life, molecules, microbes, or forms of life existence are widely theorized.

A popular first belief was that original chemical compositions originated in shallow waters of oceans or peaceful inlets. The components for life existence were present, and perhaps the energy of atmospheric lightning provided the union of chemicals to bring about living cells, from which more complex cells would evolve, forming elementary animals.

A second theory was that the rudimentary chemicals of life, the building blocks of life, came upon Earth from cosmic space, arriving on asteroids or other natural space travellers.

A third theory, currently being considered, is that the chemical elements of life existed in the hot interior contents of earth and seeped through fissures in the sea floor to be chemically joined, and to form first life. The components of cells existed, or had been chemically developed within the earth, perhaps in the core or mantle of earth, from molten magma. The magma of superheated forms of rock and minerals broke through the surface of undersea fractures or fissures in the ocean floor. From this expulsion, microbes, bacteria, yeast, prokaryote cells, or other forms of future life would exist. Some simple forms of these exist today.

Ancient living matter were hermaphrodite, containing both male and female sex organs, capable of making both sperm and eggs, so each creature could fertilize itself, continuing its species.

Insects, over a period of 400 million years, evolved into perhaps an estimated future 5 million species. This explosion in insect growth may be example of uncontrolled specie development, versus control (Intelligent Design) of species.

Some fish of 380 million years ago, the Coelacanth for one, had mammalian features of internal birthing of young and with umbilical cords. They had claspers, as do sharks and rays today, and deposited semen inside sexual organs of females of their species.

Biblical belief of creation holds a fundamental tenet that all species were created in their present forms. Charles Darwin found evidence of evolutionary changes within various species, whereby animals and birds differ slightly in response to their environment. The Book of Genesis tells that man and his female mate were created from clay and mystically given life. Many cultures portray their own versions of Creation.

Earth became relatively quiet and cooled to a livable temperature after about 4 billion years. After many other life forms developed further as microbes, bacteria, worms, cellular organisms, yeast, or other forms, developed complexity and ability to live and propagate their species.

A theory has developed that first life was in elementary plants under seas, or by water-living fish, muscles, or energy-seeking plants. Animal life developed, as vertebrate organisms, to later develop legs or other means of locomotion. From these creatures, more complex animals developed, often bearing coverage of fur, hair, or shell. New species developed with more complex features and larger size.

Amphibians developed by evolution to climb upon the land, could breathe air, and had ability to live either on the land, rocks, or in water to fill voids where life on earth could exist, with various and unique changes in their biology. Members of the reptile family formed at about 400 million years B.C.

From this complexity, animals, fish, and reptiles formed appendages and continued in complex life forms by growing legs, arms, fins, flappers, claspers, or undulating body sections to facilitate movement, some developing a coat of hair, fur, or scales for warmth, insulation, or protection.

Among the developing animals were Lemurs of Madagascar, Shoshonus of western North America, fish of the sea, mammal fish-like creatures of Coelacanth, whale, and porpoise in a continually more complex line of animals, fish, reptiles, amphibians, and mammal development.

There developed at about 10 million years ago a division from other animals in the Great Ape family, and at about 7 million years ago the Chimpanzee divided from other animals, apparently from mutations and changes in the genome. The Great Ape family consisted of the Gorilla, Chimpanzee, Old World Monkey, and the Bonobo. The Chimpanzee is the closest genome relative, and the Gorilla is the second closest to the genome of man.

Source: National Geographic magazine. September 2008.

HOMININS AND PRIMATES

Hominines are a group of bi-pedal primates that appeared in Africa about 7 million years ago. Australopithicines lived between 4 and 1 million years ago. Homo habilis appeared about 2.5 million years ago. Homo ergaster evolved 1.8 million years ago, and Homo erectus, with its various sub-species, is believed by the author to be the "common thread" in human species evolution throughout the world, about 1.7 million years ago.

At about the time from primates of 1.2 million years ago (first of the ape family) to a time when hominids split into at least three forms, including Homo ergaster, Homo habilis, and Homo erectus. The period ended with extinction of Homo ergaster and Homo habilis at about 1.8 million years ago. These species (H. ergaster and H. habilis) have been evaluated to not be on the evolutionary ladder leading to our present humanoids.

Studies by anthropologists and cultural psychologists reveal that these mammalian members, especially the Chimpanzee, exhibited familial practices of care and teaching of their young, exhibiting emotions, communicating with others of their species, building and using simple tools, engaging in small-scale warfare, and showing male/female division of domestic duties, practices that can be observed even today among the Great Ape family. Almost 99 percent of the Chimpanzee's genetic code is the same as that of the human being.

In the wild, Chimpanzees live in groups or communities. Each community is divided into smaller bands in size of 6 to 10 animals. Chimpanzees often move back and forth between different bands within the community. Chimpanzees live, even today, primarily in the forest, but also live in grasslands and desert-like areas. They sleep in trees on nests made of branches and leaves and may make a new nest each night. In recent decades British biologists, Jane Goodall, spent many years studying the chimpanzee communities and was one of the first scientists to document chimpanzee's hunting and use of natural tools.

First human fossils were unearthed only 180 years ago and anthropologists have since amassed a formidable record of our forbearers.

The author believes this development followed an uncanny process of Intelligent Evolution. It is unbelievable that development of plants, fish, animals, bacteria, microbes, algae, yeast, and the like would not follow a logical pattern of successive development without some sort of goal-oriented, intelligent guidance, in an extremely technical development, to avoid chaos in biologic design. Changes in life forms and mutations in DNA and RNA occurred to develop in the genome of new genes that would be instigated by need, naturally, and describe qualities and species as environmental conditions required.

Changes were processed by actions within the genome and to enzymes that changed man's various organs. Changes were prompted by environmental situations, but intelligence in evolution dictated the concept to be directed by the genome, and direction within the environmental requirements.

Further change in the Homo species of animals produced a variety of species that had some features of man, but this miraculous break-through was emergence of the Homo, or man-like species which developed, some changes occurring from environmental needs. But more importantly there appeared to be a direction in evolution, as opposed to chaotic development, even though useless-developed features disappeared with succeeding generations. It should be noted that the many animals of the forests also underwent evolution in a series of body changes.

The hominids increased their mental abilities in reasoning, analysis, conceptualizing, tool-making technology, and belief in an external, controlling spirit which he should recognize and honor. At about 600,000 B.C. Homo erectus was a "base specie" that advanced its abilities and migrated to various locations of the ancient world into various sub-species sprouting from the base species. Early specie's RNA and DNA produced greater complexity in the mentality, analytical, emotions, spiritual, and technological ability. With advancements in developing tools from chipped stone, and later from iron found in certain locations, these could be formed using fire-heat that had been discovered. Later mixing with copper, tin, and other metals would form bronze, a still harder and more malleable metal, useful for weapons in elementary warfare and in fashioning more complex tools and instruments to aid in his living.

LAKE TURKANA—MANKIND'S ORIGINS

Ancient Lake Turkana, also known as Lake Rudolph, is in Northern Kenya. It is irony that in this bare land around Lake Turkana, where man now struggles to survive, the past was neither waterless nor barren. Ample fossil evidence that the lake's shores were teeming with wildlife, including prehistoric elephants and three-toed ancestors of the horse, and before them some truly enormous dinosaurs such as the 50-ton Diplodocus, a fossil of which was found on the east side of the lake. Other animal fossil finds include both black and white rhino, the extinct giant otter, hippo and the extinct pygmy hippo, elephant, monkeys, wild camels, and lion.

But most remarkable, and the reason for Turkana's world-wide fame as the purported 'Cradle of Mankind', are the finds of early hominids, including remains of various Australopithecus species, Homo habilis, Homo erectus, and homo sapiens, creating speculation that Lake Turkana was mankind's origin. The cause of diaspora from Lake Turkana is unknown, but we know from fossil finds that younger groups and species were discovered at the Rift Valley, in Kenya, Ethiopia, Tanzania, eastern and southern Africa, and perhaps to the Congo.

Although Lake Turkana and adjacent parts of the Rift valley may not be the only, or even the earliest "cradle of mankind", the abundance of the fossil record here has been taken by many to support the "Out of Africa" or "African Eve" theories of human origin, which postulates that mankind began life in Africa, from where it began fanning out across the rest of the world. Whether or not mankind originated in East Africa, the path of human evolution, from the first faltering steps of Australopithecus some 4.5 million years ago, to the tools shaped by Homo habilis 3 million years ago, through Homo erectus, and Homo sapiens ("knowledgeable man"), to ourselves, Homo sapiens, can be traced through the fossils that have been unearthed along the shores of Lake Turkana, (Source: Google, Turkana prehistory).

Early Homo erectus species features included: 1) bipedal stature and locomotion, 2) possible origination in the Georgia area of present Russia, 3) probably lived sometime in Africa, and migrated through Mesopotamia, India, China, and the Far East, 4) may have populated eastern and southern Africa, 5) commenced intellectually and spiritually about 600,000 years ago, and perhaps were a forerunner species of Homo neanderthal and possibly Homo sapiens, and 6) had unique appearance, physical features that included an outward nose and downturned nostrils, mental abilities, forward-looking and stereoscopic eyesight, and ability in elementary technology of controlled fire and tools. (Source: Google)

HOMO ERECTUS

Among the first hominids, after primates, were Australopithicines. Homo erectus entered the fossil record when Java man (1.8 million years ago) and Peking man (500,000 years ago) were early names for sub-species of Homo erectus. Not all Homo erectus were in Africa, but fossils were found, not only at Olduvai Gorge in Tanzania, but also at Lake Chad, 700 miles distant to the west, and northeastward through Mesopotamia, Russian Georgia, India, and into China, Java and Indochina.

Homo erectus further refined stone tools, mainly the hand ax. They ate meat and travelled great distances as bipeds. Homo erectus was able to control fire, cook food, used fire for defensive purposes against wild animals and marauders, and to extend the available day. Fire spots have been found at Zhoukoudia Cave near Peking China, showing migration of Homo habilis and Homo erectus to China at 1.8 million years ago.

Homo erectus had a larger brain than Homo habilis, approximately in the range of size with modern humans. Their skulls had heavy brow ridges over the eyes and the forehead was low and sloping. Different sub-species of Homo erectus differed in various locations of Europe, Asia, Siberia, Africa, China, Indochina, to Papua New Guinea, Australia, and possibly to Pacific Ocean islands, and beyond to South America.

Homo erectus marked a change in culture, achievement, social learning, and teaching. He was accomplishing events on a larger scale previously unknown. Homo erectus was likely a predecessor in a chain of species leading to Homo neanderthal, Homo sapiens, and Modern Man. (Source: Great Courses, Prof. Barbara J. King)

Homo erectus overlapped in location and time with other fossil forms. For one, Australopithecus robustus was a localized hominid that did not leave Africa, but went extinct after about 1 million years. Ending date for Homo erectus is indefinite and controversial, but estimated at 500,000, 300,000, 250,000 or 200,000 years ago, or sub-species later in different parts of the world. Homo erectus species may well extend to a sub-species of Homo sapiens

Homo erectus marked a change in culture, hominid achievement, social learning, and teaching. He was accomplishing events on a larger scale previously unknown. Homo erectus was likely a base predecessor in a chain of species leading to Homo neanderthal, Homo sapiens, and modern Man. He experienced a spirituality and Summa bonum (sense of an invisible superior power).

TIME PROGRESSION OF COSMOS, EARTH, PRIMATES, HOMINIDS, AND HOMINOIDS

Prior to 6 billion years ago life had developed progressively from first Prokaryote cells to Eukaryote cells having a cell wall enclosing a nucleus of DNA and other traits. Life had evolved in the waters, of growing plants, protozoans, fish and shellfish of various species, reptiles, amphibians moving alternately from sea onto land, various species of vertebrate animals, some having fur, four legs, feet for movement, with fingers and toes to facilitate climbing. Complexity of bodies and nervous system had developed, perhaps from first, elementary worms, then to forms having developed necessary components of heart, cardiovascular system, kidneys, lungs, brain, and other organs.

ENTER THE HOMINOID

Many developments in man's evolution seem to converge at a certain period. A stupendous period commenced about 2.75 million years ago and matured about 100,000 B.C.E. Prior to 2.75 million years ago there was gradual and very slow change from when first primate life appeared on Earth, continuing to 600,000 years ago when first hominoid life appeared.

The whole associated process of change came with addition of the Homo species, institution of new species and sub-species, incarnation of soul to hominid animals, evolution of sexual differences, further development of the soul with its many basic inherent qualities, and maturing in evolution of the soul, may all have happened over thousands of years, but in the same general geographic areas in Africa of life development on Earth. The general scheme for the creation of man, and incarnation of soul described in the Book of Genesis would logically have transpired several thousands of years earlier, perhaps at 600,000 B.C.

The Book of Genesis describes a lineal family and population descent from creation of Adam and Eve. In year 1654 A.D. Archbishop James Ussher in Ireland projected from study of the various generations reported in the Holy Bible, that Earth, the cosmos, as well as Adam and Eve, were created on October 23, 4004 B.C. This author proposes that the lineal descent story as described above was placed at a time well after a true creation process, albeit at a somewhat arbitrary period or span of time in history.

Adam and Eve may well have been, not two individuals, but instead a symbolic group of hominoids inhabiting the new species involved. The sub-plot of the serpent and Eve is symbolic more than actual, introducing the factors of temptation and evil. The era of time would have been gradual, between 3.75 million years ago (time of "Lucy") and 125,000 B.C.E. (initial appearance of Homo sapiens).

Contrast in change of Homo genus from Australopithicus to Homo erectus genus was largely an increase in the cranium volume, number of brain cells, brain circuits and their functioning, to provide increased thinking ability, reasoning, and emotion capacity, in addition to obvious physical body changes. This was accomplished by the continued incarnation in successive generations, with soul direction working in conjunction with the genome and enzymes, providing a whole new dimension to mankind.

RACES

It is possible that the ancestors of Neanderthal (Homo erectus) were black skinned, but ancestors of Homo sapiens in Israel and northern Africa were fair-skinned. Intermingling of races occurred with various "mid-complexions" in Mideast, Northern Africa, and South Asian races. Very Black race people are biologic descendants of Negroid populations living in the Rift Valley. Moderation in intensity of sun rays away from Earth's equator may account for lighter-skin color humans (Caucasians).

The author believes an explanation based on logic, reason, and present-day accumulation of knowledge explains creation of the Caucasian Race. The anthropology trail of Homo species between 1.8 mya and 600,000 years ago is not largely documented.

Homo erectus first entered the anthropology realm at 1.8 million years ago and is paired with Homo ergaster of 1.6 mya. It is not clear whether Homo ergaster was a conflicting species from Homo erectus, or whether the two species integrated.

Homo erectus emerged again about 600,000 B.C., but of Caucasian Race. Possible explanation is that there was migration from Africa to Southeast Asia about 1.7 million years B.C. One group branched off to the Caspian and Black Sea and Caucasus Mountains region where it maintained a separate culture for over a million years in a climate contrasting to the tepid heat of eastern equatorial Africa and home of recognized species of that area.

The colder climate of Russian Georgia bordered a colder, southern-reaching glacier encroachment. The colder climate of 1 million years ago was cause of change in the genome of Modern Homo erectus branch, having higher emotions, great spiritual recognition, biologic changes, greater cranial capacity, and abilities in problem solving and technology, to emerge 500,000 years later as virtually a new species, but still called by historical anthropologists as Homo erectus. Modern Homo erectus was to become predecessor of Homo Neanderthal (a possible sub-species), and later Home sapiens and Modern Man.

Archeology fossils in the Russia Georgia region were not well preserved due largely to agrarian land tilling beginning about 10,000 B.C. It is known that the later Celts of Ireland originated from the Caucasus Mountains region, to proceed along the Danube River valley, to Switzerland, France, northern Italy, Germany, England, and Ireland. They also migrated northward and eastward into northern Russia, Mesopotamia, India, China, Korea, and Southeast Asia as Denisovan hominids, to eventually populate all major countries of the world.

It could well be said that glacial action was a major instigator of the Caucasian Race, as the Negroid Race was preserved at a critical glacial time at Moesel Bay in Southwest Africa at 195,000 to 120,000 B.C. and in Central, East, and West Africa.

FIRST HOMINOIDS

It is believed that first hominoids date back to a modern Homo erectus, starting about 600,000 B.C. Homo erectus had some genetic attributes that enabled him to believe in spirituality, a belief in an inner spirit, or outside superior power that controlled or influenced his life, if nothing more than to bring success on upcoming food hunting expeditions. He must have wondered about god's making of a man-child, to be produced by his female mate in some wondrous event where a person came out of a person. These are all wonders that went unexplained in the Pleistocene age.

Homo ergaster and Homo erectus had populated Africa, and spread through migration to the Arabian Peninsula, Mesopotamia, India, and as far as China and Java in Indonesia, from the end of the Pleistocene epoch about 1.8 to 1.3 million years ago. Homo erectus is believed to have split into two or more branches, one or more migrating to Europe, and the others to Mesopotamia, India, China and the Far East, including Indonesia, and acquired qualities of the present North American Indian and oriental races.

Early African Homo erectus and Homo ergaster fossils represent the oldest known early humans to have possessed modern human-like body proportions with relatively elongated legs and shorter arms compared to the size of the torso. These features are considered adaptations, with the skeletal ability to walk and possibly run long distances, and with associated dramatic changes in bones of the foot and knee structure.

The ability of Homo erectus species eventually merged with "design improvements" to a new species (or sub-species as proposed by the author) of Homo neanderthal, and a later contemporaneously developed, either within or separately, to the Homo sapiens sapiens, or Modern Man, at about 28,000 years ago.

TO AN AGRARIAN CULTURE

Homo sapiens man was still Paleolithic Man however, living in caves, recognizing familial relationships, recognizing an external spirit having powers greater than himself but physically untouchable. He had ability to reason, to plan, to analyze, to hunt, to pick berries and plants, to differentiate between nature's animals for hunting and for fishing, and to develop a social relationship in family and community.

A society developed with association of people, from both near and afar. Production of items in excess of his immediate needs allowed trading to acquire additional goods to enrich his life. Success would often breed further success, with individual wealth and power developing. Communities developed showing impressive wealth and sophistication, an important trading center was a major aspect in the early game of civilization. Various geographic and political cultures developed, each rising and many subsequently falling, often by defeat by a neighboring warring force. It seemed that urban civilization, technology, and organized warfare emerged hand-in-hand.

Starting about 12,000 B.C. (The timing varied with different geographic areas) man discovered another, easier, more sociologic, and perhaps more sensible way of living from his previous hunting, fishing, and live-by-the-moment way of existence. Plants that he had encountered in the wild could be replanted in a localized area, and seeds of plants could produce new growing plants. Animals previously encountered in the wild, some in the Canine family like wolves, coyotes, and wild dog, could be tamed, and milk-producing animals could be controlled to a domestic state.

With this new stability in his life, man could abandon his cave dwelling and construct a form of house to accommodate himself and his family in a permanent, or at least a semi-permanent location, tending his new form of agriculture and a form of controlled animal husbandry.

MALE/FEMALE SUBSPECIES?

I have often wondered whether either male, or female humans are perhaps subspecies within the species of Homo erectus. Certainly the differences of male and female are obvious and unchanging. Females have obvious abilities to birth and nurse their young. They generally are less tall than males, and are of lesser muscular build. Physicians and psychologists could list several distinctive and unique features of female from male in humans.

Timing of such division of male and female subspecies might be 1) at introduction of Homo erectus at 1.8 million years to 600,000 years ago, except that identified female footprints have been found and dated to about 4.5 million years ago, or 2) when eukaryote cells replaced prokaryote cells at about 3.5 billion years ago when sex entered the world of evolution to "engineer" qualities of both male and female into the embryo, or 3) after the original life of amoeba, bacteria, microbes, and yeast there clearly was no second sex within animal or plant, and even today there are examples of hermaphrodites that do not require a second sex for birthing, reproduction, evolution, or care of their young, or 4) sexual differences have been studied in early Chimpanzees, the female, of course, birthing offspring, but also is relegated to rearing and teaching their young, and preparing food brought in by the male.

Truly existence and division of male and female goes back at least 6 million years, but the differences are definite.

DEVELOPMENT OF HOMO ERECTUS

Homo erectus definitely shared some anatomy features of later Homo Neanderthal, Homo sapiens, and Modern Man, and moved away from prior transitional species. Dates for Homo erectus are 1.9 and 1.8 million years B.C. (exact dates are not certain). He exhibited 1) an increase in brain size, 2) a vertical shortening of the face, 3) shortening of arm bones, 4) formation of an external nose and down-turned nostrils, 5) a modern human size in height, 6) resemblance of a very robust modern human form from the neck down, and a less hairy body, 7) an elongated femoral neck, 8) longer vertebral spines, 9) a narrow air canal used for fine air control during speech in modern humans, 10) a rounded edge to bottom of the eye sockets, 11) lack of a clear demarcation between the nasal region and the lower face, 12) thick cranial bones which are flattened along the sides of the braincase, 13) a greater cranial breadth at the base of the cranium, 14) a reduced development of the frontal and parietal lobes of the brain, though more developed elsewhere, 15) prominent muscle markings along the sides and back of the cranium, 16) possibly a light skin coloration as Caucasian, reflecting origin near the Caucuses Mountains of Russian Georgia, 17) a more pronounced physical difference between the male and female sub-species, and 18) forward-facing eyes and stereoscopic eyesight.

The strength of his spiritual beliefs increased from Homo erectus to Homo Neanderthal, Homo sapiens, and to Modern Man. There are many qualities and evidence of soul. Not all of these qualities were present and strong in man at first presence. Homo erectus is believed to possess the first evidence of soul, which was probably the presence of basic emotions, that set Homo erectus different from primates and hominids. The qualities were not immediately strong in its presence, but was certainly exhibited. That, and other qualities, became stronger with each generation. The qualities of soul became more common, as the genome that determines height, growth of intelligence, or sophistication in man's physical abilities, each generation stronger than its former in prominence.

In the specie Homo neanderthal the qualities of soul were even stronger, present to the extent that man would express the presence in quality of soul in drawings and art work on walls of European caves of his living, perhaps much like modern man prepares religious symbols (such as a Crucifix) on walls of his residence home, or constructs alters within his home as location of his personal devotions. (A friend of mine of the Jain religion of India, has constructed a modern alter in his home at a former closet space.)

Homo sapiens possessed the qualities of soul, even more so than Neanderthals, with religious expression even more public, and groups of the ancient public meeting regularly to express their religious or mystical views regarding the intangible presence of a variety of gods, each god controlling a particular aspect of man's life. These aspects grew with organization of temples, synagogues, and churches, with a concept that gods would be present upon earth in a close association with man. This doctrine, with later belief in a single monotheism, became strongest in the years of Modern Man, approaching the Hebrew prediction of a coming Savior on Earth, with presence and earthly birth of Jesus as manifestation on earth of God.

THE PRESENCE OF JESUS

The presence and lore of Jesus was not wholly accepted within the Hebrew community, and at a later time Jesus was convicted and crucified for his alleged heresy statements that he was indeed the son of God on earth. (See another chapter in which it is advanced that the essence of God is within the soul of each human, and that God in Afterlife is actually present on the surface of earth but in a higher space/time dimension.) If so, was not Jesus actually telling the truth about the relationship of God and himself? This is in contrast with the extent in presence of emotion in previous species of man. The concept of one universal God expanded publically in following years with growth in religious dogma, priest, synagogues, churches, religious relics, alternative religions (such as Islam), with strong effects of genome development, even a "God gene", believed by some to be present in man's genome.

Alternative religions developed throughout the world, with presence in variety of religious thought and various rituals, not only in the former Mesopotamia, but also in widespread areas of Europe, China, Indonesia, North America (Indians) and religious practices of Central and South America settlers (even though yet undiscovered by Europeans.) The presence of genome that hungered for worship of a God or gods, was worldwide in Modern man, with the "common thread", I believe, in the "God Gene" genome of Modern Man.

HOMINOIDS AND DINOSAURS

About 350,000 years ago the system of life on Earth had increased in complexity with additional forms of species and subspecies. Hominoid creatures, generally resembling features of today's humans, had developed on Earth at about 1.8 million years ago.

Some of the previous hominids had appeared in the Great Ape primate family of Gorilla, Chimpanzee, Monkey, and Bonobo, each having a notable degree of intelligence, and familial relationships. Homo erectus had developed an ability in making elementary tools that would make their lives easier, and were able to perform activities they had not been able to do before.

First hominoids had appeared on Earth, four-legged, but later to become bipedal, and demonstrated familial traits in child rearing.

Many species, each believed by some anthropologists to be independent of each other species, and even some, like Denisovan hominid, recently discovered, violating biological rules of necessary failure in crossbreeding between species, came upon the surface of earth.

What today is known as the "Cambrian Explosion" happened about 500,000 years B. C. Many invertebrates, and other life associated with sea life, were on geologic plates of the eastern Pacific Ocean coast, and were raised up on the western continent, in the present Canadian British Columbia during compression action in plate tectonics.

At about 480,000 years ago dinosaurs came upon the face of the earth, perhaps by genome mutations, the various species lasting until about 65,000 years ago, when supposedly a large space object collided with earth in the present Gulf of Mexico, just east of the Yucatan Peninsula, its cloud to blacken the skies from sunlight, causing world-wide destruction of food sources, and causing death of nearly all reptile, egg-laying, land life. Species of Homo erectus, with his ability in problem solving, and many more species of mammals, fish, sea life, subterranean worms, bacteria, some birds and insects, all managed to survive this world-wide catastrophe.

A wide variety of mammal species filled the void left by this extinction of reptiles, and have greatly multiplied in species and subspecies to this day.

HOMO NEANDERTHALS AND HOMO SAPIENS

In the evolution of man-like bodies, it has been conventional wisdom by some anthropologists that man-like bodies evolved from early primates of Ape and Chimpanzee, to become in the future Homo habilis, Homo ergaster, Homo erectus, Homo neanderthal, and Homo sapiens, to Modern Man, which is currently called Homo sapiens sapiens (the true modern humans). Among other qualities, the skin color would have logically continued to have been black, or a dark complexion, until time for development of Caucasian originating at the Russian Georgia region of settlement.

About 125,000 to 100,000 B.C. a new genus of man appeared on Earth. The new genus has been dubbed, Homo sapiens, and had many basic anatomy and biologic features of Neanderthal, but also had some critical new features or qualities. Homo sapiens had a more complex brain capacity, with increased brain synapses that allowed abilities in mind and soul of reason, increased emotion, creativity, and many other qualities. (Please see "Qualities of Soul" elsewhere in this book.)

Anthropologists held a few years ago that Homo sapiens was a new genus, not a mere outgrowth of Homo neanderthal, How the new genus was created is unknown, but I have proposed a "Rose Petal Theory", described in my previous book, "Do We Live In Two Worlds?", Various qualities developed from a stem, like separate rose petals develop. From this, there evolved a form of divergent evolution, each petal, so to speak, representing a separate quality, or even a sub-specie of hominoid development, with its own unique features and DNA. This was followed by a phase of consolidation evolution, like a rose closes, whereby the new qualities merged, like a pod, bud, nodule, or rose hip on a rose bush stem, and from this a new genus or sub-species developed.

I propose that one of those qualities or mutations was a gene controlling change of skin complexion or color, being similar to the Caucasian race today. With this, two, and as many as four and five separate races, evolved and lived in common, from time of advent of Homo sapiens at about 100,000 B.C., to dying away of Homo neanderthal at about 22,000 to 20,000 B.C. Some isolated remnants of dark complexion people continued in remote locations on earth, of central Africa and migration to certain ocean island communities, or may have been adaptation of skin coloration dependent upon exposure to sun in equatorial areas.

Five defined races are:1) Negroid, 2) Caucasian, 3) Oriental, 4) North American Indian, and 5) Asian Indian. From these various sub-species, there developed from local environments and biologic needs peculiar to all in those sub-species.

The ruddy complexions seen in people of Mesopotamia and North Africa, including Egypt, may have evolved from sexual intermixture of Negroid and Caucasian people over hundreds or thousands of years. Change in skin coloration can be observed today in persons whose grandparents were decidedly black in complexion, but after only one or two generations of black/white marriages, offspring today appear much like Caucasian.

Did Homo sapiens specie evolve from Homo Neanderthal? Could they interbreed, a former test in biology of whether same or different species existed? Consensus of anthropology academia has in the past stated "No".

Did Homo sapiens evolve from some other species, perhaps Homo erectus? This is a possibility, and recent writings that Homo erectus, Homo Neanderthal, and Homo sapiens are in fact within the same genetic family and specie, much like a grandparent, their son or daughter, and you, are genetically related in the same species, but probably definitely different from each other.

OUT OF AFRICA?

There is continuing discussion among biological and cultural anthropologists on this subject, but most seem to favor the "Out of Africa" model. Two models arise for question in diaspora of species. One is an "Out of Africa" model that all the evolutionary action for the origins leading to modern humans originated and occurred in Africa. African sites are oldest, and modern forms of humanoid spread out from a particular place in Africa and then migrated throughout the rest of the world.

On the other side of academic discussions, the Multiregional Model, followed by those anthropologists who reject the "Out of Africa" model, suggest origins can be found of modern humans on three continents, if not also in the present Americas. The persuasion is that there was gradual evolution in all of these places that can account for modern humans. They slowly and gradually developed towards modernity in response to local selection pressures. The "seeds of evolution" and species development, had been broadcast, like seeds of grain in a plowed field, throughout the world at the same time. There is evidence of human existence in North America as early as 13,000 years B.C. (Clovis points of west and southwest United States present evidence of ancient human existence, and there is evidence of human life in South America 20.000 years ago.).

There are worthy arguments and logic supplied by both sides of the "Out of Africa" interpretations, and multi-regional anthropologists, with still no consensus to decide which is correct. They have investigated and analyzed 1) the fossils, 2) the dating of the sites, and 3) the genetics involved.

There is a mid-ground called the "Partial Replacement Model" which holds that Homo sapiens did originate in Africa and did migrate north to Asia and east through India to Southeast Asia, to China, through Indonesia, and to Australia. But along the way there was sometimes replacement with other species that had migrated before them, with interbreeding and replacement, with hybridization so that replacement of species occurred gradually and basically subsumed the genetic material of the other hominoids. This integrated or absorbed the other specie populations. An example might be in study of Homo Neanderthal, that was overrun by superior Homo sapiens, which would actually create a new sub-specie of Homo sapiens, even perhaps a new race.

In summary, of many forms of life, including fish, amphibians, reptiles, the mammals, bacteria, algae, and yeast, following the Hellish time of Earth, cooling to the catastrophe of cosmic collision at 65 million years ago, mammals developed, among other species, into great apes, chimpanzees, monkeys, bonobos, and later orangutans at about 7 million years ago when first human-like, and eventually two-legged creatures, existed as primates, part animal and part life-suggesting man. Many forms of species developed, and became extinct, including Homo ergaster. Modern Homo erectus appeared upon Earth about 600,000 years ago and was present at 350,000 years ago when Homo neanderthal species appeared, perhaps as a predecessor of Homo sapiens.

THE CAUCASIAN RACE

Still another theory is advanced by the author. Homo erectus is reported to have originated in the present Georgia area of southern Russia, based on archaeological reasoning and fossil finds. The species migrated south 200 miles, eastward through Mesopotamia, India, northern China, perhaps Siberian Russia, southeast Asia, Indochina, Papua New Guinea, Australia, New Zeeland, Fiji, Polynesian and other islands of the Pacific, and to South America. In effect, all groupings of species emanated from Homo erectus or its subspecies.

It is believed that a "modern" Homo erectus originated or evolved in the Caucuses Mountains between the Caspian and Black seas, at mountains of 11,000 feet altitude and among the world's lowest elevations at the Caspian Depression, at around 600.000 B.C. Today heavy rains in the area encourage still another geographic environment. The species were of lighter skin color and were unlike in other biologic ways from the Black race originating in Africa, many near the Rift valley and at other areas of Africa.

First Homo erectus fossils date to 1.7 million years ago at China. Modern Homo erectus, demonstrating advanced mental, emotional, and spiritual qualities, date from about 600,000 B.C. It is not certain how or when Homo erectus migrated from Russian Georgia. Recent finds of when Denisovan hominines, modern humans, and Homo neanderthals last shared a common ancestor was around a million years ago. Recent finds of Denisovan hominids date to about 50,000 B.C. Peking man dates to 41,000 B. C., Heidelberg Man to 50,000 B. C., and Java Man at 1.8 million years ago. The book "Prehistoric Man" dates Homo erectus in Russian Georgia about 1.6 million years ago, and the fore-mentioned book, "The Celts" recognizes outward migration starting about 3,500 B.C. It seems reasonable to date Caucasian Modern Homo erectus in Russian Georgia at around 100,000 B.C.

Groups of Homo erectus species radiated from the Caucuses Mountains of present Russian Georgia. A second group, separated from the first main movement, later migrated about 200 miles southward into western Turkey and present northern Iraq (the Kurds of Kurdistan), Mesopotamia, northern Africa coast, and Egypt. A third group radiated north and eastward into Russian Asia and Siberia (possibly to become the Denisovan hominids). A fourth group radiated eastward through India to south east Asia, China, Indochina, and later eastward beyond. A fifth group radiated westward into Switzerland and Europe, the Balkans, Greece, northern Italy, France, Spain, Germany, the Norse land of Denmark, Norway, Sweden, and Finland, the lowlands of Netherlands and Belgium, and westward to England, Ireland, and Scotland. From these radiations various sub-groups developed, grew, civilized, nurtured, and developed in unique geographic areas, with cultures, language, and common biologic features.

A listing of man-like, human evolution fossil origins is presented by Google-Wikipedia, including those of Home erectus, Several Homo erectus fossils were listed as found in China (4), South Africa (1), Kenya (1), Tanzania (2), Ethiopia (1), Indonesia (5), Spain (1), Algeria (1), France (1), and Greece (1).

Migration charts ("Prehistoric Life", DK Publishers, 2009) indicate a diaspora from Africa (about 1.7 million years ago) eastward, with a branch of Homo erectus migrating northward to present Russian Georgia, near the Caucuses Mountains between the Black and Caspian Seas, at about 1.7 million years ago, and others continued migration eastward to China and Indochina about 1.6 million years ago.

Few Homo erectus fossils are listed as found in Russian Georgia, although it seems apparent that the area near the Caucuses Mountains was rich with the species of Caucasian race. It should be recognized, however, that absence of fossils is not proof that some certain species had not previously existed there. For example, many persons would have died along the migration path to the east, to China, and elsewhere, but few, if any, fossils have been discovered along the migration trail.

Perhaps reason for this void in Russian Georgia fossils can be related to environment, elevation, climate, weather, geology, a more dense population, agrarian farmers, and others since, plowing grounds and burying fossils, or other reduced-preservation conditions.

The author believes Homo erectus of the Russian Georgia region was home for diaspora west, north, east, and south, propagating those diverse areas. The eastern area became origin of the Denisovan hominids in northeast Asia and Siberia, Mongolia, India, China, and to Java, about 50,000 B.C. Other Homo erectus migrated to Mesopotamia, Turkey, the Balkans, Greece, and westward to Switzerland, France, Germany, Spain, Norse land, and to England, Ireland, and Scotland of Great Britain. Eastward migration populated India, China, Indochina, and later Papua New Guinea, Australia, New Zealand, Polynesian islands in the Pacific Ocean, and to western and northern South America and Central America.

The author believes the Kurds of Kurdistan in Turkey and northern Iraq are distant cousins or ancestors of most people having origins in Central Europe, the British Isles, Russia, Asia, and other locations of predominantly Caucasians.

The place of origin of the future Celts was also the origin of Homo erectus, which branched westward to Europe, southward to Mesopotamia, and eastward to India, Indochina, Indonesia, northern

China, Mongolia, and the Denisovan hominids of Asia and oddly to Java and Papua new Guinea, enlarging the area populated by Caucasians, or sub-species of Aryan, Indo, Oriental, Asian, future North American Indians, southeastern Asia Polynesians at Pacific Islands, and to South America. This biologic strain might have also been ancestors of people of Turkey, Greece, Macedonia, Balkans, Italy, Corsica, Sicily, and Spain.

Similar biologic features can be observed even today by Kurds and Caucasians of Europe and the Celts. Both are of Caucasian race, many with dark-color hair or observable Nordic blond hair, common facial features, and clearly original and common biologic ancestral features, with a conspicuous difference from Black species originating in Africa.

THE HOBBITS OF INDONESIA

Another example of eastward Homo erectus migration involves the "Hobbits of Indonesia". In 2004 researchers working on the island of Flores in Indonesia found bones of a much shorter human species, formally named Homo floresiensis and nick-named the "Hobbit", that lived as recently as 17,000 years ago. Scientists originally postulated that H. floresiensis species descended from Homo erectus. New investigations show that the hobbits were more primitive than researchers thought, a finding that could overturn key assumptions about human evolution. (Source: Scientific American, Nov. 2010)

Our biologic, anatomy, mental, psychological, and spiritual progression had advanced tremendously when man, in various parts of the world, made transition from Paleolithic age to a cultural breakthrough to the Neolithic and then agrarian mode of civilization.

Life had advanced from single-cell amoeba to multi-cell eukaryotes having developed a cell wall, enclosing a nucleus of genome, RNA, DNA, and one or more electrons orbiting an atom center. The cells had joined to form molecules, formed life-allowing organs, other life—enhancing organs and appendages, with covering of fur, hair, or shell. Two goals in life were 1) continued existence, and 2) reproduction of their species.

Fifty-six million years ago a mysterious surge of carbon into the atmosphere sent global temperatures rising. In a geologic eye blink life was forever changed.

Earth was hot and ice free at the end of the Paleocene Epoch. With sea levels 220 feet higher than now, the Americas, not yet joined by continental drift, were smaller.

Forms of man developed from primate animals, to hominids having man-resembling skeletons, appendages, protective skin covering, and the five senses of sight, hearing, smell, taste, and touch.

From this hominid world a branch of life developed of hominoid, specifically the Ape family of Gorilla, Chimpanzee, New world monkey, and Bonobo.

The Neolithic period could be generally defined as the time in development of Homo sapiens prior to the agrarian, agricultural mode of living. We commonly picture this as days of the caveman, void of housing, and subsisting in families in day-to-day hunting for food and cooking by open fire. Clothes consisted of animal skins. In some locations art, pictured on cave walls of rock or sculpted into spiritualistic dolls, had existed from their ancient times. Tools of axe and knives for cutting were fashioned from stones, with some fashioned from chance findings of iron or copper.

RISE OF AGRARIAN CULTURE

Such was the mode of living in Mesopotamia, Egypt, and other Mideast locations until about year 10,000 B.C. At that time the mode of living changed to an agrarian culture whereby seeds were planted to provide a planned source of food. Meat from animals was preserved by rubbing with salt found in dried up sea beds from salt water. Change from a culture of hunting and gathering to an agrarian mode of living and culture of land development, even like farmers today, who cultivate their lands, participate and enjoy activities of hunting wild animals and preserving their quarry for future use. This was not an either/or decision for economic use and allocation of their time. Over many years, guided by economic decisions (a new quality in the mind of man) greater time allocation was given to agrarian activities.

Communities developed with various persons assembling tools, items that others of near and far communities would desire, the rudiments of trade. Housing provided permanence to their location and building of communities.

People in communities would find it advantageous to band together for purpose of trading, performing skills in various specialties, and for protection from outsiders, both animal and human. Communities banded together into kingdoms. The first kingdom of record in Mesopotamia was Akkad, ruled by King Sargon The Great, who lived 2,270 to 2,215 B.C., in a location now known as Iraq. Many other kingdoms arose, and some were warring kingdoms.

HOMO ERECTUS AS A BASE SPECIES

At about 125,000 years ago a new species or subspecies of Homo erectus, Homo sapiens appeared from some unresolved source. Both Homo Neanderthal and Homo sapiens existed during the common time until 30,000 to 28,000 years ago when Homo Neanderthal, as a separate species, died out, or perhaps merely was replaced by a subspecies of Homo sapiens, leaving Homo sapiens as the sole human species on earth, before speech and language were developed.

The author believes that some aspects of spiritual soul were introduced with Homo erectus at about 600,000 years ago. The various qualities of soul were expanded gradually over thousands of years by mutation and development, and appeared "full blossom" in Homo sapiens between 125,000 and 30,000 years ago, to be unified and coordinated, replete with reincarnation to humans, of that cultural time.

Recent genetic analysis indicates that early Homo sapiens did interbreed with Neanderthals, according to researchers at the Max Planck Institute for Evolutionary Anthropology in Germany. Using DNA fragments from ancient human bones, the team sequenced 60 percent of the Neanderthal genome and found that humans in Asia and Europe share between 1 and 4 percent of their DNA with the ancient Homo neanderthal. The researchers also identified genetic differences related to cognitive functions and bone development that may have helped modern man flourish, whereas Neanderthals died out 30,000 years ago. (Source: Scientific American, July 2010). Is it possible that there was "flow through" in genetics from some populations of Homo erectus direct to Homo sapiens and certain populations of Homo neanderthals were by-passed?

About 350,000 years ago the system of life on Earth had increased in complexity, with additional forms of species and sub-species such as H. neanderthal and later H. sapiens and H. sapiens sapiens (Modern Man). Hominoid creatures, generally resembling features of today's humans, developed on Earth at about 1.8 million years ago.

Some of the previous hominids had appeared in the Great Ape primate family of gorilla, chimpanzee, bonobo, and old world monkeys.

CREATION AND DEVELOPMENT OF SOUL

The author theorizes that at moment of the Big Bang, some of the positive-charge particles continued existence during a moment of annihilation. As negative-charge particles formed matter of our cosmos, most positive-charge elements evolved to an intangible cosmos in one or more of a probable 10 or 11 or more dimensions in our space/time relationships, to become the mystical "world" for soul development.

It is proposed that there was development that the institution of soul would "mature" at perhaps 600,000, 300,000 to 200,000 years B.C.E., or some other time, with soul incarnated further (reincarnation) to the human body, then known as Homo erectus.

With incarnation of soul to the human body, the mind, the soul, and conscience, to the brain of man, complete new species, or sub-species, developed what would be known as Homo erectus, Homo Neanderthal, and later Homo sapiens. Sub-species developed under each of these primary species as moderate environmental conditions required.

SOUL AND REINCARNATION

Although the moment or period of Annihilation, affecting both negative-charge particles and positive-charge elements were both present at the Big Bang, the process from creation would be much different. Unlike the vicissitudes of negative-charge elements that created and evolved into various matter, positive-charge elements evolved in a time-space universe of antimatter elements.

Being non-physical, positive-charge elements in antimatter had no worldly concern with development or evolution of its otherwise negative-particle self. Soul, being non-physical, would possess no sense of sight, hearing, smell, taste, or touch associated with animals in a world of matter. These were qualities only of animals living in a world of matter.

Evolution in the existence of soul would evolve primarily in emotions. This quality of emotions in hominoids would complement the animal qualities of matter to produce the current state of wholeness in modern-day humans. Souls from positive-charge elementary particles would be nurtured in a form of evolution, not only in emotions, but also appreciation of a higher dimension (summum bonum) to preserve the body above the body's own existence.

In a previous book by this author, "Do We Live In Two Worlds?, I detailed two existences concerning soul. In one instance there was one basic brain activity needed by an animal to subsist, to take in food, to process the food in biological conversion to gain and expend energy, excrement, movement, communicate by hearing and doing, possess a nervous system, have body maintenance activities, brain thought processes, and other processes to live, allow reproduction of its specie, and to die, with little purpose to its being.

A second listing details further qualities that soul adds to humans. At least 50 or 60 qualities, purely of the mind (or soul), have been listed and more could be added with additional analysis. These qualities are purely of the mind and soul, and not necessarily of pure brain working.

It is conceivable that scientists and technicians, at some future time, could assemble a machine in a laboratory that would sustain itself like a basic physical brain and body. However, such a machine would probably not have ability for analysis, emotion, judgment, soul, and "mystical" qualities of the mind and soul as does the mind of man today.

Because the process of annihilation occurred at a brief moment during the Big Bang 13.7 billion years ago, development or evolution of the soul had more time to develop than did items of matter and animals. First life of algae and bacteria did not happen on Earth until about 3.8 billion years ago.

Meanwhile, the "world" of soul and antimatter was not restricted by the cosmic changes unique to Earth formation. A beginning evolutionary process for soul development could have commenced at a time 13.7 billion years ago at the time of annihilation, far earlier than evolution of a solid and inhabitable period by early earth life formations, a clear advantage of about 10 billion years.

There are several events in spiritual development that all seem to tie together. First is the onset of Homo Neanderthal and Homo sapiens. Scientists project that the time span of each were different, with Homo Neanderthal entering the scene of evolution at about 275,000 years ago. H. sapiens fossils had not existed until age about 125,000 years ago. Both possessed many qualities attributed to soul. Paintings on cave walls by Neanderthals expressed knowledge and belief of a superior being that would hopefully affect success in hunting expeditions. Homo sapiens, thousands of years later, to preserve the deceased body, went further in burying their dead, sprinkled with a preservative of Red Ochre that would supposedly allow future life of the soul in an Afterlife, a thought and philosophy manifested thousands of years later in Egyptian mummification.

A second factor in the unification of body and soul deals with when incarnation of soul to the animal body took place. Some Biblical theory holds that the event happened about 4004 B.C. Logic in time of this creation event would set time of incarnation of soul, and existence of certain qualities of soul, at a much earlier time, a time to 600,000 B. C., with advent of modern Homo erectus. The actual event of Creation is far before any Biblical reports, originally developed, perhaps by Pleistocene cave men sitting on rocks, trying to devise a concept of where they came from.

SEX IN ANTHROPOLOGY

Another element to this evolutionary conundrum is the wide discrepancy of when sex came into the evolutionary picture. Some accounts of the Bible place events of the Garden of Eden, creation of man (Adam), and creation of women (Eve) at about 4004 B.C.E., without sexual union being mentioned in Genesis, although their offspring were named.

Fossil remains of Australopithicus africanus places time of the skeletal fossil of a female, nick-named "Eve" or "Lucy", at about 275 million years ago. Obviously, females existed before the 4004 B.C.E. time biblically described.

SOUL AND THE MONARCH BUTTERFLYS

An analogy is presented to show a parallel of the miraculous procession of soul, with its many existences, to that of the common Monarch butterfly. The Monarch butterfly, first of all, is birth in morphology, through an egg laid by a previous generation of that butterfly specie. The egg develops to give birth to a caterpillar having colorful bands resembling the coloration once a later butterfly will be formed. Through a still unknown process, the caterpillar transforms (morphs) to form a pupa, like a cocoon. In a matter of days the pupa breaks itself open to reveal the birth of a fully-grown Monarch butterfly, whose wings will harden in a few minutes, allowing the butterfly to take flight in a new generation. The butterfly then experiences a life on earth, and at a certain time of the year, commences a journey to southern climes of the Sierra Madre Mountains in Mexico. Pictures exist of thousands of Monarch butterflies wholly covering trees of the locale where they remain until a certain time at Spring when they will return North, and that generation will later return to Mexico in the Fall, as did its grandparents and parents.

This is where the magic begins! Each butterfly in the late winter or springtime will leave its spot of location in Mexico and set forth to return to the place of its northern origin, that wonder in memory of route from the north not being at all understood. Further, that particular butterfly may not live to arrive at its spot of northern origin, but instead stops on the way to initiate the birth cycle of laying eggs in which its parent formerly was engaged. When this happens, its next generation (the third) continues the route back to the place of origin of its grandparent, arrives there, and initiates again a cycle of egg, caterpillar, pupa, and birth. Not only is the effort exerted in the long-distance flights a wonder, but the memory process, and knowledge transference between generations, all within its insect-size brain, is miraculous, and all within its morphogenesis through the physical stages of parent, egg, caterpillar, pupa, and emergence as a newly born Monarch butterfly in a new generation.

A parallel can be visualized of the Monarch butterfly and the existence of Soul. The butterfly lays its eggs (like a human reproduction process), to form a caterpillar (birth process), change to pupa (human living process), performing a miraculous morphing (human death), to a completely different form (emergence of soul), and emergence into starting a new generation (reincarnation). The process seems miraculous, and as mystical as is the incarnation of soul to man, and transpiring over a number of generations. The process has been scientifically observed, but so far is unexplained, let alone understanding the memory transference between generations. A general system exists in the ways of the world, and it may be logical that the system of soul would parallel this system.

A second evidence of retention of long term memory is observed in the Salmon fish. Young Salmon will be hatched from the egg of some previous generation, frequently in headwaters of some stream, river, or perhaps an inland lake, leave its freshwater origins to enter other freshwater, or salt water oceans, where that individual salmon will swim freely or in a school of salmon. At a certain later time, that mature salmon will somehow find and return to the same river, lake, and small stream from which it originated, where it too will lay fertilized eggs, as did its parent, and probably grandparent before, and a new generation will enter the birth cycle of nature. The operation of this manipulation and migration in nature remains as an unknown quality, but is believed to relate to analysis of chemicals peculiar to the river of origin of that salmon.

The phenomenon relates also to why geese and summer birds fly south, even before winter temperatures drop, and when Fall and later Spring migrations take place, to former locations.

This event in nature may also parallel the miraculous memory that transfers to a new generation, such as sometimes experienced in reincarnation of soul to a new human.

AN INDUSTRIAL MICROCOSM OF MAN'S EVOLUTION

The succession of biologic changes in various species of hominid (possessing hint of man-like physical features), hominoid (possessing basic physical, neurons, brain, and mind, features later to be found in man and Modern Man), can be likened to the growth in modes of transportation, specifically the automobile, where vehicles started out as very rudimentary, and have evolved to today's marvelous state of automotive technology.

For centuries man had recognized various elements existing, and connected them together in an intelligent manner to ease his life. Horses and other animals lived that possessed power, individually or in a team, more than man himself could supply. Vehicles had developed from "carry sticks" to be pulled by humans "Indian style". Wheeled vehicles, such as wagons, were developed that were better for transporting goods and belongings of man. Combining horse for power and wheeled vehicles was a major improvement.

In the early 19th century man made further improvements in transportation. A gasoline-fueled engine for power was joined to a wheeled carriage, with a tiller bar to enable control of direction. The automobile was born (although similar creations had been made in Europe—in 1893 the first automobile built in the United States was modeled from the Duryea brother's creation), an intelligent joining and improvement over previous modes of transportation. This was like a new form, or a new "species", had been created. Henry Ford and Ransom E. Olds were giants in development and manufacturing of the "horseless carriage".

A period of time occurred in which designs and mechanical innovations occurred for Henry Ford to present the "Model T" automobile in 1908 (like an improved species of automobile), a substantial improvement over the "horseless carriage". Henry Ford developed a standardized system where many such vehicles could be reproduced, and the number of Model T automobiles grew to meet the awaiting demands of people. Other persons developed and manufactured other styles and types of vehicles, like "sub-species" in evolution of the automobile.

The Ford "Model T" series was surpassed in year 1927 by a revolutionary Model "A", and in year 1932 by a Model "B". A succession of improvements followed with intelligent improvement in appearance and mechanical changes, with effort to produce and lower costs for widespread usage and ownership. Designs and modifications took place, almost as if the manufacturer had a master plan for the future, and a development program to be followed. Such plans seemed to follow, each year presenting improvements in the "sub-species" of automobiles.

We live today in what might seem as the ultimate in efficiency, style, design, and comfort features of automobiles, but sense that the future will hold even greater, perhaps unthought-of automotive improvements. Other modes of transportation (like new "species") existed too in railways, busses, trucking, as well as in modes of communication, and computers developed for providing answers to new needs of the world.

With some change in terminology, this development of automotive engineering can be visualized as a modern-day microstructure, paralleling species evolution, in which new models (species) were periodically introduced to meet demands of life's needs. Such is the parallel with presence of new species and sub-species, in intelligent presence of new types of hominids, hominoids, Homo Neanderthal, Homo sapiens, and Modern Man, preceded by many species of hominoids as "new and improved models" of their time.

Robert Fulton, in 1787, joined boats and steam power to initiate steamboats. (He also invented the submarine).

Development of species is like development of the computer, internet, and cell phones. Change comes not all at once, but gradually, giving the occasional viewer the impression that a new series was developed instantaneously. In species development these small changes are known as sub-species. Sub-species developed gradually over many years.

INTELLIGENT DESIGN DIFFERS FROM AN EVOLUTION PROCESS

There are many questions in evolution that anthropologists attempt to answer, step by step. One major question is whether each species developed from a prior specie, or was each specie newly created?

Intelligent Design is the plan for life movement. Evolution is the step-by-step method of how life movement occurs. Intelligent Design and evolution are compatible in life changes. In an industrial setting, engineers and designers intellectually plan the changes in product (Intelligent Design), with workers and technicians performing the actions (Evolution) to affect the changes.

The mystic of the automobile industry, with multi-national operations, might be likened to the soul, with its operation, knowledge, communication, training, concepts, chain of command in organization, planning, development, and support staff, all paralleling that of soul within man.

The workers, supervisors and managers might be likened to Souls, Guardian Angels and Archangels. In life development, an unseen intelligence plans the direction, and manipulates DNA to affect the change.

Manipulation of RNA and DNA, with consequent change in enzymes and organs, is performed by soul and mind at the direction of a Supreme Being (God).

ADDITIONAL DEVELOPMENTS

With demise of many reptile species of dinosaurs, mammals expanded their presence and proliferation in many species from Saber Toothed Tigers to Mammoths and Mastodons. Great Ape species of Gorilla, Chimpanzee, Orangutan, and intelligent monkeys of over 6 million years earlier exhibited intelligence, and Intelligent Design, with their development by successors of intellectual and physical tools, raising their competitive abilities above that of other primates, from Australopithicus Africanus, personified by "Lucy", the first primate found to be bi-pedal, dated to 3.7 million years ago.

Ardipithecus ramidus (loosely nick-named "Ardi") living 4.5 million years ago, is the most complete female specimen known of the hominid groups. Several species of Australopithicus and Ardipithecus appeared on Earth. About 2 million years ago new families of Homo species appeared as Homo rudolfensis in eastern Africa, Homo habilis in sub-Saharan Africa, Homo ergaster in eastern Africa, and Homo erectus migrating southwest to Asia and Africa to eastern Asia at about 600,000 years ago and lasting until about 250,000 years ago when Homo Neanderthal in Europe and Western Africa became the dominant specie before Homo sapiens, a new and more complex species in both physical and emotion traits, appeared at about 120,000 years ago, outlasting the demise of the last Homo Neanderthals at about 28,000 B.C.

The author holds that strains of Homo erectus flowed through H. neanderthal, H. sapiens, and Modern Man to ourselves.

Meanwhile, another culture, species and several sub-species was developing in a part of the ancient world, apart from Africa, Egypt, and Mesopotamia. China was on the expansion route of Homo erectus migrating to the east.

Since inception and evolution about 1,800,000 B. C., Homo erectus had continued a relentless migration from the Asian Mideast eastward through India, Indochina, central and northern China, Papua New Guinea, Australia, New Zeeland, Fiji and other Pacific Islands, to South America. Further, Homo erectus had migrated and developed as sub-species in central China and the Korean Peninsula, (later to become Japanese).

Further sub-species had developed to form the Denisovans, a recently discovered (2010 A.D.) species or sub-species of people in northern China and Siberia, who may have been ancestors, root, culture, and population of future North American Indian, and another sub-species also through Indochina and Papua New Guinea.

Other sub-species migrated to Indochina, Vietnam, and Papua New Guinea, perhaps ancestors of the diminutive Homo floresiensis of Papua New Guinea, over an ancient land bridge, caused by low ocean water levels, to northern Australia, and eastward by raft or boat to Pacific Ocean islands of Fiji, Hawaii (as Polynesian people), and thousands of miles to the coastal area of South America, as possible forerunners and ancestors of the ancients who would evolve to be the Maya race and culture of south and central areas of South America.

The Meakumbut of Papua New Guinea, located in the mountains, may be one of the last of the cave-dwelling, seminomadic people of the earth. They are unique in their culture and symbiotic painting of their faces and bodies. (Source: National Geographic)

WHAT IT IS TO BE HUMAN

In the present status of man, the brain, (its complexity and mass), the mind, (demonstrating primarily emotion and judgment), the soul, (an extension of God to each individual person), the genome (affecting the enzymes and biology of man), extra dimensions, (some scientists say at least 11 dimensions in time/space relationships), energy, and speed of light, all exists. In Afterlife there is extra-sensory communication without benefit of the five senses, transference of thought through space and judgments (Example on Earth: The ability of Dr. Edgar Casey (1877-1945) to diagnose patient maladies from several miles distance), summum bonum,(or the appreciation of a superior, mystic power, which is the highest good as the end or determining principle in an ethical system), are all related in the essence of mankind.

Like a computer, the brain represents the hardware. The mind is like the computer software that contains the various intelligence programs that instructs and controls the hardware (the brain) to operate in a manner that provides a desired result. It is the ever living soul, with its line of communication to a superior being, that is our consciousness and conscience that guides our mind and brain to make decisions, hopefully, in a manner that provides desirable results.

The soul can affect change in the mind of man (probably by an electro-chemical process to the genome) by manipulation of the genome and subsequent actions (to specific enzymes and to the body, brain, and mind.

SOURCES

1. The Great Courses (Several courses and professors)
2. Google—Wikipedia, The free Encyclopedia (Primarily for dates)
3. Western Civilizations (College history textbook)
4. King James Bible
5. Harper Study Bible
6. Archaeology magazine
7. National Geographic magazine
8. Scientific American magazine
9. Prehistoric life, DK Publishers, 2009
10. Discover magazine
11. Nature magazine
12. Encyclopedia Britannica
13. Physics Annual Review, The University of Michigan
14. A wide variety of general reading

LIFE II—TIMELINE

The following Timeline is presented to relate various developments in perspective to other events occurring during the long era from before the start of time to the Agrarian Revolution at about 10,000 B.C., and to the time of Christ.

Period	Event
Minus time	**BEFORE THE BIG BANG** (A timeless eon before recognized time.) Islands of energy were collected into a Black Hole. There was acquisition of energy, not matter, since no matter existed in the relatively small cosmos prior to our own original Black Hole of energy. There was a subsequent Big Bang, Singularity, and Annihilation of positive charge elements. Other galaxies of stars formed nearby (relatively) in the cosmos, before and after the stars of the Milky Way Galaxy, in a similar and progressive pattern—"an archipelago of galaxies".
13.7 Billion years ago "	**BIG BANG EVENT** Moment of Singularity of first elementary particles (hydrogen and helium) and annihilation of positive-charged particles by negative-charged particles. Immense areas of energy and heat had been contained, and expanded back into the cosmos at the travel speed of light. Positive-charge elements became antimatter. Negative-charge elements remained to form matter.

13.4 billion years ago	**FIRST STARS** Separation and conversion of energy to creation of matter (Einstein's principle). Agglomeration of energy, self-incurred gravitation by rotation, build-up of heat, explosion of stars in supernovas, and conversion of energy into matter. First supernovas recede back to stars by gravity, reheats, continues chemical action of fusion, forms new elementary particles through heat and pressure. Galaxies are formed by gravitational attraction of stars and offsetting outward forces in cosmos. Stars illuminate by heat and fusion action. Much space between stars, galaxies, and star clusters.
11 "	Gamma ray burst in cosmos originated such as by explosion of a star 30 times more massive than our sun.
7 "	**COSMOS** Early supernova of cosmic stars creates new particles and elements upon star regeneration.
4.57 "	SUN Our Sun was formed when a hydrogen molecular cloud collapsed, (and re-formation followed a supernovae explosion, per author). It illuminates, caused by pressure and heat.
4.5 to 4.0 billion years ago	Sun re-forms after supernovae, leaving planets, asteroids, comets, and other debris in orbit, with chemical formation as mass increased by assortment of new minerals and agglomeration.
Unknown	Orbit of Earth (and other planets) around Sun varies, probably increasing in radius over time.

4.5 "	**HADEAN EON** Reduction to Hellish-like molten condition of hot Earth. Oldest known rocks (in Iceland and Australia). Oldest known mineral (Zircon).
4.4 billion years ago	Hadean Earth continues melt down. Differentiation in mineral solidifications and locations in Earth. Moon forms from glancing contact of large cosmic substance, gouged perhaps from present Pacific Ocean area.
4.0 to 2.5 "	**ARCHAEAN EON** (Origin or Beginning of an Ice Age) Cooling of rock surface. Continental geologic plates form and break into continent-size sections. Toxic atmosphere of methane, ammonia, sulfur, and other gases. Earliest first life on Earth in prokaryotes of oxygen-producing bacteria. First single-cell (Prokaryote) life of bacteria and archea. New theory arises about origins in Australia of life on Earth: Carbon-rich meteorites bombarded Earth for 140 million years.
4 billion years to 120 million years ago	Age of a typical "Brown Dwarf" star (in Pleiades cluster)
3.5+ to 2.5 billion years ago	**ARCHEOZOIC ERA** (Larval life, only few known fossils)
3.8 "	First life on Earth of algae, virus, microbes, and bacteria. Prokaryote cells. Split occurs between archea (future animals) and bacteria. Oxygen oxidizes iron creating iron ore compounds. First viruses.

Ancient time unknown—sometime after introduction of eukaryote cells.	Split from original amoeba, virus, and bacteria of animal and fishes in development of eukaryote cells having a cellular wall, a nucleus that contained neutrons, protons, and a basic genome, and one or more electrons in orbit of the cell, and has ability to reproduce itself to perform various functions for body and brain existence. This organism would further develop a spinal column, nerve system with a brain as a controlling center, blood and capillary system, having a heart to pump and circulate fluid of blood, intake, digestion, and excrement system, and other biologic systems of body existence. This notochord could be described as a first, explainable in its design and creation by environmental needs. Development of original viruses, bacteria, and amoeba is considered a natural, biological, science-based, and explainable in development from physical forces, the origins still not universally accepted in science and philosophy, but considered not to be mystical creation.
3.5 billion years ago	Earliest fossils of organisms (Stromolites). Increase of oxygen in atmosphere due to photosynthesis. Moon very close to Earth causes 1,000 feet-high tides with hurricane-force winds that stimulate evolutionary processes.
3.4 "	First evidence of increase of oxygen in air due to photosynthesis and oxygen output of plants. Prokaryote cell fossils.
3.2 billion ya to 542 million ya	**PROTOERZOIC EON** (Oldest forms of life) Period before the first abundant, complex life on earth.
3.2 to 2.86 bya	Little is known about this period.
2.86 to 2.8 "	Oxygen in atmosphere begins to increase due to photosynthesis.

2.5 billion to 580 million tears ago	Excess carbon dioxide causes 4 periods of mass extinction during the Proterozoic Eon.
2.1 billion years ago	Fossil algae.
1.5 "	First eukaryote cell organisms. Green algae colonies are in shallow seas.
1.2 "	Sexual reproduction first appears.
1.1 "	Tracks of multicellular animals found in India.
1.1 billion ya	Rodina exists as oldest known continent until Pangaea splits off between 750 to 225 million years ago.
800 million years ago	Supercontinent of Rodina. Global glaciation. "Snowball Earth" period. Multi-celled eukaryotes replace prokaryote cells. Complex single-celled life. Volcanoes on Planet Venus release lava and gases that triggered a powerful greenhouse effect there. Break-up of Gondwana and Pangaea.
800 to 630 "	Soft-jellied creatures in seas, shaped like bags. Worms. First sponges.
670 "	First multicellular animals (soft body), Eukaryote cells (multi-function), early plants.
600 million years ago	Chordate fishes: notochords with hollow nerve cord, gill slits.
580 "	Paleozoic Era-Cambrian Period begins. First well-preserved fossils of Nautiloids, Ammonites, Trilobites, Brachiopods, and Gastropods abundant; climate warming.

580 to 540 "	"Ediacaran biota" plant is first large, complex cellular organism flourishing in seas. Most modern phyla of animals first appear in evidence from "Cambrian Explosion" of species types found at British Columbia province of Canada. Ozone layer of atmosphere oxygen allows formation of layers blocking ultraviolet radiation, permitting colonization of the land. First fungi.
550 to 500 "	Leaf-like flora
540 to 250 "	**PALEOZOIC ERA (Ancient life)**
505 "	Fossilization forms specimens at ocean level, uplifted fossils in British Columbia, Canada later to be described as "Burgess Shale".
Unknown	Lobe of Mediterranean Sea over North Africa left deposits of Salt upon evaporation.
500 to 490	**CAMBRIAN PERIOD** (Major groups of animals first appear)
500 "	First vertebrates. Subcontinent of Rodina breaking up. Cambrian Explosion of fossil types. Most modern animal phyla appear. First chordates. Period of mass extinction at end of Cambrian period.
500 to 1 "	**PHANEROZOIC EON** (Current eon of abundant animal life exists)
500 "	First bony fishes. Insects, fungi, algae, worms, sponges, their likenesses continue to our present day.

490 "	**PALEOZOIC ERA** begins: First animals with hard parts, trees with leaves, trunk, and root system, first corals, starfish, fish-like animals. Lake Champlain and Taconic Mountains formed in New York state. Predecessors of Green Mountains in Vermont formed.
488 to 444 "	**ORDOVICAN PERIOD** (Period between Cambrian and Silurian Periods)
485 "	First vertebrates with true bones. Jawless fishes.
475 "	First organisms on land.
468 "	Bivalves, nautiloids, trilobites, starfish.
450 to 340 "	Vertebrate fishes
444 to 416 "	**PALEOZOIC ERA-SILURIAN PERIOD** (Major extinction of 60% of marine species)
437 "	Period of mass extinction.
430 million years ago	Heavy concentrations of carbon dioxide from many volcanoes cause abundance of carbon; pooling of extensive concentrations of water in seas as hydrogen and carbon dioxide mix and leave residue of carbon in rocks.
420 "	First ray-finned fishes and land scorpions. Western North America inundated by Western Interior Sea, called the Niobrara Sea. Colorado, Mississippi, Columbia rivers and their tributaries are major drainage channels. Utah salt flats are evaporated remains of a salt sea. K-T Explosion of numerous specimens of pre-historic sea life, encased at uplifted Canadian Rocky Mountains at British Columbia. Grand Canyon carved to reveal geologic time records there.

416 to 359 "	**PALEOZOIC ERA-DEVONIAN PERIOD** (Fins of lobe-finned fish evolve into legs, arms, digits, etc.)
410 "	First signs of jaws or teeth in fish. Earliest nautiloids, brachiopods, corals, crinoids, trilobites. Age of fishes (bony plated, lungfish, sharks, mollusks, amphibians), squid, octopus, cuttlefish. First forests as trees develop to absorb carbon dioxide and disperse oxygen in percentage near present day proportion.
400 "	First reptiles. First sharks, amphibians. Supercontinent of Gondwana. Lobe-finned and ray-finned fish, sharks, amphibians. Appalachian Mountains formed. First vascular plants, forests, land flora. Millipedes, tetrapods, crabs,
395 "	First lichens. First seed-bearing plants. First trees. First winged insects, bees as Lepidoptera.
363 "	Earth begins to be recognizable as sea and land. Insects, sharks, crabs, tetrapods, vegetation, seed-bearing plants, forests, ferns, land flora.
359 to 318 "	**MISSISSIPPIAN OR EARLY CONIFEROUS PERIOD** (Abundant coal beds)
359 "	Time of first true forests.
340 "	Diversification of amphibians. Large fresh-water predators in salt-free oceans. Sharks common and diverse.
318 to 299 "	**PENNSYLVANIAN OR LATE CONIFEROUS PERIOD** (Named after state with rocks of this age)
300 "	Dinosaurs appear. Siberia joins Europe, with other lands, forming supercontinent of Pangaea.
299 to 251 "	**PERMIAN PERIOD** (Large mass extinction)

299 "	Oldest beetles and winged insects. Oldest dinosaur fossils. Amphibians, reptiles. Coal forests, ferns. Oxygen proportion at its greatest proportion.
275 million years ago	Australopithicus africanus or "Eve".
251 to 200 "	**TRIASSIC PERIOD** (Major extinction events)
250 "	Permian-Triassic extinction eliminated 90% of marine species to clean slate of many species, leading to diversification over 30 million years. Beetles and flies. Greater number of dinosaur species'
225 mya	Pangaea splits as one large continent, from Rodina
217 "	First egg-laying mammals (like Platypus).
215 "	First mammals. Land masses unite into super-continent of Pangaea. Appalachian Mountains formed.
200 to 145 "	**JURASSIC PERIOD** (Age of reptiles)
200 "	First birds, flies, turtles. Breakup of Pangaea into Gondwana and Laurasia. Many types of dinosaurs, Ichthyosaurs, pterosaurs (birds) become much smaller and more mammal-like.
195 "	Mass extinction at end of Triassic Period.
180 to 135 "	First plants with modern type flowers and fruits. Birth of American Rockies and Andes Mountains, canyons of Mesa Verde are cut. Cartilaginous fish (sharks, skates, rays, sea urchins). First modern birds and insects. Single land mass of Gondwanaland starts breakup to form present configuration of 7 continents. Strata of rock in mantle is about 7.5 miles below earth surface.

150 "	Archaeopteryx bird (absence of a keeled sternum). Pangaea splits, forms future North America and Eurasia in northern hemisphere as Laurasia, with South America and Africa as Gondwana in southern hemisphere, but joined at Gibralter.
170 "	Oldest wasp. First mammals, crocodiles and amphibians. Sea urchins, sponges, starfish.
155 "	First Archaeopteryx (ancestor to birds). Marsupials.
150 to 145 "	Oldest birds of other species. Gondwana and Laurasia meet to form Pangaea (B)
145 to 165 "	**CRETACEOUS PERIOD** (Chalk period)
140 "	First flowering plants. Breakup of Gondwana.
135 "	**MEZOIC ERA-CREATACEOUS PERIOD:** Liaoningernis birds (possibly evolved from Archaeopteryx and not ornithurnine birds).
125 to 120 "	Oldest beaked bird. Some placental mammals.
100 "	**START OF CENOZOIC ERA (New life)**
100 "	First bees, crocodiles, sharks. Primitive birds replace pterosaurs.
98 to 94 "	Oldest snakes.
97 "	Snakes may be descended from giant sea lizards (marine mosasauroids that look more like a snake with tiny hind legs, that died out with the dinosaurs). (Fossils found in Israel).
84 "	Earth tilts 20 degrees
80 million years ago	First ants and termites.

70 to 5 "	**CENOZOIC ERA-EARLY TERTIARY PERIOD—MIOCENE EPOCH** (Less recent): First primates (lemurs, Eosimas). Small and unspecialized mammals. Other mammals become dominant, most major groups represented. First turtles, invertebrate. Life species and plant life much like that of today. First grasses. Rocks of Bryce Canyon present. Weather moderate, becoming warm.
68 "	Flowering plants, insects. New types, species and sub-species of dinosaurs and mammals. First fossils of primates.
67 to 65 "	**TERTIARY PERIOD:** Asteroid impact destroys many species including dinosaurs and archaic mammals in extinction of 50% of all species. Cretaceous-Tertiary extinction. Age of mammals begins.
65 to 55 "	**PALEOCENE EPOCH** (Mass extinction—K-T Boundary)
65 "	Dinosaurs extinct. Rapid dominance of conifers and ginkgo trees. Tree shrews, rodents, lemurs, tarsier, gibbon, gorilla, Old World monkey, New World monkey, increase in modern family of mammals.
65 mya	North and South America clearly in place, but not yet joined.
65 to 23 "	**CENOZOIC ERA** (New Life)
60 to 56 "	Expansion of mammal species and plants following demise of dinosaurs. Diversification of large, flightless birds. Earliest true primates. Pangaea splitting to form North America and Europe. Atlantic Ocean formed. Climate tropical. Modern plants appear.

	Examples of mammals and plants developing in warmer, carbon dioxide-rich atmosphere of former shallow, inland sea at Wyoming (USA) and western areas. Dawn Redwood tree, Ginkgo tree, Plesiadapis (squirrel—like). Ectocion (dwarf, hoofed deer), turtle, Champsosaurus (alligator), cattail plant, Meniscotherium (small, furry, cat-like), Hackberry bush, Prolimnocyon (weasel-like, striped fur), Gaulitia (snake), Hydracotherium (small zebra like, stripes), Cantius (larger, fur bearing, like bear but large rear legs and smaller forelegs. front-facing eyes and grasping hands), Copaifera (bean-bearing tree), Chiracus (like ring-tailed raccoon), Suzanniwana (lizard-like), Sycamore tree relative fo Katsura tree, horsetail plant, palm tree, Diacodexis (like antelope or small horse), voracious insects, Eomaia (proto-mouse). "When it needs to, evolution is fast". Many species became extinct. Source: National Geographic, October 2011.
58 "	Spike of carbon in air, rising temperatures, diversity of mammals.
57 "	First apes and tree-dwelling primates have stereoscopic vision. Hands evolve to grasp. Larger brain.
55 to 34 Million years ago	**EOCENE EPOCH** (Modern mammals). Start of severe global warming period lasting 34 million years caused by volcanoes, methane gas bubbling up from ocean floors, peat and coal fires. Most life on land adapted or migrated.
55 "	First whales re-enter sea from land, dolphins, porpoises rodents, bats. First horses; Modern bird groups diversify (song birds, parrots, loons, swifts). Rocky Mountains formed. Himalaya Mountains form from colliding geologic plates. Mammals diversify, some to large animals.

50 "	Australia splits from Antarctica. Beginning of period of colder climate. Shoshonis (small mammal) living in Wyoming. Whales split from land mammals.
45 "	Early monkey-like primates, apes, hominoids. India continues collision with Eurasia, forming higher Himalayan Mountains. Great diversity in species, mammals, and grasses. An Antarctica event (possibly Earth magnetic change) triggers ice age. Songbirds evolve in Australia or western part of then supercontinent Gondwana, the area that formed New Zealand, Australia, and New Guinea.
40 "	First butterflies and moths. Saber Tooth cats. First prairies of grasses. In genetics, X and Y chromosomes differentiate further.
40 to 25 "	Komodo Dragon of Indonesia.
37 to 34 "	Antarctica begins to freeze (Prior to this the area had rainforests on coastal plains.)
35 "	New World monkeys. Much grassland. Moderate cooling climate. Several "modern" mammal families.
34 to 23 "	**OLIGOCENE EPOCH** (Scarcity of additional modern mammalian fauna)
30 "	Volcanoes start on floor of Pacific Ocean to later form Hawaii ecosystem at 40 to 20,000 years ago. Volcanic eruptions in areas of Ethiopia, earthquakes, and much rifting in topography changes local climate, develops savannah, and leads to extinction of some species, giving hominids advantage of some environmental changes.
28 to 34 "	Warm but cooling climate moving toward glacial period. Rapid evolution of mammals and plants.
26 "	Monkeys in South America. Apes and Old World monkeys diverge.
24 "	Grasses and herbivores increase. Deer develop.

23 to 5 "	**MIOCENE EPOCH** (Less recent of modern sea invertebrates)
20 "	Proconsul africanus, an early Ape. Later gorilla, chimpanzee, bonobo, monkeys separate. Wide spread forests draws massive amounts of carbon dioxide, providing woody growth, and "exhaling" oxygen for abundance.
20 million years ago	Man-like, erect-walking creatures divided from four-footed, tree-and-ground dwelling animals.
18 "	Some primates, first hominids. (Apes lead to hominines and human species)
16 "	Human evolutionary line diverges from Orangutan.
15 to 1 "	**CENOZOIC ERA—LATE TERTIARY PERIOD- MIOCENE AND PLIOCENE EPOCHS**
15 million years ago	Advent of Man. Mountains of African Rift Valley lifted. Fossil of Baboon discovered, which could be one of our ancient relatives.
11 "	Grasses become ubiquitous throughout known world.
10 "	Alligators, crocodiles, snakes.
8.7 "	Some common ancestors in primates of African apes and hominids.
7 to 5 "	**RECENT MIOCENE EPOCH**

7 "	Earliest fossils of bi-pedal hominines. Great Ape family of Gorilla, earliest hominids. General dividing line between 4-legged fur-covered mammals, tree—and surface-dwelling vertebrates, primates of Ape family (gorilla, chimpanzee, monkey, bonobo). Humanoid skull discovered in Chad desert has about brain capacity of chimpanzee and human-like set of features, including human-like canine teeth, leading to question, "is this the fabled "missing link" between humanity and apes? Human evolutionary line diverges from chimpanzee.
7.24 "	Horses, mastodons, mammoths.
6 to 8 "	Approximate dividing line between primates (apes) and hominines having general human-like features and walking erect (Lucy).
6 to 4 "	Divergence of future human and chimpanzee lineage. Approximate dividing line between primates (Ape family) and hominoids having general human-like features and walking erect.
6 "	Previous earliest remains of first hominids and hominines before Lake Chad findings. Earliest remains of bipedal hominines. Remains of "Ape-man" found in Kenya's Rift Valley as upright-walking hominid the size of chimpanzees-the earliest—known two-legged hominid. This discovery may mean that "Lucy" was part of a "dead branch" of the human family tree as "Millennium Ancestor" is 2 million years older than "Lucy" (an Australopithecine).
5.5 "	Atlantic Ocean breaks the Straits of Gibraltar and forms the Mediterranean Sea.
5.3 "	First hominids as Australopithicines, Mammoths and Mastodonsr, mollusks appear.

5.1 TO 1.8 "	**PLIOCENE EPOCH** (Continuation of the present)
5.0 "	Primates of Great Ape family and hominoids. Species Elephant and Mammoth diverge.
4.8 "	Mammoths, Mastodons and Sabre Tooth Tiger. Earliest forms of apelike hominoid creatures that walked upright, had larger brain capacity, figured out how to make and use stone tools.
4.7 "	First hominoids as Australopithicines.
4.5 "	"Lucy", walks upright. Mediterranean Sea flood.
4.4 "	Earliest fossils of man (Ardipithecus ramidas).
4.4 to 3.9 million years ago	Australopithecus anamensis, although ape-like in some regards, these hominids walked upright and are the earliest bipedal hominids yet found. An idea currently favored is that several different hominids species may have coexisted in the same period, of which only one could presumably have any direct relation to ourselves. The answer, of course, is which? And how these different species of hominid interacted with each other? Or even, did they interbreed and converge.
3.9 "	Earth relatively quiet and cooling with glaciers. Reptiles, Tarsier, lemur, monkey, first Australopithicus anamnesis (in Kenya). Australopithicines, although we may be descended from one, look much like chimpanzees, which were much shorter than modern humans (less than 3 ½ feet tall), with brain size only slightly larger than that of chimps), were clever and made stone tools. Start of insects, Coelacanth fish.
3.6 to 2.6 "	**PLEISTOCENE EPOCH** (New)
3.6 "	Earliest evidence (footprints) of hominid life on earth.

3.5 "	Australopithicus afarensis ("Lucy") (from Ethiopia and Tanzania), and Australopithicus bahreighazalt (in Chad). Analysis of ancient Australian rocks reveals banded cherty rocks had life-sustaining levels of oxygen much like todays. Red coloration comes from iron rusted by oxygen.
4.0 to 2.5 "	Austalopithicus series (Australopithicines): A. anamensis, 4.0 mya in Kenya; A. afarensis (Lucy) 3.5 mya in Ethiopia/Tanzania (size of present African Pygmy race in Africa); A. bahreighazali, 3,5 mya in Chad Lake area; A. africanus, 2.8 mya in South Africa; A. garhi, 2.6 mya in Ethiopia; A. aethiopicus; A. boisei; A. robustus.
3.3 "	Meteorite off Argentina causes many extinctions.
3,0 "	Macaque, monkey. Collision of North and South Americas coincidental with a major Ice Age. Land connection between North and South America at Isthmus of Panama complete and may have caused an ice age to begin because of the forced detour of salt water, along with a climate that had already been slowly cooling.
2.75 "	A. garhi. Chimpanzee, orangutan, bonobo; first hominoid modifies stone tools.
2.7 to 1.6 "	Paranthropus series: P. aethiopicus, 2.7 mya in eastern Africa; P. robustus, 1.6 in South Africa; P. boisei, 1.6 mya in eastern Africa (had robust and massive chewing muscles and cheek teeth, ate vegetables).
2.6 mya to present	**QUATERNARY PERIOD of PLEITOCENE EPOCH** (Present)
2.6 Million years ago	Glacial ice conditions continue.
2.0 "	Homo habilis. (Will become extinct).

2.5 mya to 12,000 years ago	**PLEISTOCENE EPOCH**
2.5 to 2.3 million years ago	Sharp flakes struck to provide cutting instruments. Series of species: Australopithicus robustus (1.6 mya), A. africanus (2.7), A. Garhi (2.6 mya),. A. ethiopicus (2.7 mya), A. afarensis (3.0 mya), A. anamensis (4.0 mya). Homo habilis (2.5 to 2.6 mya) may have evolved to Homo erectus and is questionable "missing link" between hominoids and modern man. Back and forth of ice starts as harbinger of warmer climate. Ape-sized hominid brain begins to develop into a fully human one, four times as large and reorganized for language, music, and chains of inferences. Ours is now a brain (with mind) able to anticipate outcomes well enough to practice ethical behavior, able to head off disasters in the making by extrapolating trends.
2.5 to 2.0 million years ago	Australopithicus robustus (or A. africanus). "Taung child" where spinal cord entered skull from below rather than behind, indicating forward looking, rather than animal-like 4-footed posture in anatomy.
2.5 "	Australopithecus aethiopicus, differed from the robust A. robustus and A. boisei, and may be evidence that evolution not only leads to divergence, but possibly to convergence.
2.3 "	A. africanus, A. Paranthropus. Sophisticated stone tools made by Homo habilis.
2.1 to 1.1 "	Australopithecus boisei. Vegetarian diet. Variant of A. robustus. Believed to not be our ancestors, species died out.

2.0 mya to present	Homo series: H. rudolfensis, 2 mya in eastern Africa; H. habilis 1.8 mya in sub-Saharan Africa (had large brain and reduced back teeth) indicative of change from grass eating to meat eating; H. ergaster, 1.7 mya in eastern Africa; H. erectus 1.6 mya to 600,000 years ago in eastern Asia (i.e. Georgia, Russia), migrating eastward (Java Man was a precursor of aboriginal Australians); H. antecessor, 800,000 ya in Spain; H. heidelbergensis, 600,00 ya in Germany; H. neanderthal, 350,000 to 28,000 years ago in Europe and western Asia, and H. sapiens in Europe, Asia, and eastward migrations. Hominoids in Russian Georgia. Homo erectus migrates north, south, east, west. Homo habilis (will become extinct). Massive volcano dominates Olduvai Gorge. Hominid remains found at base of Olduvai Gorge in Kenya. The human foot completes its evolution as we know it, but human hand has not yet developed sufficiently to permit workmanship. Ape ancestors have now moved out of the trees. North American Yellowstone volcanic caldera eruption. H. habilis was not of a lesser mentality, but of a different culture, and of Black race, different than the various Caucasian subspecies emanating from H. erectus. H. habilis was a parallel species to H. erectus, H. habilis remained mostly in Africa, with some migration also to Asia, Europe, Australia, and Pacific islands.
1.9 "	First (or Archaic) Homo erectus in Africa. Approximate dividing time between hominids of Ape family and hominoids of A. Paranthropus. H. neanderthals are first to survive this ice age.

1.8 to 1.6 "	Homo erectus (or Archaic H, erectus) out of east Africa was definitely our ancestor. Homo ergaster in east Africa was first hominid to emigrate from Africa at least 1.8 million years ago, spreading all the way to China and Indonesia, then at some point, for reasons still mysterious. the lineage diverged, with one branch leading to Neanderthals and another to modern humans. Dividing of A. Australopithicus and Archaic Homo erectus as a base species, later to become much greater advanced biologically, technically, and spiritually. H. erectus was a tool maker, which provided the ability to obtain meat from relatively large grazing animals. First use of toothpicks.
1.7 "	African migration into Europe
1.6 mya to 730,000 years ago	Period when Homo erectus populated temperate areas of Africa, Europe, and Asia. Their remains have been found in northern China, but not in northern parts of Eurasia.
1.5 "	Stone hand axe and cleaver used in Africa. Surviving core of species at Mossel Bay in western South Africa.
1.5 mya to 100,000 ya.	Waves of early humans (probably H. erectus) migrate from Africa to China area and Europe.
1.0 million years ago	**CENOZOIC ERA**

1.0 "		**Pleistocene period** starts. Widespread glaciation. Southern Michigan landscapes owe form to Pleistocene glaciation. Culmination of mammals with extinction of many groups. Widespread glaciation. H. neanderthal and H. sapiens may be sub-species of H. erectus. Humans from Denisova Hominin (human like creature) shared a common ancestor with modern humans and Neanderthals about 1 million years ago. This is known as the divergence date, when Denisova hominine's ancestors split away from the line that eventually led to Neanderthals and ourselves.
800,000 years ago		Earliest northern European human settlement discovered in Britain and evidence that early humans were able to live and survive in the cold northern climate. Homo antecessor fossils found in northern Spain and possible human cannibalism. Homo antecessor may have been a common ancestor for both Neanderthals and Homo sapiens. Early hominids in Asia had migrated to island of Flores in the Java archipelago. Haleakala shield volcano in Maui, Hawaii appears about this time.
800,000 to 400,000 "		Homo heidelbergensis in various Europe locations.
750,000 "		Homo erectus in eastern Asia. Homo antecessor in Spain. "Stone age" in Europe. Tools made by chipping. Spoken, symbolic language. Use of fire.

700,000 "	Magnetic poles of Earth reverse. South Pole develops negative polarity. Early man colonizes northern Europe. These people had ventured out of Africa, colonized southern Caucuses Mountains about 1.8 million years ago, then venturing westward along the Mediterranean, reaching Italy and Spain around 800,000 years ago. Around this time Britain was connected to Europe by a land bridge that extended the length of today's English Channel, with lowlands at the shallow coast of East Anglia, meandering rivers depositing a thick sedimentary layer of mud and sand, marshland, and grassland providing food for bison, lions, wolves, hippos, rhinos, elephants, mammoths, and others.
600,000 "	Earth enters a series of ice ages with widespread glaciers and seas drop 300 feet. Caldera at Yellowstone Nat,l. Park in Wyoming from ancient volcano.
600,000 "	First hominoid life. Hominoids developed increase in mental abilities. Modern Homo erectus shows increased evidence of soul.
600,000 years ago	A creation (unexplainable in science but proven in existence) would arise at 600,000 years ago as soul was gradually introduced to the body of erect, walking man (Homo erectus) providing mind, emotions, judgments, reasoning, and other qualities in the mind of man, made other operations through the brain of man (a nerve-coordinating organ), the mind (as conscience and consciousness of man), and the soul of man (an entity with communication with God to realize communication with a superior being). Millions of years later René Descartes, a proclaimed mathematician and philosopher, would define "his humanness" in proclaiming, "I think, therefore I am", meaning his brain and mind become able to observe, analyze, and respond.

500,000 to 250,000 "	"Archaic Homo sapiens" found in Greece.
500,000 "	Homo heidelbergensis in Germany and throughout Old World. Cambrian Homo erectus. Diet was exclusively uncooked meat, the fats of which may allow Explosion of ocean-living life. Peking Man found near Beijing. Fossilized remains of rhinoceros with projectile (spear) wound found in England. Acheulean hand axe developed. Heidelberg Man in cold Eurasia, possibly an early African-evolved brain growth. Neanderthal and modern evolutionary lines diverged around 500,000 years ago showing that Denisova hominid is of a previously unknown human lineage from a hitherto unrecognized migration out of Africa. The merging picture of humankind during the late Pleistocene when modern humans left Africa and started to colonize the rest of the world. A time slice at a point in the late Pleistocene would reveal a range of human populations spread across parts of Africa.
450,000 years ago	Fossil bones found in quarry at Norfolk, United Kingdom of 2 Hippos, with remains of horse, hyena, fish, and rodents. Evidence of a wooden hut found near Nice, France evidently a home for Neanderthal, with use of a fireplace, and furs, grasses, seaweed, used as bedding.
400,000 "	Man-made spear found in Germany. Hunters were distant ancestors of Neanderthals (called archaic Homo sapiens). Man learns use of fire.
400,000 to 375,000 "	Illinoisan glaciation Period, to be followed by Sangamon Interglacial Period; Neanderthals may have developed speech by this time.

375,000 "	First neanderthals. New Homo erectus had migrated to Indonesia and China, Controlled fire common. Used "Acheulean" stone tools, which were better made than "Oldowan" tools of Homo habilis. Apes such as orangutan migrated from Africa to Asia.
375,000 to 125,000 "	Sangamon Interglacial Period. Many species and sub-species develop.
350,000 to 20,000 years ago	Homo neanderthal: Tall as modern man. Lived in Europe and east Asia. Brains as large as modern man. Developed "Mousterian" stone tools. Hunted large ice—age mammals (Mammoths, Wooly Bison, and Mastodons). No proof that Neanderthals had symbolic language. Absence, or disuse, of larynx would not allow speaking like modern man. Buried their dead and painted dead bodies with red ochre, which implies capacity for symbolic thought. Recent studies of DNA state that Neanderthal and hominoid species (Homo erectus) had split more than 500,000 years earlier. Both Erectus and Neanderthal species disappeared, as such, about 120,000 years ago with advent and pressure of Homo sapiens species.
325,000 "	Turtles.
300,000 "	Neanderthals found in northern Spain represent early stage in development of Neanderthals. Neanderthals lived 300,000 to 30,000 B.C.
300,000 to 250,000 "	Human habitation in northern Siberia. Britain's oldest known human remains.
280,000 "	Mastodon tooth and camel jaw found at Los Angeles, California.
250,000 to 10,000	Stone Age. Period of last two ice ages. Stone tools found in Siberia, near Irkutsk. Human brain size stopped its slow trend toward enlargement

248,000 "	Abrupt climate changes over 15 years as "great climate flip-flop".
240,000 "	First traces of human occupation in Denmark, but settlement is ended by Ice Age until about 15,000 B.C. when hunter-gatherers return.
200,000 "	Migrations from Africa: 200,000 years ago and more from eastern and southern Africa; 70,000 years ago from east and north Africa; 50,000 years ago from Arabian Peninsula, Mesopotamia, to India, Indonesia, Malaysian islands, and Australia 30,000 years ago; To East Asia, central China, central and north Asia; 20,000 years ago to southern, central, northern and western Europe and South America; 15,000 years ago from northeast Asia to Bering Straits, western and northern Canada, west, central, and eastern North America; 15-12,000 years ago to western South America. Homo erectus in Russian Georgia/Caucuses Mountains area, will migrate in four directions to populate Old World.
200,000 to 30,000 years ago	Evidence of stone tools in a cache indicate bands of humans had reached southern Arabia by this time in migration. Artwork in Babylon depicts genetic evidence of man and woman: (dubbed "Adam" and "Eve".)
200,000 to 12,000 years ago	Wisconsin Glaciation period, with 3 interglacial periods to 12,000 B.C. Homo sapiens remain as sole, man-like humans.
125,000 "	First Homo sapiens would live contemporaneously with Homo neanderthals for about 100,000 years and possibly with Denisovan hominids.
125,000 to 30,000 years ago	Dawn of Stone Age cultures and increasing technical complexities.

78,000 "	Great diversity of Mammals. Upward spike in global temperatures.
65,000 "	Arizona Meteor Crater formed. Demise of dinosaurs.
57,000 "	Great Diversity of mammals. Upward spike in global temperatures.
50,000 "	Dawn of human species
48,000 to 30,000 years ago	Denisova hominin lived in central Asia. Modern genetic sequencing proves that Denisova hominin is distinct from Neanderthal and modern humans.
34 to 28,000 "	Warm but cooling planet, moving towards Wisconsin glacial period. Rapid evolution of mammals and plants. Homo sapiens are sole human species.
30,000 "	Last glacial maximum (Thaw started 18,000 to 15,000 years ago).
28,000 "	Extinction of H, neanderthal. Increase in oxygen.
20,000 to present	**HALOCENE EPOCH**
20,000 "	Disappearance (extinction) of Homo neanderthals. Widespread forests draw massive amounts of carbon dioxide and increase of oxygen. Evidence of human culture in South and North America.
18 to 15,000 "	Last glacial maximum (thaw started from Wisconsin Glacier about 18,000 years ago).
17,000 "	Homo florensis ("Hobbits") fossils found in Papua New Guinea.

15,000 " (approx.)	**QUATERNARY PERIOD**
15,000 "	Quaternary ice age recedes and period ends. Current Interglacial Period begins. Rise of human civilization. Migration from Asia to become North American Indians and Eskimos.
13,000 "	Clovis points of 13,000 years ago found at Clovis, New Mexico. Culture in South America may have been result of migrations from Southeast Asia, Papua New Guinea, Fiji, Hawaii, and South Pacific islands.
12,000	Agrarian culture period starts in Mesopotamia and Egypt/Africa
11,700 "	Last ice age (Wisconsin Glacier) ended. ("Little Ice Age" will occur 1,400 to 1,850 A.D.)
11,000 years ago	Grasses become ubiquitous. Wooden building in South America (Chile); first settlements in Argentina's Man on Santa Rosa Island off coast of California; human remains found in Yucantan.
10,000 "	With divide of Paleolithic/Neolithic, culture changes, a new Agrarian culture of agriculture and animal husbandry rises. Stone age continues, to slowly die. Approximate start of Agrarian culture.
8,000 "	Kingdoms start in Mesopotamia and Egypt with advances in culture.
	We live today in the Phanerozoic Eon, Agrarian Era, Holocene Epoch, and Pleistocene Period.
3,500 "	Copper Age. (Varied with geographic area.)
2,500 "	Bronze Age. "
1,200 "	Iron Age "

| 30,000 / 20,000 years ago B.C. to 1,519 A.D. | **ANCESTORS OF MAYA PEOPLE IN SOUTH AND CENTRAL AMERICA:** |

Ironically, very little definite anthropologic information is available for this American time and place. Points of location and dates have been identified, but there is much void in information between the data. This scenario is presented here as possible and logical for the ancient time in America:

Recognition of ancient people of the Americas involved two or more groups. The most well-known are the ancient Indians from northern Asia who crossed over a land bridge at the Bering Sea Straits in hunting and pursuing food animals, in several waves between 15,000 and 10,000 B.C. They travelled southward, through Rocky Mountain valleys, eastward to central Canada, and along the west coast of North America by crude boats and rafts. Some, later to be called "Inuit", travelled eastward along frozen lands and icy waters, living from arctic sea life.

Some anthropologists believe, and the author proposes, a second group of people, possibly Homo erectus, migrated, by land and sea, from southeast Asia, Siberia, China, Korea, Papua New Guinea, Australia, New Zealand, Fiji Islands, and other islands of the south Pacific Ocean. (It is believed that discoverers and first settlers of the Hawaiian Islands originally migrated from Fiji. James A. Michener, in his book "Hawaii", suggests that first people of the Hawaiian Islands originated from the islands of Fiji.)

Species of man, resembling this early Maya man, may have therefor existed from the eastward migration of Homo erectus, to the Far East, Papua New Guinea, and Pacific Ocean islands at an early time, perhaps about 30,000 to 20,000 B.C. They eventually landed and settled at the South America west coast of present Peru and populated Central America prior to the Aztecs. With possible dates, the stones of the Easter Islands may prove to be evidence of a stopping point for the early migrants.

The Aztecs were a culture that possibly originated from the Alaskan crossing, migrating to northern Mexico or the central mountain and plateau areas of the present United States. Through merging of the cultures in the 13th century A.D. the Aztecs may have become ancestors of the later Mayan people and their culture.

Evidence of humans in the form of Clovis points, some found with bones of killed animals, has been found near present Clovis, New Mexico, dating to about 12,500 B.C. and in other locations in Texas, Colorado, southwest Wyoming, and Utah.

It seems illogical that, except for the Aztecs, origins of these South American people were associated with those of the North American Indians. Stature, height, appearance, religious practices, constructed pyramidal step temples, sacrifices, age of migration, and perhaps skin coloration were radically different from the ancestors of ancient North American Indians.

Inca, Olmec, and Toltec may be developed subcultures of the ancient Maya. A branch of the ancient Maya would have migrated both south in Peru and Chile and northward to the present Central America, at the Yucatan Peninsula, and at present Mexico City. A series of kings, gods, and cultures would evolve. At A.D. 1,324 a large Inca religious community of Tenochtitlan was constructed on an island at present Mexico City, and numerous temples constructed in now Honduras, Guatemala, Belize and San Salvador. Later voyages of exploration and settlements were made up the Mississippi and Ohio rivers as far as southern Ohio, Cahokia, (in Illinois), and Minnesota.

	Elsewhere on Earth
4004 B.C.	The time of Biblical Creation, as calculated in year 1654 B.C.E., by Archbishop James Ussher in Ireland, projected from the various generations stated in the Bible, that Earth, Cosmos, stars, Adam, Eve, the serpent, and Garden of Eden were created on the afternoon of October23, 4004 B.C. There are many different accounts of creation, also told in the cultures of Pagan groups, Babylonia, and Assyria.
China; 2100 B.C. to 220 A.D.	Civilization and rulers also appeared in China. Ancient governments and rulers before Christ: 2,100 to 1,600 B.C. The Xia Dynasty-the first dynasty of China. 1,600 to 1,046 B.C. The Xia and Shang Dynasties were reported as founded by Black tribes living in ancient China. 1,045 to 256 B.C. Kingdoms were divided into smaller states. 221 to 206 B.C. The first Chinese empire. 202 B.C. to 220 A.D. The "Golden Age" in Chinese history.
2270 B.C.	First kingdom: Akkad in Mesopotamia, headed by King Sargon The Great.
1179 to 1173 B.;C.	Ramesses III rules in Egypt. Philistines settle on the southern coast of Canaan.
1150 to 1140 B.C.	Egypt has the last remnants of its empire in The Levant.

1125 to 1103 B.C.	King Nebuchadnezzar destroys the Elamites kingdom.
Late 11th century	Egypt disintegrates into feuding principalities.
1080 B.C.	Egypt loses control of Nubia. Egyptian empire is finished.
Between 1079 and 1007 B.C.	Young David, a future king, kills Goliath of the Philistines during reign of Saul.
1030 B.C.	Philistines start to expand into the interior of Canaan. Battle of David and Goliath.
1020 to 914 B.C.	Some leading kings of Israel and Judah: Saul 1020 to 1000 B.C., David 1000 to 993 B.C., Solomon 960 to 931 B.C., Jeroboam (in Israel) 931 to 910 B.C., Rehoboam (in Judah) 921 to 914 B.C.
960 to 931 B.C.	Solomon reigns as king of Israel and begins construction of Hebrew Temple.
934 to 612 B.C.	Neo-Assyrian empire.
931 B.C.	Northern clans of Israel disagree with Solomon's son, Rehoboam, and anoint Jeroboam as their king. Israel disintegrates into northern (Israel) and southern (Judah) kingdoms.
883 to 824 B.C.	Assyria becomes a great power once more.
850 to 600 B.C.	Phoenicians colonize the western Mediterranean area.
800 B.C.	Carthage in Africa founded by Tyre on shore of eastern Mediterranean Sea.
8th century B.C.	Pharaohs in 22nd and 23rd Dynasties.

753 B.C.	Rome founded: Romulus plows furrow around Palatine hill to mark boundary of new city. (Possibly a tale or myth.)
744 to 630 B.C.	Zenith of the Assyrian empire.
732 to 730 B.C.	Late Period when Egypt became Province of Greece and Persia.
721 to 705 B.C.	Sargon II expands Assyrian power into southeastern Anatolia.
685 to 547 B.C.	Lydian kingdom at location of Turkish provinces,
465 to 30 A.D.	Greek culture with advances in philosophy, government, mathematics, medicine, and religion. **Many personalities, cultures and events occurred during the rich 500 year period leading to the time of Christ:** Aegean and Hellenic civilizations, Homeric age, Sparta, Athens, Troy, Pericles (government), Pythagoras (mathematics). Xenophanes (theology), Protagoras (Sophists), Thales of Militias (Greek founder of mathematics), Anaximander (biology), Hippocrates (medicine), Herodotus (historian), Thucydides (scientific history), Phidias (sculpture), Epicurus (Epicureanism), Zeno (Stoicism), Copernicus (astronomy), Euclid (geometry), Hipparchus (trigonometry), Archimedes (calculus and physics), Socrates, Plato and Aristotle (philosophy and logic), Geris Gracchus (representative government), Etruscan people, Roman and Greek gods, Xerxes, Darius, Alexander The Great and the Persian Empire, Pompey, Julius Caesar, Carthage, Hannibal, Octavian/Augustus Caesar, Buddha, Confucius, Marc Antony, Cleopatra VII, and many others.

Approx. 2 or 4 B.C to A.D. 34, exact dates disputed.	**Birth and time of Christ** following Old Testament prediction for coming birth of a Savior. Variety of cultures, religions, philosophies, various warring countries.
	SUMMARY OF MAIN CREATURES LEADING TO MAN-LIKE, HOMO SPECIES
	4.0 billion years ago-Earliest first life on earth in prokaryotes of bacteria. 3.5 bya-Earliest fossils of organisms (Stromolites). 2.1 bya-Fossil algae. 1.5 bya-First eukaryote cell organisms. 1.2 bya-First sexual reproducing appears. 800 million years ago-complex single-celled life. Eukaryote cells replace prokaryote cells. 670 mya-first multicellular animals (soft body). 600 mya-Soft-jellied creatures in seas, Chordate fishes, notochords. 580 mya-Fossils of nautiloids, ammonites, trilobites, brachiopods. Ediacaran biota plant is first large, complex cellular organism. 500 mya-First vertebrates, first bony fishes. 490 mya-First animals with bony parts. 485 mya-First vertebrates with true bones (jawless fishes). 475 mya-First organisms on land. 468 mys-Bivalves, nautiloids, trilobites, starfish. 450 mya-Vertebrate fishes. 437 mya-Period of mass extinction. 420 mya-First ray-finned (later arms, legs, etc.) fishes and land scorpions. 410 mya-First signs of jaws or teeth in fish. 400 mya-First reptiles. 395 mya-First lichens, seed-bearing plants, trees, winged insects.

359 mya-First true forests.
340 mya-Diversity of amphibian reptiles. Sharks common and diverse.
300 mya-Dinosaurs appear.
250 mya-Permian-Triassic Extinction eliminates 90% of marine species.
217 mya-First egg-laying mammals.
200 mya-First birds, flies, turtles. Many types of dinosaurs.
195 mya-Mass extinction at end of Triassic Period.
180 mya-First plants with modern-type flowers and fruits. Birds, insects.
155 mya-Archaeopteryx bird (absence of keeled sternum)
150 mya-Oldest birds with keeled sternum.
125 mya-Oldest beaked birds, placental mammals.
97 mya-Oldest snakes, lizards.
70 mya-First primates (lemurs, etc.) as small and unspecialized mammals.
68 mya-Asteroid impact destroys many species of dinosaurs.
58 mya—Spike of carbon in air, rising temperatures, diversity of mammals.
20 mya—Great ape: Split to Gorilla, monkey, chimpanzee, bonobo hominids.
15 mya-Advent of Paranthropus man: Fossils found in Olduvai Gorge of Tanzania, Africa. "Eve" as great (+) grandmother of all humans. (Exact dates not documented.)
11 mya-Grasses become ubiquitous
7 mya-Earliest fossils of bi-pedal hominines. (A. Paranthropus.)
6 mya-Divergence of future human and chimpanzee lineage. "Lucy".
5.3 mya-First hominids as Australopithicines. Mammoths and Mastodons.
3.5 mya-Australopithicus afarensis (Lucy) and Australopithicus family.

2.7 mya-Paranthropus family of species.

2.3 mya-Homo habilis (will become extinct). Large stones struck for cutting.

2.0 mya-Homo series of species (Homo habilis, Homo erectus).

2.0 mya-Hominoids in Georgia. Homo erectus migrates south, east, west.

1.9 mya-Homo neanderthals are first of human-types to survive Ice Age.

1.5 mya-Stone hand axe and cleaver used in Africa.

1.5 mya-Waves of early humans migrate to Mesopotamia, Asia, Far-east.

1.0 mya-Homo erectus is only survivor species. H. neanderthal and H. sapiens may be sub-species of H. erectus. Widespread glaciation.

750,000 ya-Homo erectus in eastern Asia.

600,000 ya-Homo erectus show first signs of soul.

375,000 ya-First H. neanderthal.

200,000 ya-Homo sapiens, Live at same time with H. neanderthal to 28,000.

78,000 ya-Dawn of Stone Age cultures and increasing technical abilities.

30,000 ya-Last glacial maximum (thaw starts 18,000 to 15,000 years ago).

28,000 ya-Disappearance (extinction) of H. neanderthal. Increase in oxygen due to abundance of plant life in equatorial areas.

20,000 ya-Evidence of human culture in South and North America, Pre-Maya.

17,000 ya-Fossils of Homo florensis (Hobbits) found in Papua New Guinea.

15,000 ya-Migration from Asia to become North American Indians and Eskimos

13,000 ya-Clovis (cutting and piercing) points found in New Mexico, USA.

12,000 ya-Agrarian culture period starts in Mesopotamia and Egypt/Africa.

	8,000 ya-Kingdoms start in Mesopotamia and Egypt with advances in culture Time of Christ—Physical man much as at present. Basic, elementary beliefs in mathematics, philosophy, religious, law, customs, culture, natural sciences, government

Sources:

1. The Great Courses (several courses and professors)
2. Google—Wikipedia, The Free Encyclopedia (primarily for dates)
3. Western Civilizations (college textbook)
4. King James Bible
5. Harper Study Bible
6. Archaeology magazine
7. National Geographic magazine
8. Scientific American magazine
9. Discover magazine
10. Nature magazine
11. Encyclopedia Britannica
12. Physics Annual Review, The University of Michigan
13. Prehistoric Life, DK Publishers, 2009
14. A wide variety of general reading

WONDERFUL WORLDS

PART III

BRAIN, MIND, SOUL, GENOME, ENZYMES, AND CONSCIOUSNESS

ADVENT OF THE SOUL

Planet Earth formed 4.5 billion years ago, 9.2 billion years after the Big Bang, initially from concentration of cosmic gases originating from the Big Bang, and later aggregates from the cosmic debris. The gases and aggregates were compacted into bands and stratified by the increasing gravity attraction of the spinning earth, with heavier and greater mass and density of matter, such as molten iron and nickel, becoming the core of Planet Earth.

Elements forming lesser mass became lesser concentric layers of matter, some of which would become on the earth surface during the seemingly endless volcano eruptions following magma.

Later, first strata of near-surface material and aerobic soil would support plant, microbe, amoeba, bacteria, yeast, and other early-life forms, or support matter(energy) for continued life. This included material contents of the later ocean floor, one suggested possible source of complex life on earth, by hot excrement from fissures on the ocean bottom, and chemicals of the hot excrement compound meeting the oxygen, hydrogen, and other chemical elements of sea water, allowing even more complex life forms than the first living organisms.

From a basic group of cells, there eventually developed amoeba, yeast, microbes, and virus, simple early single-cell animals, pre-human animal forms, primates, many evolved species, and finally a human being—us. Archaeologists have unearthed several specimens of prehistoric mankind in successive species, each being more refined or advanced than its predecessors in its abilities, size of brain cavity, skeletal development, as well as evidencing different living habits and traits. The creatures of species that would lead to man had evolved, by some kind of direction, either natural or by intelligence, from algae, bacteria, and an elementary worm at the dawn of solid, cooled earth, to an animal of unbelievable complexity today in its body, brain and mind.

The point in time marking the initial event in man-like species (hominid) is believed to be over 4.5 million years ago. A stage in transition would later be a developed species that is today designated Australopithicus Paranthropus. Later, about 3.0 to 2.5 million years ago, another species developed, Australopithecus afarensis, and named "Lucy" by anthropologists, as one of the first upright walking species. This specie had all features of the previous forest animals that lived largely in trees, but sometimes on the grassy ground, now standing and walking upright in a bipedal mode.

These creatures of the forest had evolved over 4 billion years through many and varied species, from first cells on a newly-cooled earth, to complex living animals, and to self-sustaining animals, fish, and sea life, with molecules of matter in a universe ruled primarily by negative-charged cells.

A parallel evolution of positive-charged elements had evolved in a world of antimatter e.g. antiprotons, a mirror image of each element on earth, with some elements existing in multiple dimensions. (Some authorities in study of particle science list 9 to 11 different dimensions of space outside our observable world of three dimensions plus the fourth dimension of time.)

A following list detailing qualities of soul describes the qualities of soul, contrasted with basic qualities required only to sustain life and reproduce their species that existed before the advent of soul. Addition of soul to the animal body of man provided a wide variety of qualities and formed the mind of the individual, each individual of the Homo family now having qualities of emotion and in greater ability to think.

Listing only a few qualities enabling thought and exhibiting a variety of emotions are: ability to visualize a supreme entity, showing abstract reasoning, showing love, honesty, mercy, grief, compassion, free-will, morality, vision, wisdom, foretelling, and a myriad of other like emotions.

This mystical moment, or perhaps an era in time, could be related as parallel to the Biblical event concerning Adam and Eve and termed as the Creation. This mystical creation is only one of many creations told by different cultures in change and development of the cosmos of earth. Other creations involve geology, firmament, waters, atmosphere, reproduction systems, biology, races, and other world changes. The Biblical Creation could realistically instead be described as the incarnation of soul to the body of man. Other factors of creation in spirituality joining the animal-like body are evidenced such as recognition of a personal god, temptation, free-will, child production, love, concern, and other emotions.

INCARNATION

In a Biblical sense, this would be the act of incarnation of the soul to the previously existing physical body. Physical changes described in the Biblical Creation incarnation are commonly thought of as the union of soul and mind to the body and brain.

The author proposes a radical change in understanding of God and soul in this regard. Incarnation is not merely the joining of soul to body and brain. Soul is proposed as actually being a portion of God, of the Supreme Being, and therefore a portion of God is incorporated into the person of each human being.

God is not merely an entity existing "up there" in the cosmos of Heaven, but is existent, here and now, within each human being, as soul of that person, and on the surface of Earth. The soul becomes paramount in understanding God, because the soul is a part of God.

This possibility, or insight, may very well be that soul is a portion of God, imaged if not by God, by one of his representatives, such as an angel. It is believed by many that we each have a Guardian Angel hosted within our soul, and that may serve as our direct contact with the world of God.

We have the mechanism within ourselves, within our soul, mind, brain, and physical existence, for most events to formulate actions necessary for problem solution through our own mind, soul, and brain, acting much as God would, through element of the soul acting as a "personal representative" of God.

The mind and soul of the individual demonstrates unusual power and strength in thought and problem analysis. The soul has ability to interact and communicate, if necessary, with the soul of other individuals to accomplish desired actions, all oriented to activity necessary to grant the request of prayer. For example, in a prayer involving human medical surgery, the individual soul of a patient may work in unison with the soul of the operating surgeon to assure necessary and skillful actions.

In problems or events involving physical or mental change, the soul has the ability to modify body cellular change through DNA activity that will result in activating necessary enzymes for corrective change in body organs.

In problems involving social matters, the soul has the wisdom and ability to communicate with soul of others, with an objective of corrective change.

The soul may also communicate directly with the essence of God to correct and take action beyond the powers of an individual soul.

In the wisdom of the soul and mind, not all request of prayer will prove out in that request of prayer, but will be modified within a total desire and plan of the individual's soul and of God.

PRAYER

We pray to God, but the corrective action is within the power of the individual's soul, mind, biological, psychological (soul to soul communications) essence within the body and mind. The soul and mind act within the individual in a truly miraculous, yet logical, manner in accomplishing necessary acts or corrective actions affecting the individual.

This process is necessary because of the momentous number of people and souls in the world (nearing 6 billion and increasing), complicated with events changing moment by moment in each person. Stated here is explanation of the way that prayer request operates with the wonderful existence of soul within the person.

The soul and mind within ourselves "listens in", so to speak, to the prayer request and attempts to devise and carry out a corrective action, and with soul to soul communication to others if necessary, to actuate the prayer request. I think here of the adage, "a problem well defined is half solved", which applies here in presence of the mind. William S. Burroughs (1914-1997), an American novelist, stated, "Your mind will answer most questions if you learn to relax and wait for the answer". As we were taught since we were children, "The Lord helps him who helps himself".

In a sense, we need not pray to God for certain events and desires, but merely to communicate within ourselves, to our immediate soul and Guardian Angel as a representative of God, because as will be explained later, soul is the entity that makes mystical events occur. Communication and actions of the soul, a microcosm of God, from the genome of each body, and intra-soul communication, is the act that affects mystical proceedings, including health, mind thoughts, divine intervention, intuition, conscience, thought transference, compassion, morality, wisdom, emotions, and other qualities.

The will of God, manifested in the mind and soul, is challenged by the biological, free-will thinking process of each individual. This free-will thinking (it is theological thought that free-will thinking was present in the Adam and Eve epic) will sometimes disagree with the desires of a God-oriented soul, thereby resulting in thoughts that are contrary to the thoughts and pressures (our conscience) of soul and God.

There are shades of Goodness, and shades of Evil. No person will be perfect and in full agreement with God, all of the time, because of this free-choice interaction between will of God and acts of man.

(As an aside, It was demonstrated during the 2011 Masters Golf Tournament that one contestant in particular had a difficult putt. To hole-out would have put him in the competition lead of the tournament. The golfer obviously said a silent prayer that his golf ball would go into the cup, thus assuring him a better chance in winning the tournament, with a substantial monetary award. The putted gold ball did not drop into the cup.)

(It has been stated before that the soul and mind collaborate to answer prayer. A logical question then is "why was the prayer not answered positively?" A possible answer is that "insignificant" request in prayer are not given equal consideration compared to more meaningful requests, and perhaps not where monetary reward is a factor.)

POSSESSORS OF SOUL

Many religious leaders have told, since their own first spiritual knowledge, that we all possess a soul. We wish to explore where the soul came from, what it is, what its purpose is, and what happens with soul when we die. The world of matter is a containment from which we cannot see into other dimensions. It is believed however, that the opposite, viewing into the present world of antimatter to souls from other dimensions, is possible and perhaps common.

Empowerment, here, is in the ability to conceive and visualize what is not obvious to our five senses. We are empowered to realize the existence of soul, its purpose, and perhaps where and how soul came from, and where soul will go in the future after worldly death.

It is perplexing that so-called open minded scientists, writers, and even law courts insist on procedures of natural evolution, without regard to intelligence in evolution. It is possible, and often necessary to have both, intelligence (to plan), and natural evolutionary processes (to implement), to effect the change.

The farmer plants his crops and they mature by natural growth, but it is his intelligence and ability in evolution, when the farmer provides the intelligence of what to plant.

The automobile is assembled by many workers, but it is the intelligence of creative designers and engineers who plan the image and guide the process in order for it to be manufactured.

Evolution provides the working for the size and shape of bird's beaks (per Darwin), but it is intelligent evolution that guided whether the bird would even have a beak at all. Intelligent and directed manipulation of genes is essential to how evolution was even able to transpire.

Nothing in the human mind makes sense except in light of soul. Common sense, plus straightforward logic and reasoning, and a little questioning of received wisdom, equals empowerment. Empowerment is the ability to conceive and visualize what is not obvious to our five senses. We are empowered to realize the existence of soul, its purpose, and where and how it came from, and where soul will go in the future.

Evolution of man through the centuries was parallel with the directed evolution of biology. Changes in man's biology arose from needs in existence and function, just as Charles Darwin revealed that changes are due to environmental needs of the species, but implementation of the change was a product of the genome and biology.

This theory of soul solves many mysteries that arise in our existence, but obtaining adequate, tangible proof awaits further advancement in combined study of Elementary Particles, Physics, Philosophy, Mathematics, Logic, and related fields of study.

There were many species and sub-species that developed between the 6 million years ago occurrence of the intelligent Ape Family (Gorilla, Chimpanzee, Monkey, and Bonobo) through Homo erectus (believed to be an early possessor of "basic soul") attributed at 1.8 million years ago, and a modified Homo erectus species at 600,000 years ago, that carried advancements in emotions and other qualities.

With advent of soul within man (in Homo neanderthal and Homo sapiens) the brain became physically more complex and functional in specific areas of the brain's operations.

There was more than "simple merging" of soul into the animal body, but also involved physical change in man's brain and consequent abilities. There was interrelationship of physical body and brain to one's soul, or spirit, and of mind.

1. The existence and complexity of DNA and RNA are determinant of physical development.
2. The existence and complexity of DNA and RNA are determinant of mental activity. The existence of soul is observed with the existence of man.

SOUL

Archaeology has presented a confusing picture of Homo erectus. Two creatures were honored with the same name of "Homo erectus".

At 1.8 million years ago Homo erectus species entered the progressive development of Homo in an erect-standing creature in northeast Africa. This species of Homo erectus supposedly went extinct about 1.6 million years ago. These creatures lived in parallel with Homo ergaster (1.7 million years ago), primarily in China, Indochina, and Java.

A second related creature, emerged about 600,000 years ago, perhaps an extension of the earlier Homo erectus, but now possessed with more distinct qualities of soul, as enumerated below. His origin was perhaps in the southwest Asia region, now known as Russian Georgia, between the Black and Caspian seas, near the Caucuses Mountains. His skin color was different, in being lighter. He also possessed other physical traits, and was categorized as Caucasian race.

His emotions included Summum Bonum, or a recognition of a Superior Power within himself, and demonstrated his faith in artwork drawings on walls of his cave-residence, and in molding or carving spiritual dolls from clay, representing fertility of his species.

The modern Homo erectus species of 600,000 years ago lasted at least until arrival of Homo neanderthal, about 350,000 years ago, with no significant interim species recorded in the Mideast history before this time. Modern Homo erectus possessed increasingly distinct and developed elements of soul.

For ease in identification, we could call the earliest creature "Archaic Homo erectus", and the later creature of 600,000 B.C. as "Modern Homo erectus". Modern Homo erectus grew in his emotions as his species continued through time. His emotions developed further, but slowly.

It is known that about 350 million years ago a new species (or sub-species), called Homo neanderthal, emerged in the Homo species progression. The author proposes that Homo neanderthal, and also later Homo sapiens, were actually sub-species of the older modern Homo erectus, and continued the qualities of emotions, including the recognition of a Supreme Power, together with incarnation within the human body of an essence we call "soul".

Developing the soul constituted an extension of God, and founded the further creation of mind, which became a controlling force in man's existence, including judgment and direction in living.

From that time on, the elements of the existing brain, soul, and mind would interact cooperatively and in unison, together with the not-yet known DNA and RNA, affecting change in body and the neuron (brain) system, allowing technological improvements in organized speech, language, and recording of events, together with a numbering system of counting, addition, subtraction, multiplying, and division. A society had developed in mankind during this Neolithic life.

From this source and species, mankind developed through sub-species of Homo neanderthal, Homo sapiens, and Modern Man, each succeeding sub-species more complex and complete than its predecessor sub-species.

QUALITIES OF THE SOUL

Scientist have shown that there are certain biological changes that take place in humans, and some other mammals, that are common to emotions. The emotion of pleasure in animals, for example, is associated with a rise in body temperature and heart rate. To a less precise measure of feelings, behavior and body language are associated with certain emotions. It may therefore be scientifically possible to validate the rise of an emotion by measuring biological changes.

Some scientists believe that consciousness arose when animals began to experience physical pleasure and displeasure, and evolved further along this value, with other emotions developing later.

The processes in the total functioning of the human being can be divided into two categories. The first will encompass those body and neural processes that are common to all animals, most fish, and some insects, generally the notochord and chordate families of life. These processes are devoted primarily to sustaining life and are basic to existence.

The second are many qualities of a deeper and individualistic nature. They are qualities of the mind, more complex, variable, and are unique to the individual organisms expressing these qualities. They go beyond functions of organs and systems that serve only to sustain life.

Examples of the functions to maintain life and qualities of life of the soul are given on the following pages.

To illustrate, one can think of a draft horse, or some other form of farm animal, as a fair example of a pure division between these and the more complex qualities of the human "animal". These are functions purely of the body and brain, and are absent of the functions of the mind.

As far as we know, we see no qualities in these animals such as love, vision, mental telepathy, creativity, empathy, consciousness, intuition, morality, emotion, or recognition of a supreme good, although it could be argued that "man's best friend", dogs, cats, and certain other pets and wild animals do exhibit various qualities and degrees of affection, planning, cunningness, emotion, creativity, and even thought transference, such as is trained in Leader Dogs for the blind.

The second category involves functioning and processes beyond the electro-mechanical-chemical-muscular-neural system functioning of the brain and body. They involve sensations (emotions) that are wholly intangible and not reproducible by some rote, computer-like communication process.

Some qualities are required only to maintain the animal body to function in maintaining the animal life. Other additional qualities are found that maintain the essence of soul in a human body.

Two lists of many of these qualities are presented. I suggest that among these are the qualities of the soul after physical death of the body has taken place. These are qualities of the souls in Heaven. (Perhaps they too, are the qualities of the God substance, and God himself.)

When the animal (man or other) expresses these qualities, it is expressing the presence of soul. Perhaps the degree of development (maturity) of the soul is defined as the extent that these qualities are present.

Findings in science, and those in metaphysical existence in theology, comprise an undivided wholeness of soul with body, brain, and mind. A relationship exists between thought and "feelings" and emotions, which is a deeper level of consciousness. Thought is an electro-biological operation of the brain, mind, and soul in processing impulses to actuate response. That response may be modified by judgment, which might also be a learned electro-biological operation of the brain, or it might be modified by a feeling that is associated with our deeper consciousness or intelligence.

Feelings, or emotions, on the other hand, are ethereal and non-systematic impulses that is different from the mere thought processes of the brain. It will take considerable scientific effort to prove this relationship in a scientific method that requires validity and reliability.

A definition of consciousness, which includes feelings, is: "The ability and performance to think, to analyze, to visualize, to show concept of self to other beings, and to plan ahead".

FUNCTIONS NECESSARY FOR LIFE

(VARIOUS EMOTIONS ARE THE KEY TO LIFE AND HUMAN EXPERIENCE)

BASIC FUNCTIONS OF THE BODY AND BRAIN

These qualities require only molecular systems that will keep an animal alive and functional. These mechanisms build tissue, power the body, assist in form and motion, have memory, elementary analytic ability, and mount defense against threats such as virus. Such may be found in farm draft horses, worms, or elementary animals.

The molecular mechanism might be viewed as near mechanical-like, with an observable cause and effect, or perhaps viewed like a computer that will give predictable results from specific stimuli. The organism may demonstrate certain elementary, esoteric qualities of the mind that are detailed in Qualities Purely of The Mind:

Communication ability (hearing and doing)
Nervous system (sensing and reaction)
Motivation and sensing to muscles: voluntary and involuntary

Sensory and nerve impulse transference to the brain (sight, hearing, taste, touch, smell)
Mental computation and information analysis (basic thought)
Maintenance:
Digestion (intake, nourishment, excretion
cell replacement and organ life retention)
Blood and circulation systems
Respiratory and pulmonary functioning:
Nerves, neural system, brain, reflexes, other organs and systems
Guttural throat sounds
Motor abilities (muscle movement)
Knowledge: accumulation, storage, and processing
States of Mind (chemically altered or stimulated)
Thoughtless (involuntary) muscular operation and reflexes
Cognitive (e.g. food=eat or life-extension thought)
Persistence

QUALITIES AND EMOTIONS PURELY OF THE MIND

The qualities of emotions listed here are those of the mind, and are beyond functions of organs and systems that serve only to sustain life. The qualities are ethereal in that they are intangible, personal, unpredictable, and mystical. Many are involuntary and untaught. None of these qualities found in humans would be found, say, in an impersonal computer hardware system, robot, or simple animal:

Abstract reasoning and planning (Creative, theoretical planning, aesthetic, visionary, metaphysical)
Affection
Altruism
Anger
Anticipation
Appreciation of aesthetics
Approximate (to estimate)
Artistic appreciation
Awareness of self

Awareness of needs of others
Astonishment
Audacity
Beauty (appreciation of)
Bonum Summa (recognition of a Supreme or Superior Power)
Charity
Community concern
Compassion
Conscience (intuitive proper action or direction)
Consciousness (awareness)
Concern
Cooperation (mutual advantage)
Creativity (reflection)
Curiosity, Wonder, Suspense
Deceit
Déjà vu (experienced before)
Embarrassment
Emotions (feelings, sensations)
Empathy (feeling for others)
Envy
Expression of art
Faith (trust), Hope, Peace, Bliss
Fear
Foretelling
Forgiveness
Free Will (belief in, choice)
Generosity
Guilt, Sense of justice
Grace, Poise
Grief, Sorrow
Hate
Honesty values
Insight
Intellectuality
Intention and planning
Introspection, Judgment (whether to extend or adhere to tolerance limits)

Intuition, Joy, Cheer, Triumph, Elation
Kindness
Love (personal)
Love (societal)
Mental telepathy (thought projection and receiving)
Mercy
Morality, Ethics (knowing and practice of right from wrong)
Nostalgia and memory recall
Parental responsibility and association
Personality
Pride
Reflect God in humanity of Earth
Righteous indignation
Shame
Societal awareness
Spirit communication
Spontaneity (sudden idea)
Sympathy
Think metaphorically
Understanding
Vision (to experience sense)
Vision (to imagine)
Wisdom (best judgment)

THE SEQUENCE OF SPECIES ON EARTH

From the Big Bang, in a period of 9.7 billion years after the Big Bang, and 4 million years ago, there followed elements and successive particles from matter and antimatter of the Big Bang, of cosmic gases, molten earth, earth cooling, change from non-organic to organic molecules, amoeba and elementary cells, multi-cell organisms, development of functional organs in animals, refinement and coordination of animal and plant organ systems, leading to worms, sea plants, mollusks, fish, amphibians, dry-land animals, primates, to man-like animals, such as "Lucy" of an Australopithicines species, various Homo species, and finally to Modern Man. The sequence is as follows. (Dates are approximate and recognize first appearance):

Ape family of hominids (Gorilla, Monkey, Chimpanzee, Bonobo)	6.0 million years ago
A Paranthropus	Dates undetermined
Australopithicines	5.3 million years ago
Ardipithicus ramidas	4.5 million years ago
Australopithicus anamensis	4.17 " " "
Australopithicus Afarensis ("Lucy")	3.6 " " "
P. aethiopicus	2.7 " " "
A. garhi	2.6 " " "
H. habilis	2.6 " " "
A. boisei	2.1 " " "
H. rudolfensis	2.0 " " "
H. erectus (Archaic)	1.8 " " "
H. ergaster	1.7 " " "
P. robustus	1.6 " " "
Homo erectus:	
" " (in Europe)	800,000 years ago
" " (in Indonesia)	700,000 " "
" " (more modern species)	600,000 " "
" " (in Germany)	500.000 " "
" " (Peking Man)	500,000 " "
Homo heidelbergensis in Europe	500,000 " "
Homo neanderthal (first appearance)	350,000 " "
Homo sapiens in eastern Africa	160,000 " "
Homo sapiens in western Africa	120,000 " "
Homo sapiens in South Africa	113,000 " "
Humans (in southwest Asia)	92,000 " "
" (in Eastern Europe)	46,000 " "

Humans (in Mongolia)	40,000	" "
" (in Papua New Guinea)	40,000	" "
Denisovan hominids in northern China, Melanesia, and Java)	41,000	" "

EVENTS IN MAN'S PROGRESSION:

Cave Art (European—in 200+ caves)	38,000	" "
First fossils of Cro-Magnon man (in Romania)	35,000	" "
Human settlement of offshore islands of southwest Pacific Ocean	33,000	" "
Modern Man: Homo sapiens sapiens/ Cro-Magnon in France	28,000	" "
Man paralleling European Sulterian Culture at Chesapeake Bay, No. America	22,000	
Homo neanderthal becomes extinct	20,000	" "
North American Indians cross Bering Strait (assproximate)	15,000	" "
Palo-Indians in western & central North America (Evidenced by Clovis points)	12,500	" "
New Agrarian culture in Mesopotamia	10,000 years ago	
Existence of Troy in eastern Turkey	3rd Millennium B.C.	
Settlements in Fiji Islands of South Pacific Ocean	3.500 B.C.	
Egypt Old Kingdom	2,686 B.C.	
First Olmec culture (South America)	2,500 B.C.	
First kingdom (Akkad) in Mesopotamia)	2.334 B.C.	
First Maya culture (South America)	2,000 B.C.	
Rome founded	753 B.C.	

Some Books of Old Testament (Talmud) written foretelling of a Savior coming	600 B.C.
Persia expansion/Alexander the Great	550 B.C.
Hellenistic Greece	323 B.C.
Rome defeats Carthage Empire	146 B.C.
Roman occupancy of Israel	63 B.C
Roman Empire founded from Roman Republic	44 B.C.
Birth of Jesus (approximate)	4. B.C.

SOUL EXISTS

This is an explanation of the soul of man, not as a religious dogma or portrayal, but as a realistic dissection of soul, what are its elements, perhaps what might be the origin, how the immaterial soul, from the material of matter, became a part of man, its purpose, and its future.

With annihilation of positive-charge particles during the moment of Annihilation, this leaves the question of what happened to the positive-charge units as the negative-charge units became units of matter, which we today can see, touch, and physically determine.

Proving the existence of soul is like attempting to prove the existence of God. Neither is tangible. They are not physical. We cannot physically pick up a substance of it, see it, feel it, or describe its dimensions. We can only describe its qualities and understand its composition when the qualities are present and when they are absent.

Modern man possesses these qualities. The author has described some of these qualities of soul, basically emotions, and presented them here. The qualities are not physical, not matter, but identifiably present. In fact, it has been questioned over thousands of years of interpreting, "What is God, and what is man?" In the Book of Genesis, finally codified from oral understanding about 600 years B.C., it is stated that man was created in the image of God. This statement may be an amazing truth in that soul is a portion, or reflection of God, and that God, rather than residing "up there" in the sky, is actually in residence within the soul of each of us—that soul and God are synonymous. A portion of God exists within each of us, manifested in the soul of each of us.

These qualities exists as our emotions, as our consciousness, as our intuition, our sense of values, our belief in a superior and guiding power, our ability to determine proper thoughts and actions from those that are improper and against unjustifiable actions. They are the difference between Modern Man (or Homo sapiens) and prehistoric animals and hominids. They are qualities that were joined through genetic processing, to bi-pedal and erect walking, hominoid animals of hundreds of thousands of years ago, way prior to existence of Homo neanderthal and Homo sapiens.

God is manifestation in soul as a minute quality, like a drop of water in a large container. Soul is present within our body and mind, but in a dimension additional to the four dimensions we recognize in our everyday living of height, depth, width and time.

Some theorists, in studying "String Theory", number eleven different dimensions in antimatter existence. Others state an unlimited number of dimensions exist. These dimensions can presently be only theorized, identifiable only by their qualities and strangeness, 600,000, 350,000 and 120,000 years ago, and all gradually refined genetically to our present existence as man.

Man has modified his body, mind, and abilities over Neolithic existence of cave-living style, various stages of technology and culture development, to agrarian and community living style, and unfortunately over periods of experiencing differences in human values with empire building, warfare, and subjugation. Soul is the altruistic heart of man, more permanent than his heartbeat, continuing in existence even after physical death.

Perhaps the origin of soul can be explained. An explanation goes back to the moment of the Big Bang when all elements were composed in the Singularity prior to that moment. Various elements and forces possessed electrical charges, some negative, and an equal portion positive. There existed a process of Annihilation in which negative-charged units eliminated most of the positive-charged units. This leaves the question of what happened to the positive-charged particles as the negative-charged particles became units of matter, which we today can see, touch, and physically determine.

It is reasoned that positive-charged units formed antimatter. These units of antimatter formed the universe of antimatter, which reasonably is the basis of soul.

Soul evolved, as did Matter on earth and life, and was joined by the physical matter-body of man, at time that the present species of man came into existence from hominoids, at approximately 600,000 B.C. with Homo erectus, or at about 350,000 B.C. as

Homo neanderthal evolved, or 128,000 B.C. upon advent of Homo sapiens, and likely as a continuum of Homo erectus in all these periods. Universes of matter and antimatter evolved in parallel, each in a separate universe.

It is difficult to comprehend why considered science does not make greater attempt to understand and have success in the recognition of existence, nature, and qualities of soul. It seems nearly like the popular "elephant in the living room syndrome", whereby a situation exists, but no scientist seems willing to talk about it. Study of the soul does not, it seems, meet the requirements of reproducible proof designated by the time-honored "Scientific Method" of proving validity and reliability, and thereby remains outside the margin of what will be scientifically researched. Perhaps change in this regard should be broadened to encompass the indefinite qualities of antimatter.

We, as a populace, seem to recognize that soul is present, but we are reluctant to define it, to analyze it, to speculate, let alone determine its origin and development to its modern-day existence. Why is it that "good people" recognize that spirit, or soul, exists in people, but fail to go on to explore what that is, to define the qualities of soul, when it is present and when not present, in humans, and possibly some animals, but not in machines. It seems that science has its biases. Science seems to fail in recognizing the continuation of soul even after the death of a mortal body.

We proposed in an earlier chapter the qualities of soul. We have defined soul in terms of its characteristics and its qualities. We don't know the purpose in life of soul, whether it is the same as the body and mind, or whether it is a separate assemblage that parallels the body in life existence. We think of soul as a religious object, seemingly forgetting that soul existed in mankind long before any of the world's present religions came into existence.

One function of soul is to evolve and develop the body and mind of humans, and perhaps another function is to further develop itself as soul. Soul can manipulate the various genes of our genome that in turn directs enzymes to affect proteins of various organs of the body. This "bio-engineering" was done in the past to evolve and direct development from ancient amoeba and viruses to develop, one step at a time, what is now complex organs of the body, in heart, lungs, digestive system, blood, transference of air and elements for body use, and other organs of the body and mind that make up our present status of anatomy, neural, and other body systems.

We somehow recognize that soul exists beyond the life span of the human body, and that soul exists in an Afterlife, not knowing the quality or purpose of soul in this Afterlife, if that indeed exists. Many religions, not all, have developed soul to be a religious entity, which upon death will be rewarded or punished depending upon evidence of "goodness" in the lifetime of the living body. The author believes soul has a purpose: It exists before birth of the human body. It exists during lifetime of the human development, and it exists after death of the human.

Even in sexual matters, soul becomes present from the first moments following conception and directs the developments of cell growth, culminating in a human being at the moment of birth. Development of the unborn fetus is a product of soul. This revelation may clarify thinking of persons regarding abortion and right-to-life controversies. In abortion the infant body dies, but the soul does not. The soul is "returned" to the parallel world of Afterlife.

The scope in functioning of the mind in prehistoric, Paleolithic, man may have been extremely limited. His thoughts would likely have been simple, focusing only upon immediate goals of survival, and not long range complexities, quite animalistic in comparison with mind of later species.

Soul, in conjunction with mind, is the link between the animal human body and the Supreme Power. To make the understanding tangible we can make an analogy in life: It is like a Chairman of the Board of a corporation, acting through a President, and through various Vice-presidents, each in charge of a function area, and through supervisors and workmen that cause an organization to function. This similarity is to our God, Christ, archangels, angels, prophets, and through various categories of angels.

Today great discussions take place surrounding the question of Intelligent Evolution, or Planned Evolution over time. An opposite view seems to follow a belief that untold growth and variations developed without plan through endless mutations, with non-functional pathways cancelled through disuse. It has even been proposed that cosmic rays, such as Gamma Rays, caused mutation's disuse, causing a joining of many smaller bones and body configurations into our present foot, hand, fingers, and other items of anatomy to our present configurations. We think of the absurd analogy commonly stated that an infinite number of primates using, an infinite number of typewriters, could produce the current library of great books of the world, is unlikely.

The soul, conversely, is believed to be creation of a second form of existence, formed from the antimatter particles found in science. Science in the past century has only started to explore and define the qualities, actions, and nature of particles, both minute and intangible, in antimatter. The world of antimatter is believed to be defined, in not our three dimensions of width, height, and depth, and also of a fourth dimension of time, with other previously unimaginable dimensions, some think may number 9 or 11, or others think are unlimited in scope.

In defense of free-will versus predestination, we can assume that the brain was the primitive form of thinking by pre-historic man. He had no language with which to communicate his ideas. As such, when we revert to our sub-conscious mind, such as when we sleep and dream, our dreams are often filled with decision points. These are free-will decisions. If our mind's natural state was pre-destination, we would dream of events that will come true through pre-destination, which is invariably not the case. That said, the Bible relates Daniel having and reporting dreams to King Nebuchadnezzar that in fact did come true, perhaps evidence of pre-destination as revealed in dreams.

That soul exists in modern man is quite well accepted by thinking and philosophical-minded people today. Further evidence of soul exists in different formats, some in parapsychology. The author and others have reasoned that, elements of soul were created at the very birth of the cosmos, following the Unification, Singularity, Big Bang, and Annihilation, a genesis of the soul in antimatter qualities, evolving and developing to a point that allowed later first incarnation into animal bodies and forms that preceded Neanderthal, Homo sapiens, and Modern Man.

The "mechanics" of incarnation was made physical with discovery of the "God gene" within the genome. The God gene introduced "Summa bonum", or recognition of a Supreme Being and the supreme or highest good, from which good acts are derived, into the genome system, and allowed reproduction of this presence, like other qualities of mankind are reproduced in following generations. The God gene allows the connection between man's physical, or matter being, and a non-physical, or antimatter recognition, the essence of soul within the human. Examination and verifiable existence of the God gene is subject of study by research geneticist in collaboration with philosophers, which will be a momentous breakthrough in genetics and philosophy, as well as in religions, and soul study.

I have listed many of these qualities of soul n a previous book, "Do We Live In Two Worlds?" and are repeated in this book. Some aspects explored here are complex and perhaps irrational in our current status of understanding. However the concepts seem compatible and conceivable. Perhaps, someday, they will prove to be an avenue to a more universal thinking of oneness of body, brain, mind, and soul.

The soul may even recirculate from Afterlife, to join a future new-born person (reincarnation), or it may transfer from one existing body to another during those respective lifetimes. Some cultures seem to believe that the whole purpose of life is to refine the soul, to be worthwhile in relationships to fellow man (and other souls) and to be ethical in lifetime relationships. This is the "Karma" of one's self and the quality of one's "goodness".

EVIDENCES OF SOUL EXISTENCE

I believe there are many points of evidence that soul does exists:

1. In paleoanthropology there was observable difference between the hominid of the primate family and the later hominoid of traditionally modern humans and their ancestors. A main difference was that of the hominoid who possessed, or who

started to possess, recognition of a superior entity, and a sense of symbolism shown in cave wall art, or the relationship of his actions in respect for that Superior Being.

2. In evolution, hominoid further developed qualities, including death burial and artifacts accompanying burial, to aid the deceased in later life, the Hereafter. These evidences have been found in anthropology and archaeology of burial sites.

3. Mankind developed a continuing relationship with powers of a super-life (soul) body, with his personal actions that would affect his welfare in an Afterlife. Clear evidence is found in Egyptian death and travel to the "Underworld". This was the beginning of organized religion, whereby a set of thoughts supporting the existence of gods, and procedures to pacify those various gods.

4. Prayers have been offered and rewards were given with desired action appearing, either by coincidence, or as perceived as a direct result of the prayer given, to influence a greater power. Ancient Egyptian culture was particularly notable in this soul/Afterlife relationship. Egyptian death culture was somewhat practical in preparing the deceased body (mummification) and supplying the deceased with all niceties that would be enjoyed in his "Underworld". Soul of the deceased body was anticipated to leave the tomb at night, with name markers on the tomb to guide return of the soul in the morning to his body burial site. (Perhaps this thought is forerunner of modern-day gravestones.)

5. A concept of return to secular life following Afterlife developed with belief of future earth existence in one's true soul life span.

6. Judeo, Christian, and Islamic religious cultures all developed to define a monotheism and a well-considered set of standards to be met in order to attain existence as a soul in Afterlife. The Egyptian concept of "The Book of the Dead' was quite parallel to the Mosaic Ten Commandments. The book listed all the things the Pharaoh did that were good, and events of forbearance that Pharaoh did not do, points to be used for admittance to the Egyptian Underworld. Moses, having

grown up in the household of an Egyptian Pharaoh, would be well aware of the contents of "The Book of the Dead" and may well have used this as an outline for his own writing of The Ten Commandments, embellished with a bit of "pharaoh-like creative drama".

7. Mystical events have been reported that would substantiate existence and alleged proof, of soul within man. I do not necessarily accept terms of "spooks, ghostly presence, apparitions, visions, specters, transfigurations, or ghostly beings" as infantile expressions, or that they are an impossible existences as unseen antimatter forces, or that they are unreasonable.

8. Strong doctrines in church dogma further defined an existence of soul in man and in Afterlife, with doctrines developed for monotheistic man to follow. Prayers are given after a person's decease requesting good treatment of one's soul in Afterlife.

9. Mystical events have occurred in both short and longer historic past to support power of prayer, evidence of communication with the dead, return to life after death, reincarnation, out of body experiences, telecommunication beyond natural experiences, a communion with a superior, monotheistic power (or God), an unshakeable faith in religious beliefs and dogma, and testimonials from various persons who professed to have returned to life after death, describing existence in, or enroot to, Afterlife.

10. Existence today, perhaps between many additional dimensions in nature, of inter-dimension communication, miracles that remain unexplained unless as result of actions by a superior soul, angels, or the like, testimonials of reincarnation, physical recurrence of events, déjà vu experiences, ultra powers in religious matters, and prayer. Various Christian, Jew, Islamic, and other religions with historical backing, have thoughtfully defined soul and Afterlife in their own terms.

11. Verified anatomical and psychological occasions (medical dissection) in differentiation of the physical body, the brain, and an intangible mind and soul, the mind being an ethical,

moral, emotional, and a just-guiding entity essence within man beyond the manipulative powers of the brain.

12. The soul is near-universally accepted as a presence within all Homo sapiens. Evidence of soul within other animals, plants, sea life, and other earth life is undetermined, philosophy varying within different cultures on earth. Native American Indians admired the Shaman who had developed powers to communicate with souls in another world and to recognize soul within tangible sources (rocks, winds, mountains, etc.).

13. The idea remains a question whether soul results from a mystically-placed physical origination in the mind of man. It remains undetermined whether origin of soul is a logical or illogical insertion for man's reasoning, or was a direct mutation of thought manifested in the genome, and mystical process by a supreme power.

14. There is logical reasoning that existence of soul is a cooperation of the make-up or design of the genome system, the brain, and mind of man. Soul presence within genes of man is no more mystical than having a gene that determines, say, the color of one's eyes.

15. There are numerous Biblical citations for the existence of soul, where Jesus Christ himself refers to his soul and souls of others, as well as sharing of his soul with the existence of a monotheistic God, the father of his soul. It is suggested herein that the soul lives forever.

16. There is Biblical reference of Abraham having face-to-face communication with Yahweh (God). That meeting would logically have been communication of Abraham's soul with the essence of God. Soul was present in Abraham as a medium between God and the mind of Abraham. There is reasoned theory that soul existed, and developed to increasing degrees, starting with ancient species of Homo erectus, Homo Neanderthal, and Homo sapiens, culminating today in Modern Man.

17. Existence of soul in Afterlife is reported Biblically in the Book of Revelation as drama after death. With soul there is Biblical

promise of an Afterlife and presentation to Jesus Christ after death.
18. Incidents of the real presence of soul are demonstrated in the life and teachings of Jesus Christ and others.
19. It is possible to enumerate and describe the qualities of soul (See Qualities of Soul in another chapter).
20. One can reason, in observing the actions of mankind, the existence of an internal, non-physical entity, complimentary to the brain, defined by the mind and its functions. It is believed, as presented in this book, that soul is a microcosm of God, that a portion of God exists within each man, women, and child in the form of mind, as opposed to brain or heart.
21. Spiritualism is the essence of soul activity. Spiritualism, as opposed to religion, is one's belief regarding his god, existence, and powers. Spiritualism pre-dates any present religions on earth. Spiritualism will provide guidance to one's morality, ethics, and conscience.
22. There has been a wealth of spiritual mythology throughout history in ritual centers, and findings in archaeology of precious burial artifacts for an appropriate existence in Afterlife. Soul has really nothing to do with religions. Soul existence is a function and creation of evolution, rather than creation of some religious view spoken from a pulpit or by some ancestral priest. It is to be a practical entity inhabiting our bodies, or awaiting our death.
23. One can sense, feel, and comprehend that his life-being is more than being merely a biologic "machine", without emotions, operating android-like, and that he possesses unique values in his make-up. We sense that we are more than an animal of the forest living only for subsistence and procreation. We have experienced an epiphany, a transformation, a feeling of inner cleansing and renewed appreciation of God and Jesus Christ, and what in his mind each stands for and represents.
24. The parallels of the Monarch butterfly and soul (see another chapter) regarding memory transference between lives is significant, that perhaps a "system" currently exists

in soul development to provide this function of thought and memory transference between souls. A similar pattern exists in the life panorama of Salmon fish, whereby adult Salmon will miraculously return to the backwater stream from which each had originally been hatched, to give birth to a new generation of its species. Migrating geese and other birds also demonstrate memory transfer by the migratory travel route and return chosen by themselves and their progeny. A system currently exists in nature and life for transference of knowledge transcending death and birth.

25. Like a learned son who has grown greater and more qualified than his own father, Homo neanderthal, and especially Homo sapiens and Modern Man, are sub-species of Homo erectus, but have grown greater than their "father species" of Homo erectus of 600,000 years and 1.8 million years ago.

26. Anthropology studies have found ape-like and chimpanzee-like qualities of brain in intelligence, but not intrinsic qualities that would simulate soul. This suggests a period of time in development of man when qualities of soul were not present, but became present at some later period of time. Soul has not existed in primates or hominines, but became present in hominoids to a limited extent, early, as in Homo erectus. Incarnation of soul to Man was not a product of evolution, but was purposely endowed as a separate act.

27. Nevertheless, forces of science appear to disregard any acknowledged fact for existence of soul. A portion of that reluctance seems to be a basic difference in acceptance by the science community and in obeying rules of the Scientific Method, which perhaps should be enlarged to encompass study of antimatter, with those peculiar rules of probability rather than repeatability.

The scientific method requires substantiated and repeatable proof in research, with repeatable results in experiments. With soul proposed as existing in an additional dimension, it is presently near-impossible to devise such an experiment having repeatable and provable results. It is a given in dealing with factors in antimatter that results are a

28.   Soul is reasoned to have none of the physical senses of humans of sight, hearing, taste, touch, or smell, but instead communicates by a form of thought transmission and memory in the mind. Incarnation of soul to body of living man is interpreted as a device in spiritualism to provide soul with a secular living experience, perhaps to experience free choice.

29. The first seeds of soul were sown in the first moments of the Big Bang. In this event there were positive-charge elementary particles and approximately equal numbers or quantities of negative-charge particles. In the resulting conflagration a process existed, known as Annihilation, whereby negative-charge elementary particles and forces caused elimination of most of those of positive-charge particles. Negative-charge particles went on to constitute matter in our cosmos. Positive-charge elementary particles, in separation (Annihilation) constituted a new world, a new cosmos, that today is recognized as antimatter and a world and dimension of soul.

30.   Knowing right from wrong: The soul, together with the mind, combine to judge whether a certain thought or action is right or wrong in the view of our guided conscience. The soul, acting as our conscience with spiritual ties, is the senior voice in this determination.

Neanderthals possessed soul to a substantial extent. Evidence is their existence, dating from 325,000 years ago, with demonstrated qualities of soul including cave art, small statuary, and painted sea shells in jewelry that showed belief in a superior power that was called upon to provide aid in hunting and living, plus afterlife for himself and family. The developed species of Homo possessed familial love of close family members and concern for persons in their community of living. They demonstrated emotions by their concerns for self and others. Their degree of demonstrated emotions, directed to their genome system, was basic, and derived from their predecessor species of Homo erectus, dating from about 600,000 B.C., who were also progenitors of future Homo sapiens who, too, possessed even greater emotions and other developed qualities of soul.

Neanderthals and Homo sapiens were competitors in living, Neanderthals eventually losing out to Homo sapiens, but they were not competitors for possession of soul. The symbolic portrayal in life of man, women, existence of evil, and later identification of the superior power that we today call God, were features far beyond an age of the proposed 4004 B.C. projection by the Irish bishop, James Ussher (1581-1656 A.D.). The features were gradually defined and increasingly prominent with each generation, differing from their predecessor species of hominoids.

As science in various disciplines develops, it may offer proofs that support principles of faith, as well as existence in added dimensions in science. Only what we can measure will presently, it seems, define our view of success in spirituality.

As in other specifics of the human body and brain, it would take thousands of years, and multiple generations, to duplicate integral features of man from first incarnation, of soul. First features of soul may well have started with advent of Homo erectus, 600,000 years ago, following through with Homo neanderthal at about 350,000 years ago, and later Homo sapiens. Development of soul was slow, but it is believed to be in full blossom as mature Modern Man occurred approximately 30,000 to 20,000 B.C., with all basic features of man having soul, with recognition of pagan gods who supposedly directed and were responsible for any and all aspects of existence, whether it be of man, mountains, weather, love, life, or death.

Archaeologists have found evidence of man having symbolic reasoning in Africa, Europe, Asia, India, Papua New Guinea, Java, South Sea islands, and the Americas, to be later refined in culture to Modern Man. A specific time to identify such change is difficult, and so far impossible, to pinpoint.

The earliest recognized days of the Maya rule in South America was about 400 to 600 B.C., but with evidence of even earlier existence there, and at least 12,500 B.C. in North America.

Significant events were occurring elsewhere in the world besides in the Maya culture:

1. This was near the time of the Babylonian captivity of Israel (587-537 B.C.).
2. This was near the end of the Egyptian Pharaohs in Egypt (about 350 B.C.), leading to a period culminating at Queen Cleopatra VII in the first century B.C.
3. This was near the time of Socrates in Greece (469-399 B.C.).

Did ancient people of Mesopotamia, of the Near East, Egypt, Israel, or Judea, migrate eastward, past India, to China, Southeast Asia, and Australia, (perhaps Denisova Hominin species) to land on the east coast in South America, perhaps settling Pacific Ocean islands on the way? It is possible that early civilizations in western South America originated from Southeast Asia where height of man was generally short, skin pigmentation was lighter, and access could be obtained by sailing east from Southeast Asia, or from South Pacific Polynesian islands.

Similarities in facial and body features of Maya do not follow those of the Old World of Mesopotamia. Modern day Mesopotamian descendants, though small in stature, do not exhibit the thick lips, bulbous nose, and flat face of Maya persons in stone carvings. (The flat faces may have been portrayal by portraying the three-dimensional faces on two dimensional surfaces.) The thick lips and bulbous nose may be characteristics of people from Ethiopia, Abyssinia, or neighboring lands like Kush described in the Bible. Temple structures, like pyramids, were common to both Maya and Nile River locations.

SOUL AND MAN

It is believed that soul first began its long, slow integration with the animal body in hominoids, starting about 600,000 B.C. Prior to this stage in evolution the hominid of pre-man, the Great Ape family of Gorilla, Chimpanzee, Old World Monkey, and Bonobo, were basically animals of the forests in their composition. In pre-hominoid (hominid) their purpose in life was oriented entirely to continued existence as a living object and to reproduction.

Homo erectus exhibited first features of soul, probably by recognition of a superior entity. For the first time a new specie of the former animal imaged a symbolism of his existence and developed a questioning in his thinking process, which would soon become his mind, of why he differed from other animals of the forest with whom he associated. He perhaps related success in hunting to the powers of some external power, which he should recognize, honor, and which would hopefully also improve his luck for success in hunting.

Although reproduction of his species had been only a physical event before, he now related the sexual event with birth of a child, and growth from infant, through youth, to adulthood. He enlarged his thinking to an affection for others, both within his immediate born family, and also with acquaintances who would mother his child, resulting from after a sexual encounter.

He developed emotions other than mystical, affection, and love, such as responsibility for the welfare and protection of his mate, family, and offspring. Through anthropological studies scientists have determined that a division of familial duties existed, even in the early stages of chimpanzee social development, not unlike that which exists today in a family. The male chimpanzee exercised duties of protecting and obtaining food for the familial group, whereas the female exercised food division and allocation to offspring, with caring for the young.

He developed ways of demonstrating and physically portraying his ideas in basic art forms of pictures, images, and dolls, imagined tales of earlier existences, relationships in family and in groups. Procedures were developed in death of family members to preserve the deceased body by adding red ochre, a magical red earth type material (ferrous oxide) that somehow would preserve the body of the deceased, hopefully to renew his life from the dead, a form of Hereafter, and continued existence different than when he was living on earth.

Emotions also gradually enlarged in scope of hope, self-esteem, friendships, ideals, benevolence, and other traits that set off man as different and advanced from other animals.

Each step became an advancement in evolution of the soul and mind, different from the mere survival tactics of animals, whom he recognized as different from himself and others like him.

He developed additional emotions of affection and responsibility, and recognition of a superior being. He recognized that he had a heart, internal organs, and brain, but something affected his desire to do what he somehow felt was "right", a morality and first manifestation of a conscience, with reasoning or association of problem factors, arranged to formulate some form of livable problem solution.

Codification of vocal sounds and body movements led to simple, controllable language with ability to communicate to others his ideas, desires, thoughts, and actions in living. He also recognized his thoughts, many of which were not communicated to others, being his own and original mental pictures within himself and recognition that this occurred in his mind. Undoubtedly he experienced strange portrayals (dreams) when sleeping, portraying a different and strange world from what he experienced in the morning upon waking.

Writing and counting developed to record experiences and needs to express differences, quantity, and enlarged ideas.

Man intensified and enlarged in his mind and soul a sense of possession, purpose of food, use of tools and weapons, a sense of ownership, that items were his and not to be used without his permission, possession and individuality being another factor of the mind and soul. He also possessed certain boundaries of land for his own or his group's purpose of hunting, with recognition of possessions of others. Violations of ownership, in his sense of possession, resulted in emotions of resistance and conflict. Possessions enlarged in a social sense, that he and other "families" possessed and used areas of land and meat from the hunt together, and resisted outsiders from occupying, or attempting to occupy "their" community land. Leaders of the enlarged groups were recognized and asked to adjudicate their relative possessions, to give counsel and also that there was some kind of understanding that tribal leaders would protect him and his family in exchange for his service to that leader.

Again, each step represented further evolution of the mind and soul, gradually leading to a culture more complex than that of ordinary animals. Soul is a function of a parallel universe operating in an added dimension of space/time. It is mystical, but on the verge of being explained

Beginning possibly with Homo Neanderthal, man showed his symbolic art, perhaps 60,000 years ago, on cave walls where he lived in the southwestern part of present-day France, somehow connecting himself, with animals to be hunted by him, and presence of a supreme being that would assist or direct his actions in a successful hunt. Other art forms of statuary, dolls, and icons were formed by hand, relating to fertility of Neanderthal females, with appeal to the invisible Supreme Being to bring forth a young man-son.

Fire, a strange and unexplained event, started by natural lightning to ignite wood, could be transported to domestic spots and controlled to promote warmth and heat to better prepare food for eating and preservation. He later learned the technology of starting his own fire when and where he desired.

Myths were developed and became oral history, one of which grew to that told of Adam, Eve, Creation, sacrifice, and temptation. Timing of these myths is considered to be late in the development of mind and soul, exhibiting a developed complexity far different than that of original presence of mind and soul.

Man was then at a social state of having family, recognizing the existence and power of a superior being, consciously, and with reason, deciding right from wrong in mental choices. Men possessed a near fully developed soul, and would attend the disabled members of their species, dating back to Homo heidelbergensis of 500,000 years ago, and endeavored to explain the purpose and significance of that soul, with possible existence in an Afterlife, with even thoughts of reincarnation, or new birth after death. (Source: Discovery magazine, Jan/Feb2010).

Man would evolve with significant changes physically, emotionally, in mind, soul, and a more comprehensive outlook, socially and theologically. The species changes would be identified later by anthropologists as Homo erectus, then Homo neanderthal, and eventually Homo sapiens, forerunners of Cro-Magnon and Modern Man, with mind capacity and soul development generally comparable to that existing today.

Man, with soul and a more complex mind-set, would change in culture from a primitive hunter of animals and food, living in caves, to constructing crude houses to live in, with new-found fire, into an agrarian culture of planting, harvesting, and trade, having mutual advantage in trading agricultural items, with intelligence to preserve and store meat from hunting, and plant items for use in a future day. Complexity in his mind, driven by his soul and needs, formulated his developing culture and beliefs.

Religion developed and he had adopted the concept of soul and life after death into their own dogma because religions sought answers to metaphysical and human dilemmas. Zoroaster of Persia (Now Iran and surrounding areas) may have been the first to hinge the eternal destiny of an individual on his or her worldly behavior. For Zoroaster, humanity moved in a direct, linear path towards a cosmic conclusion in which good ultimately triumphs, evil is annihilated, and Paradise is established on earth.

At time of beginning of recorded genealogy, the soul and mind were near fully developed through its evolution 1.8 million years and 600,000 years ago.

A history of culture, theology, society, government, warfare, social intermixing, technical and purposeful development, understanding, communication, negotiation, and ideally just living together in a communal society would develop over the succeeding 10,000 years, and to today. We are today the product of more than 600,000 years of soul and mind development within a changing physical body and species, instigated and controlled largely by the genome, by intelligent evolution, or directed change of body, mind, and soul of man.

Intelligent Design is the plan for life growth. Evolution is the step-by step method of how life movement occurs. Intelligent Design and Evolution were seen as compatible in life changes.

In an industrial setting, engineers and designers today intellectually plan the changes in product image (intelligent design), and workers and technicians perform the actions to affect the changes (evolution).

In life development, an unseen intelligence plans the direction, and manipulates DNA to effect the change in enzymes and then organs. Manipulation is performed by soul and mind, through an electro-biological system, changing the genome at direction of a Supreme Being (God, or by the mind following the direction of God).

Soul had origin in the annihilation of positive from negative elementary particles during the Big Bang of 13.7 billion years earlier, the positive-charge antimatter even paralleling life-forms developed from electron-based matter. The world, universe, and space/ time dimensions of soul exists separately, yet concentric to, the physical body of mankind as illustrated in piecing together the evolution of mankind.

The mechanism for change lay in operation of the genome, perhaps at the guidance of a supreme being exerting the plan for an intelligent evolution. The intelligent evolution developed through eons of life, with a joining of dimensions from the three and four of known matter, to additional dimensions that are in the realm of soul and mind in antimatter.

Physical and biologic changes resulted from mutations in the genes, but those individual mutations were often first directed by a greater intelligence, even though some mutations developed from needs by the environment.

In likening to industry today, machines form a product, but intelligence caused the machines and system that allowed it. Genes caused and affected physical and even mental changes, but formation and actuation of certain genes was a product of intelligence.

SPIRITUAL RETURN, REINCARNATION, AND VESTED ABILITIES

In reincarnation, the reincarnated soul may be motivated and driven by the qualities, skills, abilities, and soul formed and matured in a previous incarnation. This motivation is subconscious, but is revealed in the present person's "natural abilities" in art, science, philosophy, finance, exploration, military command, acceptance of challenges, and like "natural qualities". This ability is demonstration of an "unconscious memory" that has evolved to one person, or mind, from a soul, not necessarily his or her own ancestor. It is memory that is not consciously thought of or consciously recalled.

This concept of mind "subconscious memory", and abilities between different life spans, may relate to the transfer of knowledge and abilities between generations of Monarch Butterflies, salmon fish, geese and birds, and other creatures of nature discussed elsewhere in this book. (See memory transfer between generations of Monarch Butterflies.)

Medieval and modern persons who exhibited excellence in some field, had exhibited even a greater excellence with their body and brain which that person's abilities had exhibited previously. Medieval artists and writers such as Leonardo, Michelangelo, Raphael, Shakespeare, and others performed extraordinary research or imagination regarding events of history that were reported in paintings and writings of 1,500 to 300 years ago and before, in order to depict the various religious and historical scenes and events they portrayed. These artists were Masters of concept and composition, and of course technical aspects of art. Many subject events are in excess of those recorded in the Bible or in recorded history. These depictions have been largely accepted by the public as actual truths over the many years.

In this, there is a remembrance, through the subconscious, of a previous life. The subconscious ability(s) will repeat in further incarnations. Some persons (namely General George S. Patton of World War II fame, and many others, both famous and common, believed firmly in details of a previous life and life strategies, in a field of expertise of warfare and living. Many persons report experiences of

Déjà vu, a momentary thought that you have personally experienced a certain act or moment in time. Déjà vu may be an example of precognition in that a person's mind "jumps ahead in time" in the mind's anticipation of occurrence of an event, and then, when it actually happens, you properly sense that you have experienced the event before.

A living person may exhibit multiple abilities resulting from enlargement of his developed field of abilities and/or reinforcement of abilities resulting from multiple past incarnations and reincarnations. A person may do very well in government or business diplomacy, negotiation, finance, problem solving, memory recall, or other attributes that "come easy" to the modern-day person. In contrast, sometimes a person with practically zero "natural abilities" that some person realizes is not their "forte". In absence of basic pre-historic lives, any modern-day developed abilities in past spiritual lives would be an improvement, and will be cumulative to his present soul's "natural abilities".

A living person's propensity to accept, reject, or be neutral concerning his or her, say, belief in God, may also be cumulative to following soul periods, reinforced from strong and multiple maturing of spiritual beliefs. Belief in God and spiritual existence, for most people of Earth, has become, to varying extent, a natural quality, or way in life. Perhaps this is a factor that "calls" a person to seek a career in the Ministry or Priesthood, or other service to the Church—a "drive" in his mind, generated in his soul by a like "drive" in the mind of a person who lived in a previous life.

A modern-day person may also demonstrate an "unreasoned" fear, such as fear of water, the dark, the unknown, betrayal, or some other unpleasant experience that occurred in the life/soul of a person in a previous life. This probe into cause of a modern-day person's affliction has been duplicated, and cause revealed, by hypnosis, bringing out knowledge of an experience in a previous life that relates to the modern person's fears. Studies and techniques in psychology exist that support the phenomena.

Some geneticists have proposed a "God Gene" in our genome. This theory will either 1) replace the theory of a God Gene, or

2) indicate a development in the genome, from experience, to the genetic make-up of exhibiting unusual strength in that field.

Looking forward, abilities developed in the present life of operating within the realm of today's electronic mechanisms of computers, cell phones, understanding of science, philosophy, medical affairs, and the like, will be a further "natural ability" of future incarnations. This will be partially from nurture, but also from "natural ability" as described here, developed in basic related fields of knowledge by soul in a previous life. It is well-known that many persons exhibit natural abilities of his or her parents or ancestors.

In ancient history, nearly any one who thought intensely about anything was considered a philosopher. The philosopher conjured in his mind all subjects of a nondescript nature.

A first "break out" from this general body of philosophy was making a separate study of mathematics of which many ancients, including Pythagoras, were pioneers in the specialized study.

Another "break out" was Astronomy in which separate concentration was made of the sky and stars, with attention paid to Astrology as a fortune-telling device.

Still another "break out" may have been as a religious leader, priest, or shaman provided counsel and guidance in matters of a mystical sense.

Other concentrations from the "love of wisdom" further "broke out" in concentrated study of physics, anthropology, sociology, linguistics, medicine, and psychology. Even today specialized study is continually being amplified to sub-subjects for study, one of this century, being particle science, growing from Newtonian physics. In fact, universities today award a degree recognizing the total field of philosophy in the Doctor of Philosophy (Ph. D) degree, that might recognize one of many fields of study.

A pure Philosopher in the 17th century, Rene Descartes, reasoned in the field of duality, or whether brain functions and mind are separate, stating, "I think, therefore I am". In the exposition of this book, I personally feel that I have thoughts in a number of disciplines, and therefore will call myself a philosopher, in a general sense. I sometimes wonder if I am, in some respects, like my Grandfather, as well as of my direct parents, or even like someone else in past life, and I have attributes from soul, like that person of an earlier time.

It follows, if the reader feels that some fields of knowledge or skill "come easily" to him, perhaps this is reflection of the abilities that some person in a previous life exhibited exceptionally well, and brought to you in this life as a reincarnated soul. And it also follows that if there are some things in this life "that you just don't get easily" (you feel like a complete dud), perhaps this also is reflection of that absence in quality in soul of some person in a previous life.

For eons, spirit-recognizing mankind has prayed to their god to affect change, or stability, to their relationship in life. The author proposes that the change is effected by the soul.

As an Octogenarian, and After devoting considerable time and effort, including writing, reasoning, and analyzing my thoughts regarding soul, more directly over nearly the past 30 years of my 84 year life, I propose change in a major thought that should be further reasoned.

That thought is that there is day-to-day, operational God in the long-thought, conventional, or usual concept of a Supreme Entity, but that, to a great extent, our God is within our own soul, and our soul is evidence of our presence of God. God and soul are one and the same, at least in part. Regardless, there still exists God as the Supreme Being over all, but something reflects God into the mind and soul of each person.

Regarding prayer, when we have prayed to our God, we actually are requesting our soul, existing within the confines of our physical being, to take physical, biologic, mental, or metaphysical action to accomplish whatever is our prayer.

In Buddhism and Hinduism, religious Karma is sometimes described in reflecting actions of soul. These Asian religions hold that an individual's consciousness, or an event, will direct good Karma or bad Karma, and this will direct whether good things or bad events will prevail in one's future. Karma is defined as "the sum of a person's actions in previous states of existence, viewed as deciding his or her fate in future existence—a destiny of that person's future life or existence.

Such a change in concept could account for the need of multitudinous oversight of over six billion persons on earth, with their soul acts changing moment by moment. It is a more understandable concept to believe that all persons are not continually overseen by some central source (the God or Supreme Entity), but are overseen individually by their own responding souls.

This proposition changes from a difficult-to understand—concept involving billion of people, all being effected at various moments of time, to an understandable concept of one soul relating to one person, and direct effect of that one soul on the one person.

This recognition of soul being the key to human activity, is reflecting recognition of God. This has been so for over 5,000 years, and is paramount in this presentation. The soul exerts direct effect upon the biology, mind, and human efforts of both individual and civilization, acting collectively upon all people.

I prefer to think that physical things were created in a biologic way, and the mystical things are the ways of God. "Render unto Caesar that which is Caesar's, and to God that which is God's" (Mathew 22.21, Mark 12.17, and Luke 20.25). There is purpose and organization in the decision of acceptance into God's substance, and into the hierarchy of angels and saints. There was purpose in God sending to earth the souls of various prophets and of Jesus Christ to communicate and inform for productive liaison between beings of earth and existence in Heaven. Death is truly, after all, a part of living.

This view that the soul of Jesus Christ was of the same substance and part of God, will go far in reconciling the historic question whether Jesus was God, was the Son of God, or was the Spirit of God. Acceptance of this theory that souls of all persons are part of God, the answer to the question would be that Jesus Christ was all, God, Soul, and Spirit.

A PROOF OF SOUL

In a previous book by this author, "Afterlife", George Underwood was a fictitious person who had been suddenly killed and his soul had gone to Heaven. George believed he could prove the existence of soul for earthbound people. He did not need to prove the existence of soul to himself—he was "living proof" of the existence of soul, of himself, and others he met in Heaven.

To understand the concept of positive-charge elements (or Antimatter), we must think of the opposite of all things. As stated, the opposite of negative charge, is a positive charge. The opposite of light is darkness (an absence of photons). The opposite of forward movement is movement in reverse. The opposite of time progression is negative time progression (i.e. going backward in time). The opposite of speed of light is negative (or even zero) speed that light travels. (Speed of light, although not reversed, has been greatly reduced in a laboratory, for a fraction of a second, to near zero). Reversal of speed of light would be tantamount to time reversal.

The opposite of gravity is a force pulling away from Earth, or some other physical body (centripetal force), to the extent that downward forces of earth in revolving around itself (centrifugal force), exceeds the attraction of gravity, thus forcing an object lying at rest, to rise.

In the absence of or reduction of gravity, astronauts in space vehicles must "steer" their space vehicle in mid-course corrections via ejection of gasses from various ports on the space vehicle. Also, without gravity, or with sufficient forward velocity to overcome (exceed) gravity, space vehicles would proceed out of orbit of Earth, in a straight line in space, driven by their forward linear projection.

The opposite of sound (vibrations of air) is the movement of atmosphere air in one direction exceeding the velocity of sound waves coming from the source of sound, or in an opposite direction. (A person in a pressurized room will find it difficult to hear sounds from outside.)

George Underwood, introduced earlier, called upon his high school algebra, a basic knowledge of high school physics, Albert Einstein's universal formula, knowledge of general science, elementary particle science and quantum physics, and with the relationship of matter and antimatter, to develop a line of reasoning.

George was well aware, as was the entire scientific world on earth, the formula advanced and proven by Albert Einstein of $E=MC^2$, where E is energy, M is mass, and C is the value for the velocity of light (which is about 186,000 miles per second). When the value for velocity of light is squared (multiplied by itself), E (energy) is of a tremendously large value.

At this point I would like to describe a different type of energy. Unlike an explosion type of energy, such as evolves from our Sun, or power of energy in dynamite, Latent Energy builds slowly in building a giant sink or reservoir. Latent Energy is trapped in frozen ice glaciers of the world. The ice first required energy to become frozen, like making ice cubes in a home freezer requires energy. The energy remains trapped until the ice thaws. As ice thaws, the trapped latent energy is slowly removed. As latent energy transforms from thawing ice, that energy becomes available as energy to produce different forms on earth, such as motivating man and animals to commence moving about, and to use that energy for thought purposes by man, to create thoughts, and invent from those ideas various ways for man to perhaps have an easier life, or to develop machines or technology systems that are new to his being.

Latent energy is reminiscent of a bear in hibernation. Prior to winter hibernation a bear will eat profusely, converting the energy (plants, fish, animals, roots, etc.) to fat (a sink of energy within his own body). During hibernation that fat energy allows the bear to survive. The energy is gradually used.

But trapped energy, latent energy, is not used in a glacier until the glacier begins to thaw, like the ice cubes from the freezer thaws when we want it to thaw, to cool a drink, or whatever, releasing energy gradually to counteract heat or warmth of the liquid in the drink.

That latent energy of glaciers, upon thawing, permeates the atmosphere and into every particle of being on Earth, and brings into existence new energy to be used on earth in the many forms of energy usage, to energize animals, plants, and humans, and to warm the earth. Man develops ideas from his brain, and implements those ideas in technological improvement.

Again, an example to illustrate: A person, in an un-energized morning mood, may swallow a vitamin tablet. That tablet acts upon his biologic system to generate an increase in energy within the man. There is a transformation of energy here, to cause man to activate his latent energy into productive energy.

I believe Albert Einstein's formula, with transposition, can be used in this new sense to esplain: $E=MC^2$, where E is latent energy, and M is mass of, say, a glacier. C^2 (speed of light squared) remains constant. With C^2 held as a constant, Energy becomes proportional to the Mass.

Application of the formula, transposed to produce values for different transpositions, is as follows: (1a) $E=MC^2$; (1b) $M= E/C^2$; (1c) $M/E= C^2$; or (1d) square root of $M/E = C$. In other words, with the speed of light (a constant), the mass (the glacier), when divided by energy (available) results in a dramatic force. Repeating, with the speed of light a constant, the mass is proportional to the energy. (The greater the mass, the greater the latent energy: the larger the glacier, the greater is the latent energy trapped in the glacier, and conversely, reduction (thaw) of the glacier results in release of energy into the world).

This proves out to logic. When the temperature of Earth becomes frigid, historically there is very little accomplished in growth, or even subsistence. This was the situation from about 30,000 B.C. to 10,000 B.C. during growth, and later thawing of the Wisconsin Glacier. Contrary, at about 10,000 B.C. glaciers began to thaw, and even slid down mountain sides and ravines, with melting water acting as a lubricant.

This is about the time when man changed from the Pleistocene period to an Agrarian culture of agriculture nurturing and animal husbandry. As time progressed, man became more active in society, leaders developed, warfare started, and technology increased. Culture became more complex, religion became more defined, great names in history performed, as Pythagoras, Hippocrates, Socrates, Plato,

Aristotle, Cyrus, Alexander, and many other greats, progressing to the leaders, in technology and in systems leading to today.

Despite claims and evidence of Earth Warming, the average temperature of Earth is now progressing in the cosmos where temperatures will be declining in its cycle. In the distant future, glaciers will again form, as they have always done in all of history, sapping energy, from world processes, and progress in the world will again become static.

Enlarging from the above explanation, this change in phase of energy may also be seen from the change in glacier content on earth. As the glaciers melted, mass of the glaciers reduced, and latent energy was released to the world. This increase in energy affects many events including advances in technology, ideas, invention, and culture on earth. In this period of release of latent energy, man did collectively advance, in other fields of technology medicine, philosophy, earth knowledge, antimatter, and devised means to place man on the moon, and explore, among other means, by devices of outer space exploration of planets, asteroids, and Sun.

Man will devise experimentation at CERN of elementary particles and their actions in change. Changes will be made with ground-breaking (and mind-boggling) advances in culture, society, and matters of a social nature including increase in knowledge of soul and its ramifications, all such notable advances have occurred over the past 15,000 years, coinciding with start of increasing earth temperatures (melting) in global warming, minute (very small) changes at the start 10,000 years ago, but great in magnitude of world concern of late. The "seeds" of the growth in technology were first planted when the Wisconsin Glacier first started to melt 20,000 and 15,000 years ago. As glaciers will come to be melted completely (from energy sinks), even more latent energy will be released to the world, earth, culture, and to civilization.

Again, when glaciers reform in the distant future from decreasing global temperatures, energy on earth will decrease in latent energy and mankind will become static in available energy of matter and activity. But don't worry–this cycle will take thousands of years, as it has cycled many times in the past as recorded in history, measured by depth probes into ice and earth.

With the Earth in a freezing, glacial cycle, it will make little practical concern thousands of years from now, for example, whether the universe is expanding or contracting, or similar academic questions!—It will be cold! Like in the past, at the southwest portion of Africa during a past ice age, certain islands of civilization may exist at certain parts of the globe during otherwise ice-freezing temperatures.

The human body and soul seemed to have worked together and coordinated best when temperatures on earth were in a range of temperate variation, when trend of temperature change was varying either to higher or lower temperatures depending on the phase in the cycle of temperature change. These changes in culture and anatomy are detailed in records of anthropology.

This author was first introduced to the relationship of culture and species advancement with moderate temperatures in a class of geography at college. It was related that most changes and advancement in culture took place in temperate zones of the world, which proves out in observation.

It is only a short step to associate changes of a world-wide scope, to time when Earth was exposed to moderate temperatures, due to Earth's location and moderate temperatures in the cosmos. This change is due to location in the cosmos of the solar system, including Earth. In travel of the Milky Way Galaxy, Sun, and earth through relative warming and colder locations within space, the relative temperatures of Earth changed in accordance with relative locations of cosmic radiation that resulted from the Big Bang.

There are many events of earth, our culture, and our body functions that would not have happened if it were not for the presence of soul. First, as was mentioned earlier, is that animal life would not have advanced to human status from animals of the wild. Soul promoted emotion, reasoning ability, empathy with others of our species, recognition of some power greater than that able to be expressed by humans, and increase in a wide variety of qualities of soul, perhaps even in knowledge of antimatter, extra dimensions, and soul.

Second, man would not have advanced from a Pleistocene age of hunter and gatherer to an agrarian, communal, and skill-sharing, economic status.

Mary, mother of Jesus, was created, of course as a soul and body, as were Adam and Eve. Mary had a physical body. Unexplainably, the creation of Jesus within Mary was similar to the creation of Adam: They were both originated from God with a body formed without human fathers.

The human body and soul seemed to have worked together and coordinated best when temperatures on earth were in a range of temperate variation, when trend of temperature change was tending either to higher or lower temperatures depending on the phase in the cycle of temperature change. These changes in culture and anatomy are detailed in records of anthropology.

It is only a short step to associate changes of a world-wide scope, to time when Earth was exposed to moderate temperatures, due to Earth's location and moderate temperatures in the cosmos. This change is due to location in the cosmos of the solar system, including Earth. In travel of the Milky Way Galaxy, Sun, and earth through relative warming and colder locations within space, the relative temperatures of Earth changed in accordance with relating locations of cosmic radiation that resulted from the Big Bang.

Matter can neither be created nor destroyed. This is a basic theorem in Physics. But matter can be added, such as when earth grew in matter, as it collected, by gravitational pull, matter from space to matter of earth. With increased matter there was increase in a different form of energy, Latent energy.

Instead of being trapped in a substance, or early Earth, with growth in quantity of matter, either by accretion or change in distribution of matter in and on earth, such as occurs in explosion of supernovae, the phase of matter will change from latent energy to real energy

Enlarging from the above explanation, this change in phase of energy may also be seen from the change in glacier content on earth. As the glaciers melt, mass of the glaciers reduce, and latent energy is released to the world. This increase in energy affects many events, including advances in technology, ideas, invention, and culture on earth. In this period of release of latent energy, man did collectively advance, in other fields, of technology, and devised means to place man on the moon, and explore, by fly-by-devices of outer space, planets, asteroids, and Sun. He will devise experimentation at CERN of elementary particles and their actions in change. Changes will be made with ground-breaking (and mind-boggling) advances in culture, society, and matters of a social nature including increase in knowledge of soul and its ramifications, all such notable advances have occurred over the past 15,000 years, coinciding with start of increasing earth temperatures (melting) in global warming, minute (very small) at the start 10,000 years ago, but great in magnitude of world concern of late. The "seeds" of the growth in technology were first planted when the Wisconsin Glacier first started to melt 20 and 15,000 years ago. As glaciers come to be melted completely, even more latent energy will be released to the world, earth, and to civilization.

Again, when glaciers reform in the distant future from decreasing global temperatures, energy on earth will decrease into latent energy, and mankind will become static in available energy of matter and activity. But don't worry—this cycle will take thousands of years, as it has cycled many times in the past in recorded in history, measured by depth probes into ice and earth.

All animals have some degree in qualities of soul. Among these are emotions (as well as learning and memory), that are controlled by the hippocampus part of the brain. Question: Is the activity of the hippocampus greater in humans having soul in its Homo sapiens form, than in non-humans? If so, this would indicate that the hippocampus is a scientific attribute in development of soul—soul and science are dependent upon each other. Existence of soul is a function of biology, brain, and spiritual acts.

It is often said in science that some theories can never be proved. They can only be disproved. Because study of the paranormal in the laboratory is presently often not repeatable, and therefore in violation of the scientific method which requires repeatability, verification, reliability, non-arbitrary representation, and without bias or prejudice in testing, some theories, such as in spirituality, are difficult to prove.

Four steps are required of the Scientific Method:

1- Observation, description and isolation of a phenomena,
2- Formulation of a hypothesis to explain the phenomena,
3- Use of the hypothesis to predict the existence of other phenomena,
4- Performance of experimental tests of the predictions by several, independent experimenters, and properly performed experiments.

If the experiments bear out the hypothesis, it may come to be regarded as a theory of law of nature. If the experiments do not bear out the hypothesis, it must be rejected or modified.

The scientific method requires that a hypothesis be ruled out or modified if its predictions are clearly and repeatedly incompatible with experimental tests, no matter how elegant a theory is. Theories which cannot be tested because they have no observable ramifications do not qualify as scientific theories. The theory may be discarded as a description of reality, but it may continue to be applicable within a limited range of measurable parameters. For example, the laws of classical mechanics (Newton's laws) are valid only when the velocities of interest are much smaller than the speed of light.

In a field where there is active experimentation and open communication among members of the scientific community, the biases of the individual may cancel out, because experimental tests are repeated by different scientists who may have different biases.

With experimentation in the world of antimatter, there are few events that are reproducible. Events can only be anticipated in probability, and these violate the principles of Scientific Method. It may be worthy of change in strict adherence to principles of the Scientific Method, in order to gain experimental knowledge in fields of antimatter and higher dimensions.

WHAT IS SOUL?

Soul! What is Soul? Soul is one of the most difficult entities to describe. Soul is a non-physical part of the human existence within the secular body. It is non-matter, perhaps even antimatter, existing in a fifth, or greater, dimension, additional to the four dimensions of which we are familiar of height, width, depth and time. (Why were weight or mass never included in describing basic dimensions?) Soul may also bridge to additional dimensions of existence.

Soul is a factor complimentary to our brain and mind. It is harbored in the mind and its scope is greater than merely providing animal existence to our being.

Soul has many dimensions, which are described in another chapter, but generally provide emotions and insight to man. These include inquisitiveness in the mind of man to recognize a power greater than some mechanical or biologic process for physical existence, such as common to life on the early Earth that had basic cellular and molecular life, and such as in vertebrates, sea life, fish, amphibian organisms, hominid creatures, or even early hominoid characterizations of future man.

Soul was perhaps introduced to human-like characterizations in early entities of mankind as early as about 600 million years ago in Homo erectus species. Some basic features of soul may even have been also introduced via the genome 1.8 million years ago. This writing does not counter that God ultimately "made" flesh and spirit, but questions the timing as presently held in Jewish, Christian, and Islamic doctrines. Biblical timing holds that creation of man is fairly recent (10,000 to 6,000 years ago, or in year 4004 B.C. as calculated by the Irish Bishop James Ussher).

This exposition attempts to show that the development of flesh, as known today, and the development of mind, soul, or spirit, transpired over thousands of years, starting perhaps at the species Homo erectus at 600,000 years ago, with gradual additions of soul qualities during life refinement and development through stages or species of Homo neanderthal, Homo sapiens, and Modern Man. Culmination of soul development came at time of Christ to reveal the oneness of soul in man, with God. Exploration and development of this lore has taken place more recently over the past 2,000 years.

The progression of man in species succeeding or incorporating Homo erectus was a planned, progression, to take place over 600,000 years and more, as the time would reveal. Who could predict the future of man and soul over the millenniums? As shown here, man and soul in development has been dramatic over historical time. Will it be any less in the future? St. Augustine, Saint Thomas Aquinas, and others studied and proposed the existence of soul and its relationship to the body.

There is Biblical evidence and authority for the essence of God, with several references to the nearness of the kingdom of God, and "God created man in His own image" (Genesis 1:27).

For years certain philosophers have advocated the existence of a parallel world "on the direct opposite side of the Sun". They may have been partially correct, in anticipating a world of antimatter, in an additional dimension, parallel to our world of matter, but unknown to our secular life.

The elements of soul are believed to have become present at the Big Bang, during an event of annihilation, when negative-charged elementary particles cancelled counterpart positive-charged elementary particles, leaving negatively-charged particles as matter and positively-charged particles as antimatter. Antimatter became the universe of protons, as matter became the universe of electrons. Soul in antimatter had about 13 billion years more time to evolve and mature than did man in matter, allowing a much more advanced culture in soul existence.

First soul was not fully and mystically created or incarnated to the human-like body, but qualities of soul were introduced gradually to the body through genetic addition to the genome system, physically by what some scientist refer to as "the God gene"

As such, the God gene of the developed genome provides for qualities of the soul, much like the genome provides and directs protein and control of various physical features, organs, and abilities, as well as keenness of body and brain.

Soul is unique to man and absent in animals of the forests, although it seems to be later recognized to a minor extent in domesticated animals, including animal pets and trained animal species of the world, a quasi-civilized universe of earth life. Soul indicates tendency to provide emotion and to receive and react to emotion.

Soul is a guide for morality or desire to do what is personally desirable in light of one's development and one's accepted standards of thought and behavior. Morality is personal and may or may not agree with popular social mores.

Soul allows today for love and affection of infants and family members, much as anthropologists reason that first elements of soul occurred six million years ago. This element of love and affection was something new and different in the genome-derived traits from the independence shown by other animals of the forest, and was a key as hominoid types slowly developed towards mankind. Soul is sometimes a word used in life as a synonym for spirit. Although very much related, one meaning of spirit is an attitude, often highly emotional, an eagerness to pursue a certain activity or goal.

Spirit is also a word to describe an entity within the human body, derived from an exterior or inward-felt idealism. Spirit is also a term to describe a superior entity evidencing within the mind of man. Spirit may be the essence of another entity, such as the spirit of God. Soul is an extension of that spirit as it is found in man.

Biblical writings indicates that God created Man in his own image. Mankind has somehow interpreted this statement that the body of man is an image of the spirit of God. It is illogical to interpret the statement that the physical body of man is like that of God. God is a non-physical spirit while man is a physical being. Man's soul is like the spirit of God. At the Crucifixion it is said Christ gave up his soul to the spirit of God.

It could be interpreted that spirit of God and soul of man are identical and that God's spirit (or at least a portion) exists within each and every person on earth in the form of man's soul. Soul acts through the mind, the genome, and consequently our enzymes, the brain, and organs, and has ability to communicate with a Superior Being, within the emotions and reason of the soul, and with souls outside our own human body, all to facilitate change and judgment of the soul with mind.

Soul within man has developed today to include some features that will be present as soul in Afterlife. Not possessing a human body, there would be no ability to hear, see, talk, touch, or smell, have brain activity (but would have mind), or other bodily functions. These physical attributes would have been left in the earthly grave, tomb, or crematorium. Such abilities remaining are those of mind and soul, including ability in non-verbal communication, déjà vu, to foretell, anticipation, specters, and sense of a greater spiritual power. Product of brain power, certain memory (short of instinct), and organ functioning will be absent.

The soul will experience an existence following secular death, exhibiting a primal love and affection in emotion, in mind and spirit, of his or her soul. Memory of worldly skills and knowledge will be absent as functions of the brain in the deceased upon secular death.

To answer a question in my own mind, dinosaurs of 350 to 65 million years ago would not have possessed any genetic element of soul in their cannibal and aggressive, reptile and mammal manor of living. Soul seems to be reserved for only the Homo species of mankind.

The "common thread" that runs through this book is the concept of soul. It is proposed, what the reader will realize that the very origin of soul was derived from the Big Bang when positive- and negative-charge elementary particles separated, one from another, in a process called Annihilation. Negative-charge particles emerged to form matter, with the electron being the common, unifying quality. Following that division, positive-charge elementary particles proceeded to respective forms of antimatter, and existing in alternative dimensions from the four dimensions we know today of width, height, depth, and time.

DESCRIPTION OF SOUL

For a more complete explaination please see Qualities of Soul in another chapter.

Presentation here is perhaps a clinical study, not necessarily a religious or religion-oriented pursuit of the existence, creation, purpose, and future of soul. It proposes what immaterial soul is, as opposed to biology of the body and brain.

This presentation proposes how and when the human, primate, or early animal organisms acquired elementary qualities of soul and mind in addition to the historic animal body. It proposes how the soul, through the mind, affects actions of humans. It proposes an unending future of soul, with reincarnation to future earthly bodies. It proposes that soul affects the very culture of different civilizations around the world.

This presentation shows effects of how our traditional understanding of spiritual history could change, recognizing changes in knowledge in medical and anatomical science over 2,000 years since Books of the Old and New Testaments were written by well-meaning, but not fully informed writers (by modern standards) of that time. It proposes evidence, acquired mostly over the past 300 years (which incidentally coincides with study and findings in modern-day psychology), that clarifies history, changes, and status of peoples over past centuries.

Scientists studying the cosmos, Earth, cultures, human body, and philosophy, have detailed secular changes, but have been lax in recognizing the existence of soul in the body of humans, even animals, and perhaps plants of the earth. Shamans of Native American Indians recognized soul even in the winds, rocks, and nearly all matter of Earth.

This presentation also proposes how myth can often be overridden through explanation of possible alternatives providing logical and rational happenings in historical events, recognizing knowledge and technology available today. In short, a clinical (analytic) study of soul can advance understanding of the human existence.

It is a thesis presented here that an individual's soul is key to evolutionary development of that individual and socially related classifications of like individuals in a biological society. This phenomenon is important to understand evolutionary development in a society. There may be theological implications regarding the existence or non-existence of soul in the human.

Dedicated research in Life Sciences oriented to soul could add knowledge to the following:

1. The existence of soul,
2. The purpose of soul,
3. Interaction of soul on genetics of RNA and DNA,
4. Interaction of genetics on production of proteins and enzymes related to specific organs,
5. Influence of associated enzymes and proteins on development of individual organs of the body,
6. How development of organs affects evolution development,
7. How this process continues even to today,
8. Effects of developed evolution and historical evidence.
9. Is there life after secular death, and in added dimensions?
10. Is there reincarnation from a spirit world again into the body and mind of Modern Man?

Study is multi-discipline:
Numbers 1 and 2 above are studies in Philosophy and Theology; numbers 1 and 3 are dual cooperative studies involving Philosophy and Biology; numbers 4 and 5 are studies in Biology and Microbiology; and 6, 7, and 8 are studies in Anthropology. Numbers 9 and 10 involve associated study in all fields.

IN A NUTSHELL

The soul exists within the psyche and body of each of us. The existence of soul can be identified by its qualities. The soul is dynamic within the body and manipulates the RNA and DNA system of the biological body, possibly spurred by bio-electrical energy, thereby causing changes in the body's biological functioning.

The author holds that the soul is a small extension, or microcosm, of the Supreme Being of the soul that acts in a manner as if an agent of the Supreme Being. Manipulation of the RNA and DNA is usually for the physical and mental good and health of the body, but bad health can result from bad manipulation from the mind, or by mutation by the soul to its responsibilities, or by some intended action or inaction. Genes of DNA are controlled by an active intelligent force we know as the soul of man.

The soul, therefor, is the prime mover, under direction of the Supreme Being (or His soul representative), that initiates changes in RNA and DNA in the genome of the body as well as in the mentality, and is the controlling element to the individual body's existence and direction.

INTELLIGENT DESIGN

Intelligent Design is believed real, regardless of a legal court ruling in the state of New York. Through series of planned changes by mutations throughout the ages of evolution, mankind has developed in a direction away from that of other animals. Various forms of man-kind species have existed, some leading all the way to Homo sapiens, or Modern Man.

The soul is the vehicle for unification of humans with the divine. There is history of human experience in direct mystical communication with God or Holy figures in the divine. Notable persons reportedly having direct communication with God were Adam, Abraham, Moses, Noah, Christ, Muhammad, and many lesser individuals, people striving for personal enlightenment throughout the centuries.

Existence of soul dates back in history of mankind at least to the advent of Homo sapiens to perhaps 1.8 million and 600,000 years ago, to men and women of the Pleistocene period, especially Homo erectus, who may have possessed elementary qualities of soul. The mystical soul would have been recognized as an existence after death in a continuing life of soul in some other world or dimension beyond secular earth.

The soul is consciously inactive. You realize that it is present, but you cannot feel it. It is not like a decaying tooth that every so often gives a reminder of its presence, but more like an innocuous surprise that occasionally reminds one of its non-physical presence, and you're glad that it is there.

TWO PHILOSOPHIES AT TIME OF CHRIST

Two major philosophies were common near the time of Christ. First, the stoics made a bridge to Christianity that there was possibility of a spiritually present and knowledgeable God having a rational plan and order of things, being present in the world, revealing himself through His works, and working on the physics of reality in a divine way to realize divine purposes.

Second, also present was the Gnostic world view. Gnosticism is based on the knowledge of transcendence, or excelling higher than expectation through interior and intuitive means through the medium of mind, as reasoned that the world is imperfect, that earthly life is filled with suffering, that transgressions committed by the first human pair, Adam and Eve, brought about the fall resulting in the corrupt state of the world that followed. The religion of Gnosticism recognized existence of only one universal God. John The Baptist was considered a personification of the Old Testament Elijah.

Gnostics thought God did not create the world, but brought forth the substance of all there is in the world, visible and invisible. The basic Gnostic mind-set refers to intermediate deific beings of angels and souls in Heaven, who exist between the ultimate true God and ourselves. Yet humans are generally ignorant of the divine spark, or spiritual component resident within them, and death releases the divine spark from its biologic person.

Not all humans are spiritual to bring about spiritual freedom. They are thus in a predicament consisting of physical existence, combined with ignorance of their origins, essential nature, and ultimate destiny. They needed help and looked primarily to Jesus as the principal savior. They did not look to salvation from sin, but from the ignorance of which sin is a consequence.

Gnosis provides a way for an individual to dispel ignorance, and the potential in Gnosis can provide salvation. Morality, an inner integrity, is guide to human conduct and attitude. This transcended knowledge must come to men while they are still embodied on Earth, rather than in an Afterlife, lest they be trapped, with reincarnation denied. Gnosticism is both a religion and a psychology. As test, if the follower's reactions to this doctrine are of a positive order perhaps he is a Gnostic himself.

It might be said that, "New Testament II" has already been written by the events of religious history that took place from after 100 A.D. to the present. Religion was redefined in this 2,000 year period, different from that of Christ's time and the immediate years following.

In the Maya culture of South America it has been expressed that soul existed in forms of "good soul" and "bad soul". Good soul described good omens and actions, and evil soul controlled bad omens, unpopular, and not-tasteful acts. The omens exist even today by the Maya, continuing to be controlled by good soul and evil soul.

In more traditional existence of soul, both good soul and evil soul exist and affects behavior of mankind. Manifestation of this is seen in behavior of some living persons who perform heinous crime, and in contrast with others who seem to possess absolute qualities of sainthood. (Source: Descent into the Maya Underworld, National Geographic magazine, November 2004.)

Many of these thoughts ring true in today's view of soul, ethics, and morality.

CONCLUSION

The author turns now to an object or goal of this writing. He has explored the subject for at least 30 years. Those subjects are:

1. Does soul exist within humans?
2. How can soul be described?
3. When and where did the union of soul come to be within the animal body of the historic species that have inhabited the earth?
4. What was the purpose of soul creation, and
5. What is the future of soul in both living mankind and after death?

PHYSICAL CONSIDERATIONS

The author also explored possible time and location(s) of the event of first creation. Of course, the Biblical story was considered, although no geologic time or location for the event has reliably been found. The location of the Garden of Eden was hypothesized to presently be where, or perhaps in the past before flooding of the present Persian Gulf, where the Persian Gulf meets the delta of the Tigris and Euphrates Rivers, that may now be under water. It was surmised that in ancient geology, the waters of the present Arabian Sea were blocked completely, or possibly by only the riverbed of the joined rivers. At some ancient time, it is reasoned, an earthquake would have opened up the mountain chain that bridges both sides of the present Straits of Hormuz. This would have allowed waters of the Arabian Sea and Gulf of Oman to flood the lowlands under what is now the Persian Gulf.

The site of the Biblical Garden of Eden as proposed would have been inundated and lost to geographic history. Many earthquakes occur at the edge of these geologic plates of Earth. One major geologic plate lies under the subcontinent of India, well recognized as having travelled over eons ago from southern waters, perhaps from near present Antarctica, northward, and causing uplift of the Himalaya Mountains as it collided with the continent and geologic plate of Asia.

Another geologic plate would be under the present Mediterranean Sea. The edges of such geologic plates, geologically rubbing against each other, either horizontally or vertically, would cause severe and disastrous sub-surface land and seismic rock movement, causing frequent and historic earthquakes in the region.

One of these earthquakes may well have further opened the mountain range, now on either side of the Straits of Hormuz. (The tremendous wash of water travelling up the two rivers could well have flooded that land, creating the Great Flood told Biblically).

Alternate evaporation of waters, and a resulting consequent repeated rain, could have caused continuation of the Great Flood incident for the reported 40 days, similar to the

WONDERFUL WORLDS

continuous flooding by the Mississippi River with alternating rains at the states of Iowa and Illinois in the recent past.

A specific location may have been near the ancient city of Ur. This event in time and location is presented as a possibility, but reluctantly evaluated as not supported in records.

The elements of soul are believed to have been created at moment of the Big Bang when annihilation occurred, separating elementary particles described in Quantum Physics into particles of matter and particles of antimatter. The positive-charged particles of antimatter, I reasoned, became the original essence that would evolve into the quality of soul.

Life was created on Earth, although the cause and biology is far from a consensus among Palo physicists. Elementary life increased in complexity, the author believes, by intelligent and planned evolution, to more and more complex organisms, to recognizable, four legged, small and large, fur-bearing animals. Some of these animals became bipedal, with two arms and opposing thumbs to aid in arboreal living in trees, and later to stand erect on flat, high grass savannah lands of Africa.

Speciation, or study of separate species, developed to produce primates and hominids of Great Ape, Chimpanzee, Monkey, and Bonobos. These species developed but never formed a basis for later Homo species and modern man. A branch of these hominid species developed in Homo ergaster, that lived on earth from 1.9 million years ago for about 100,000 years. Later, about 1.8 million years ago, a new species, more complex, bipedal, and hair-covered, a man-like creature came upon the earth, with evidence of greater intelligence and development of elementary stone tools. These were called Homo (man-like) erectus.

Homo erectus species are reported to have existed until about 350,000 years ago, when again a new species, of Homo neanderthal, appeared on earth, and Homo sapiens appeared at about 120,000 years ago. Some anthropologists recognize Homo neanderthal as a forerunner of Homo sapiens, although they were supposedly each of a different species and unable to reproduce outside their own species, which is held as a biologic fact. Others believe that the qualities of

Homo erectus continued through the recognized species of homo neanderthal and Homo sapiens.

WHY SOUL?

Man having soul within him explains many things and answers many questions.

Finding hard evidence of events occurring in a parallel universe that houses soul are hard to come by, but here are some indications of a parallel universe and activities of soul:

1. Soul is a guiding force of a person's ethics and morality.
2. Soul makes a person more than a biologic automaton, or different from animals of the forest.
3. Certain men having exceptional soul have been shown to have a physical radiance (Undocumented), that is commonly in depiction of saints.
4. Upon death, the body weight of man reduces as his soul departs from the body (Undocumented).
5. Soul departing from the physical body upon death is a basic process described by various religions including Christianity, Jewish, Islam, ancient Egyptian, and others.
6. Existence of soul is basic to the concept of reincarnation or spiritual return to Earth.
7. Some persons of history are documented to have remembrance of events that existed at a previous time and in a previous life before reincarnation.
8. Soul is sometimes suggested as a synonym of spirit, with soul an intangible existence, sometimes displayed outwardly as an element of a person's emotional make-up. Spirit is sometimes portrayed as a reflection within man of the essence of God within him.
9. Existence of soul, and man's belief in this existence, is a fundamental dogma in Christianity for continuation of existence after mortal death.

10. Existence of soul is portrayed in metaphysical actions sometimes formed as specters, apparitions, ghosts, or ghostly presence, and acts on earth by unseen forces.
11. Soul of deceased persons can influence acts and thoughts of living persons. (observed but not documented)
12. Soul is a subdivision or microcosm of God, which is believed as a theory, that a division of God exists in each man, and soul is that embodiment in man.
13. Soul exists as an additional dimension of time/space and parallel universes.
14. Soul was created in a particle annihilation phase at the Big Bang as separation of positive-charge elements, from negative-charge elements, that became substance of matter. Those positive-charge elements of soul matured over time to exist in antimatter today.
15. Soul, as a spector, exists on the surface of Earth, but in an additional dimension. Specters and apparitions are sometimes seen by living persons.
16. Soul is a spiritual subdivision of God.
17. Soul, as opposed to time-honored pictorials of having human biological qualities after death and in Afterlife of human senses, does not possess a body after death and in Afterlife, nor have senses of sight, hearing, speech, smell, or touch. "Sight" by soul and "hearing" in Afterlife is a remembrance of physical appearance, within the mind of soul; communication, without voice or hearing, is by thought transmission between souls to one another. There is no smell or touch senses by souls of deceased persons.
18. Souls of deceased persons are evaluated in Afterlife, upon mortal death, and assigned to a place, existence, or function in Afterlife.
19. A possible explanation of astrology and fortunes: Souls newly "processed" in Afterlife appear to be assigned to certain "Astrology Groups". Souls have a mutual relationship common to all souls in an Astrology group. Souls appear to have a somewhat pre-determined life upon reincarnation. When a soul is reincarnated (born) at a certain date, from a certain

Astrology group, his or her traits and fortunes are similar to other souls born from that group. Grouped reincarnation provided like traits of all from that group. Astrology is one of the oldest of life's tales.
20. There are limited numbers of souls, repeated into mortal life by reincarnation.
21. Jesus Christ, if not actually God, had a portion, the essence of God, within him. However, all humans also have a portion of God within them as souls.
22. Jesus Christ, and other Biblical persons, i.e. Ezekiel, did not physically rise and leave earth, but their souls, sometimes even with a glow, did leave the body and removed to another place in Afterlife. The body of Jesus Christ may well have been found after a "second death" and was placed in a family tomb, as has been suggested in anthropology studies.
23. Resurrection after mortal death, as reported Biblically, was by soul to soul communication from Jesus to the soul in apparent death of Lazarus, Elijah, and others. Lazarus and other bodily persons may not have been "brain dead" as medically defined in present times, and were revived after their medical state of shock was nullified in the physical body.

It seems possible to predict that at some time in the future, perhaps 50 or so years from now, there will be more communication with other dimensions. That communication will reveal:

1. Existence in those newly understood dimensions, and
2. How to procedurally progress to make that understanding valid, real, repeatable, and reproducible.

There are presently events of communication with other dimensions:

1. Some notable persons have reported memories of a previous life. Among those are World War II General George Patton, and a Hollywood personality, Shirley McClain. Other less famous individuals have privately reported mental experiences of existence at previous times. It is nearly impossible to experimentally or scientifically document such previous existences.
2. There is evidence of communication with a mystical world in:

   a) Dreams, having unexplained topics and outcomes. Some are interpreted by analyst as having reflection to real events or conditions.
   b) Séances claiming contact with deceased persons and testimonials of its realism.
   c) Ghosts, specters, spirits, or apparitions from an alternative world or parallel universe in performing mystical, or physical, acts on Earth.
   d) Individuals have been documented to have had afterlife experiences and return to secular life, commonly reporting an intense light to be seen in a long tunnel, supposedly originating from Afterlife.
   e) Individuals are documented to have had experiences in foretelling. These include documented events concerning Jean Dixon regarding the visualized future assassination of President John F. Kennedy, and Edward Casey regarding numerous medical evaluations of patients being several miles away. In the Old Testament a young Daniel is said to have correctly interpreted a dream for King Nebuchadnezzar. Others claim to have had mystical abilities.
   f) It is possible to set one's "inner clock", as some people realize, to arise at some predetermined time at the next morning, or possibly to "think about" some subject at

a certain future time of day, or upon presence of some stimulus. These are qualities of the mind.

g) Some people express a guidance of their, or others, earthly actions, by prayer, through God, a Guardian Angel, or mind expression. In many cases prayer is largely self-contained within the prayerful individual. The conscious mind makes conscious statements and requests in prayer, supposedly direct to God.

h) These concepts of additional dimensions allow existence and change in humanity, individuality, culture, and our future. It explains how we came to be humans, different from early life on Earth, animals of the forest, hominids, hominoids, and through a progressive variety of species, to Modern Man.

i) The author suggests that Homo neanderthal and Homo sapiens are actually sub-species of Homo erectus, refinement bringing spirituality to our present-day concept of humans, reasoned by logic, realism, and understanding, gained from a wealth of sources explaining the cosmos, Solar System, Earth, man, mind, and spiritual development.

j) Corrective action is often formed in cooperation of the soul, mind, genome, or exterior communication. Communication of soul or mind, sometimes involves other individuals, requiring soul-to-soul communication by thought transmission.

It is theorized and proposed that the conscious brain communicates with the mystical soul and mind. The soul and mind formulate a practical and desired action, replying to the person's prayer, and resulting, insofar as practical, logical, and physically possible acts to bring about corrective results. The act may be within the confines of the prayerful person, or it may involve another person, requiring a mystical communication, soul with soul, of that person.

As stated earlier, "Nothing involving thought makes sense unless the soul is involved".

WHEN AND HOW DO SOULS AND GENES INTERACT?

The era of when some differences of species due to sex commenced would seem to be when the X chromosome first appeared in the DNA of the specie. The X chromosome is particular to the male of the specie, and absence of the X chromosome indicates the female of the specie, or it could mean there was no sexual difference at all (species were homosexual, or had no opposite sexual attributes, a phenomenon found even today in some exotic circumstances of elementary plant life, amoeba, and bacteria existence).

Without any difference between male and female of the specie, the living organisms would have needed to be hermaphrodite, in which the organisms were able to reproduce themselves. There are instances, even today, where a sexual process is not required to foment reproduction of the specie.

History of mankind could conceivably be traced to the first presence of the X chromosome. As stated, there was a time when there was zero difference between the many organisms of species. Today wheat in the field has grains that are capable of reproducing new plants without any sexual event. Earth worms too, reproduce without requirement of male and female. Bacteria, algae, and microorganisms reproduce without sex, growing from single cell to multiple cells, or from self-generated duplication. There was a division point in time whereby the X chromosome became present, and in fact essential, for reproduction of the specie, unless the specie was hermaphrodite. Today, 99.9 percent of all species require union, of one sort or another in sexual activity, in order to reproduce, all contingent upon presence of an X chromosome in the gene.

It is still undetermined whether or not sexual differences existed in all dinosaurs of 350 to 65 million years ago, although Dinosaur eggs have been discovered with young dinosaurs half out of the shells.

It is proposed that soul may choose to make change through the genome of the body. This affect would be actuated by perhaps electro-chemical force on the gene, which would cause subsequent action on enzymes, causing protein change in organs, or otherwise in the body, brain or mind of man.

Research in biochemistry may soon prove or disprove this relationship between the soul and the genome. The question will arise of just what does cause changes in the genome that result in mutations? Scientists have proposed that gamma rays can cause mutations. A theorem of existence is that that there is a cause for everything. It is illogical that advancement will be made strictly by a series of accidents, or mismatches, and agitation within the genome. The author believes science will determine in the future a causing force that is present.

The soul, acting bio-electrically on the genome, is instigated by intelligence of the soul and desired direction in characteristics of the person. Changes may be in health, appearance, personality, physical attributes, or a variety of other factors.

The condition of the DNA describing the chromosome, affects the way that the genes interact with chemicals inside the cell. These interactions determine when, how much, and what types of enzymes a cell's DNA generates. Consequent proteins are the building blocks of an organism, so the way and time a genome expresses itself is as important as the DNA itself.

Gene therapy involves sending a corrected set of genetic instructions into the body cells. To do so, scientist or doctors insert a copy of a healthy or desired gene into a vector, an agent such as a virus that is capable of carrying the gene into the body. Once there, the gene can insert itself into the cell's DNA throughout the body and start duplicating DNA and normal or desired proteins.

Gene Therapy is still experimental, in its early stages of experimentation, but recent research shows its potential to battle many diseases of the body and brain, and even trials to re-grow amputated limbs.

During the birth and life development of a person's genes (a single cell may contain a string, if unraveled, nearly 6 feet long and containing up to 3 billion base pairs of DNA). Many variations or groups of genes from perfect development can take place. Certain genes in certain situations, and at different times, will be switched on or off, controlling life-affecting changes in various forms of cancer, heart disease, autism, diabetes, schizophrenia, asthma, and other diseases that cause life shortening.

These variations can manifest illness, cancer, and other malfunctions in operation of both brain and body. The change in intensity of the geo-magnetic field, possibly around one billion years ago, may thus have made it possible for life to expand from the seas to dry land. This medical phenomenon took place in ancient times, as well as during today, sometimes set off by stress, of which early man had much, in just surviving.

Genetic changes also account for racial pigmentation. A gene, designated as the "tan gene", was responsible for part of the skin color differences between species and sub-species. This suggests that the basis for the difference is genetic, rather than environmental, although environment can also affect genetic change. Organelles within cell walls include protons, neutrons, DNA, RNA, proteins, enzymes, and mitochondria (mainly to extract energy from oxygen. Mitochondria have their own DNA. Chloroplasts extract energy from sunlight through photosynthesis. All plant-like organisms have chloroplasts.

Source: Great Courses, Big History, Prof. David Christian, San Diego University, "Thresholds of Biology", 2008.

PROKARYOTES AND EUKARYOTES:

Life on earth, whatever the source, began with single cell organisms called prokaryotes. These were simple, single cell organisms. A number of today's different species could be 10 million to 100 million years old. (For example, certain house flies or other insects are identical and survive by the millions.) Of the approximate 1 million species, 750,000 are insects. All species share the DNA molecule, even though differences occur in color, shape, function, and scale. They share the same basic chemical process and basic mechanics of DNA.

Life of prokaryotes started about 3.8 billion years ago, with capacity to evolve, adapt, and create an astonishing number of different forms. This distinguished life-forms from non-life metals, minerals, and rock. The variety of living organisms reflected a slow exploration by life of all possible ways of getting energy from the environment. This long process of change, evolution, and proliferation of life, is a history of life on earth—a complex history of life dating to archaic times.

Prokaryotes exist today in yeast, algae, and bacteria. Viruses are even more simplistic, which had shed their capacity to generate energy on their own, and must live from their attachment to other life forms.

EUKARYOTE CELLS:

Eukaryote cells developed about 600 million years ago. The first multi-cell organisms, prokaryotes, existed for only about 15 per cent of time of life on earth. Eukaryotes have a fatty membrane, a surface that is semi-permeable to keep most important contents inside, but does allow chemicals to flow inwards for nutrition, and waste products to be excreted. Within the Eukaryote cell there are free-floating molecules of DNA. RNA molecules appear with DNA to get instructions for their next function inside the cell. This is how the reproduction process works. Inside a cell is an extreme in activity and violence, especially by proteins and enzymes (a type of protein), that builds and rebuilds molecules. Doomed, worn-out proteins proceeded to a status called "proteasome", where they were stripped down, and components used to build new proteins. Earliest prokaryotes probably received most of their food from chemicals on the sea floor, near ocean floor volcanoes, from the mantle and center of earth, such as methane, and had been protected by depth from ultraviolet rays.

Photosynthesis evolution is found in all plants or plant-like organisms. This enables organisms to tap into an entire new source of energy. This is how some prokaryotes had learned to live near the surface of the ocean where they could utilize the sun's energy and ingest the light.

Chemical ingredients for photosynthesis are carbon dioxide and water. Energy for all reactions comes from sunlight. Products of photosynthesis are sugary molecules, such as glucose, which can act as stores of energy, generated in the core of the sun, released by hydrogen fusion and combined with another element, oxygen.

Chlorophyll manages the entire process. As plants are consumed by other organisms (animals), this captured energy ceases.

Date of photosynthesis is determined from oldest microfossils of about 3.8 to 3.5 billion years ago of algae (like modern cyanobacteria). Cyanobacteria-like bacteria created coral-like structures called "stromatolites", huge crowds of organisms that die and build even larger structures, evidenced, for example in the great coral reef off the east coast of Australia and elsewhere, originated during the Archaean seascape.

As photosynthesis produced oxygen, free oxygen began to build up in earth's atmosphere by 2.5 billion years ago, seen in rusted bands of ancient iron in geologic formations. For most eukaryotes, oxygen was poisonous. Oxygen is a violently reactive chemical that dissolves other fragile chemicals (an "oxygen holocaust"), killing off a large number, making a change in the atmosphere for beginning of a new eon, the Proterozoic Eon, from about 2.5 billion years ago.

Eukaryote cells appeared sometime during the Proterozoic Era, more than 1 billion years ago. This division between prokaryotes and eukaryotes is regarded as one of the most fundamental and important divisions of all between different types of living organisms. We are all constructed entirely of eukaryotes.

Evidence exists for the presence of prokaryotes within eukaryotes in the organelles, and are more complex in various different organelles inside cells.

Some of those organelles have their own DNA, suggesting that they once existed quite independently in a process of symbiosis. Eukaryotes include mitochondria that can extract energy from oxygen, and they have their own DNA and chloroplasts which can extract energy from sunlight through photosynthesis. All plant-life organisms have chloroplasts. Eukaryotes would make possible the creation of new types of organisms that were even more complex than themselves, adding to the complexity of life on earth.

Merging of various organisms within eukaryotes anticipates the later creation of multi-cellular organisms. DNA of eukaryotes is protected by the nucleus, important because DNA located in the nucleus can preserve the genetic code.

Mitochondria are special organelles that once lived as independent organisms. Eukaryotes flourished in an oxygen-rich atmosphere. The appearance of eukaryotes marks a significant increase in the complexity of life.

It is possible that eukaryote evolution was a response to the appearance of oxygen.

SEXUAL REPRODUCTION:

Sexual reproduction first occurred about 1 billion years ago, associated somehow with the appearance of eukaryotes. The prior prokaryotes regularly exchanged genetic material. They reproduce simply by splitting into two identical individuals or "clones". But in eukaryotes two organisms, two eukaryote cells trade some genetic materials before reproducing, and the offspring shares genetic material from both parents. Offspring of eukaryote cells were no longer simply clones of the parents. Sexual reproduction introduced greater variation between individuals. Individuals started to vary from each other more than in the prokaryotic world.

Natural selection seizes on variations between individuals that eventually lead to change and to new species. With sexual reproduction in eukaryote cells, evolution speeds up, and has done so in the past 1 billion years. Natural selection helps organisms search constantly for new ways of exploiting the natural environment, so evolution speeds up.

A RANDOM THOUGHT

At various times in the history of culture, different persons were introduced to earth-living who had exceptional natural attributes in particular aspects of life, of culture, science, philosophy, exploration of new ideas, or other fields. Many accomplished men seemed to live all about the same grouping of time. Some great men seemed to appear in time clusters. Persons, with souls, produced needed changes at the time.

It was somewhat like a modern-day basketball coach who sends into the game certain players who possess certain strengths, of high percentage in making 3-point baskets, or another, perhaps of more than average height who can rebound to gain possession of the ball, or another who is particularly skilled in stealing possession of the basketball, all skills needed at certain moments of a game. Their presence goes to produce needed development in outcome of the game.

So too, history reveals the presence of persons at certain times having great qualities in meeting problems of a time, or offering solutions to existing questions.

One wonders why certain persons were born on earth, grouping at certain times. Strong and able persons form a long list of notables. Only a few include Socrates, Plato, Aristotle, Hippocrates, Pythagoras, Jesus, Charlemagne, Constantine, Copernicus, Martin Luther, Michelangelo, Rembrandt, Leonardo Da Vinci, Raphael, Titian, and a wealth of other great painters, various great music composers, and certain popes, kings and rulers. The coming to earth of Jesus was predicted by many Hebrews of the time.

Is this more than just coincidence? Does the presence of one person having outstanding abilities be merely symbiotic to attract development of others also having outstanding potential? In art, many master painters existed in the 1400's, 1500's, and even through the 1800's. Why did this concentration of talent occur in art, science, medicine, and other aspects of culture? Does planned reincarnation or Intelligent Evolution have anything to do with this?

CONSCIOUSNESS, BRAIN, MIND, AND SOUL

The mind, as opposed to the brain, is equally as difficult to explain and define as is the soul of man. If we were to relate to athletics, it might be said the players are like the brain, that actually executes the desired play, but it is the coach, like the mind, that must consider "the big picture', based on all knowledge, emotions, and abilities of his and the opposing team (components), to result in a desired outcome. In travel, the brain is like the automobile, an electrical-mechanical assemblage of parts. The driver is like the mind that will plan and direct the tour, and modify the travel as needed. The brain is the vehicle, and the driver is the mind.

One reason why study of the brain and of the mind is so difficult is because we tend to combine the two, rather than to separate them.

It seems that many persons confuse the brain and the mind. The two are separate, but coordinating components, working together to produce mental results, even more complex today than they were only a few thousand years before. The mind and the soul are also interrelated in operation and goal achievements.

The brain is an electro-physiological organ that is primarily concerned with processing a multitude of operations that require interactions within the body, to raise an arm, to manipulate a pen, to take a step by moving a leg and then another, to process what is seen by the eye, and to process what is heard through the ears, or to taste various factors of sweet, sour, pleasant or unpleasant, and numerous actions of our bodies. The brain is like a computer, a complex hardware agglomeration of parts, the human components each sensitive to particular aspects, such as sight, hearing, judgment, emotions, recognition, vocabulary, remembrance, and a wide variety of other sensations.

The mind, on the other hand, is a quantity that provides judgment, reason, logic, sense of self, summon bonum, emotions (of all types), interpretation, calculation, methods, memory, and reactions to senses.

The mind is working in conjunction and in sympathy with the brain, but thought is a function of the mind, and not the brain. The brain executes the electrical impulses and synapses through axons necessary for transmission of energy through the brain. The mind composes the thoughts and compositions that facilitate speech and recognizable ideas, thoughts, sentences, and create writing.

The mind, separate but intertwined with the brain, is a reflection of multiple points of view, of past experiences, of association and melding of views and experiences. It also contains "home" of emotions, theories, concepts, and the like, that "jell" to form decisions and attitudes, and communicates that to the operative brain. The brain, in simplicity, is like an electronic machine, complete with electrically generated gates, neurons, and nerve endings leading to enzymes, muscle connections, and to the mind and soul.

Functions of the mind are frequently attributed to the brain. This is as wrong as presuming that an automobile, with its component systems of electrical, mechanical, physical, and chemicals could proceed down a road without a human driver that provides thought, judgment, knowledge, reason, logic, memory, interpretation, calculation, method, awareness, and reaction to various stimuli. It is comparable to stating, "that a computer is only a computer" without recognizing the individual components of software and hardware, let alone the new logic programs that are constantly changing the total ability of the computer system. "Imagination is the most essential piece of machinery we have if we are going to live the lives of human beings. Your imagination is your preview of life's coming attractions". (Albert Einstein).

To relate to a perhaps delicate family setting, roommates in a college setting, or perhaps husband and wife, study of brain and mind is reminiscent of "drawing out" one child, family member, or person to somewhat scientifically analyze and understand the physical, mental, and emotional attributes, or positioning, of the member, when we are really trying to determine "how the family or group works" and arrive at a compromise plan in its sociologic, family membership emotions, historical aspects, personality strengths, and other intangible factors that confront a person. The two, brain and components of the mind, views and counterviews, are definitely related, but separate entities.

Of analogy again, in performance, faults of the brain in execution are not necessarily faults of mental management, and vice versa, that errors in mental processing are not necessarily reflected by the operational or execution of the combine. The mind is a collection of all past remembrances and experiences, emotions, past reactions, intuitiveness, learning, analysis, our judgment applied, morality, ethics, beliefs in godliness, familial aspects, love, attraction, responsibilities, life goals, past infractions from correctness, and many other thoughts. Each of these can be symbolically traced in a "mind map", as if various "person residents", having unique personalities and unique concerns, all "come out of their houses", so to speak, to discuss an event.

These aspects of mind are flavored with the electro-physiological functions of the brain that will cause reactions in the body, evolution, comfort level, and decision making.

Autistic children have a problem of "wiring" in their brain, but at the same time, some are superior in their reasoning, problem solving, and logic in using their minds. They have problems of implementing their thoughts through their brain.

These factors can be placed, as if a physical map of the brain, again reacting in certain parts of the brain to affect responses. The two, mind and brain, acting in co-sympathy with each other, determine our actions, decisions, and thoughts.

Qualities of the mind have been refined during successive species of man types over thousands of years to produce our present state of mind and brain functioning cooperation. Studies show that mind characteristics develop mostly after birth. (Source: Great courses, "Origins of the Human Mind"). The mind then begins to form its unique qualities as an individual.

A proposed theory: The soul reincarnation takes place soon after the moment of conception, and expands in its qualities from that moment on. Mind and brain function are two separate, but co-dependent, features of humans. Again, the brain executes, but the mind composes thoughts and compositions. In evolution, the brain develops ability to develop more complex energy routes and capabilities, but it is development of the mind that allows greater skills in composition and arrangement of thoughts.

The increasing complexity in the state of mind parallels the increasing complexity of the soul, again over thousands of years, to the present state of soul as a complete inner-self. Consciousness is a necessary status of the mind.

A writer (in Discovery magazine, Dec. 2010), has proposed 12 steps to consciousness:

1. The mind in the brain: Organisms within the brain make programs out of the activity of special cells known as neurons. Neurons are sensitive to changes around them; they are excitable.
2. The conscious symphony: Conscious minds result from the smoothly articulated operation of several brain-component sites.
3. Mind Maps: The patterns, or maps of the mind, represent things or events outside the brain.
4. The beginning of consciousness: Portions of the brain's cerebral cortex show activation of neurons that are result of the momentary activity of some neurons, and the inactivity of others.
5. Consciousness in motion: Brain maps are not static, but change from moment to moment to reflect changes that are happening in the neurons. A consequence of the brain's mapping is the mind.
6. The body in the mind: Map making brains have the power of introducing the body as content into the mind. The mapped images of the brain have a way of permanently influencing the body.
7. Sensual Windows on the World: The brain's mapping incorporates the body senses and operation of the body in operation.
8. Emotions are complex programs of action carried out in our bodies.
9. Consciousness is a state of mind--if there is no consciousness, there is no mind.
10. Personal memories in consciousness are sum-totals of our life experiences.
11. Systematic discovery of events in human existence was possible only after development of full human consciousness.
12. Consciousness prevailed in evolution of species.

With our complex brains we have evolved the ability to project the process of consciousness into a completely different dimension. It is like an organization whereby the workers execute performance of skills, timing, "intra-coordination" with fellow workers, and even quality of their intangible spirits and mutual goals, but it is the executive and supervisory management that makes the strategic decisions that result in best productivity.

SPIRIT, SOUL, MIND, BRAIN, AND GENOME— HOW THEY RELATE

A few definitions are in order at this point to clarify the mind, the soul, the spirit of man, our brain, and the genome.

SPIRIT

The spirit is soul and mind of man. There are two definitions I can offer of Spirit, but the clearest way to define spirit is as an intangible, comparable perhaps to athletics where the spectators feel an emotion for the well-being and hoped-for victory of their favorite team. The spirit is also in the mind of the participating players, There is mind-set that their team, the entity of which they are one, will be victorious in the game (or of life in this analogy). They have perhaps been energized by the head and leader, the coach and coaching staff, to call up emotions that they have as individuals. This will produce greater-than-normal performance, to be tested in the various aspects of the game (again comparable to the "game of life").

A second definition of Spirit, again an emotional, non-physical entity or idea of a power, greater than that of the individual, that provides the power and ability for the individual to characterize the origins of that power and ability. Many refer to that spirit as the spirit of God, as the idea, the concept, that motivates them to a goal in life. The power is so great that it is a force, non-physical, that drives the emotions of the individual, perhaps similar to that force that derives the familial affairs of a prodigal adult to actively participate in daily work to provide sustenance and well-being for his or her family. It may also be similar to the non-physical force that drives or forces an individual to accomplish the best his ability allows in economic, social, or spiritual life. A recent University of Michigan Clinical Study found that attending a non-denominational spiritual retreat helped patients with serious heart disease feel less depressed and more hopeful about the future.

That idea—that spirit provides the energy and ability to communicate and motivate the individual to performance, perhaps previously unrealized, in a symbiotic action.

The spirit is the essence in personification of an idea, that may be even larger than life, an idea that men would die for. The spirit weighs the divine against the everyday, and finds one ensconced within the other.

SOUL

The soul, again, is non-physical, but is real, and more than an idea. It is like a presence, an essence within the body. The soul is sustenance and a reflection of a superior power of God and a sub-unit, so to speak, of his God. The soul "works with" (in sympathy with) the mind, the brain, and the genome system to affect the well-being, or change in organs of the body, emotions, and qualities that cause the brain by nerve impulses received and transmitted. It is perhaps like a body, or essence within the body.

The soul has a presence in Afterlife, and previously from the actual derivation of elementary particles of the Big Bang and physical creation of the cosmos, and exists within the body from reincarnation, being united with the physical body normally in the human birth process, sometime between conception and human birth. It is the guiding force to that individual, like a Guardian Angel for moral guidance of the individual. Orientation of soul could change during a person's mortal lifetime, from perhaps a non-believer in God to one having sincere beliefs in the existence of soul.

The soul will remain with that individual for his or her lifetime, and will return to Afterlife upon secular death of the individual. The soul is an integral part of the Kingdom of God. The soul maintains a communication with God. The soul, non-physical, has ability to communicate in a mystical manner, to communicate with souls of other individuals, without benefit of voice or hearing in speech. The soul "works in sympathy with" the mind, brain, and genome of the individual person on Earth to accomplish its goals and actions in life and existence. The soul may be assisted by the individual's Guardian Angel, sent from Afterlife to guide the physical person during his or her lifetime. The soul and spirit are connected but separable. Qualities of the soul are enumerated elsewhere in this book.

In Hebrew unwritten oral lore or history, the "Guff" is a term used to refer to the repository of all unborn souls. Literally, the word Guff means "body". The Hebrew Talmud essentially says there are a certain number of souls in heaven waiting to be born, and the Messiah would not arrive until there are no more souls in the Guff. Neither the Hebrew Scriptures, nor the New Testament, states there is a storehouse of unborn souls in Heaven. The Bible is not explicitly clear on when or how souls are created, but the concept of Guff does not agree with what the Bible does teach about the origin of soul, a belief that God creates each human soul at the moment of conception, or that the human soul is generated along with the body, through the physical/spiritual union of conception. Reincarnation is believed probable.

THE MIND

The mind also is more than just an idea, as described at the spirit description above. The mind is an actual essence within the body. The mind provides various emotions, judgment, ideals, beliefs, morality, willpower, morals, and personality, with recognition of summum bonum (belief in a power beyond that of physical life). The mind works in conjunction with the soul, brain, genome, and body through released electrical brain impulses, and enzymes originated by the genome that produce protein and desired action within cells of the physical body and brain of the individual.

The mind is akin to the computer software that comprises an ingenious action program that computer hardware chips will direct flow of electricity to gates that will actuate into desired results.

The mind is like the complex software design and action. The brain is like the mechanistic computer chip hardware, actions proposed by the soul (perhaps like you, the operator), changes occurring by bio-electrical impulses within the body. The mind controls the concept of self, identity, personality, consciousness, reason, emotion, judgment, values, goals, and other qualities.

THE BRAIN

The brain, as compared with the mind, is like the computer hardware, or an action system. The brain consists of a miraculously complex system for sensing, memory, communication, and action processes, much like some complex, electronically actuated factory machine.

Given certain stimuli, usually from the five senses of sight, hearing, smell, taste or feel, the brain will process communications throughout the neural system, resulting in stimuli to generate actions in muscles, organs, or the various biologic systems within the body. The brain, in its essence, is an electrical "bio-machine", both receiving and stimulating organs and systems of the body, reflecting like early basic evolution of the body matter-oriented mammals. The brain has been estimated to have as many as 100 billion neurons, with accompanying synapses. The brain, much like a machine, requires maintenance, through nourishment of food and mental exercise. Cells of the brain will die and become inactive upon human death.

The brain originated as the the nerve cord of fish and animals, required as an information transmission pathway to a center, and developed with the need for nerve stimuli coordination, and evolved to its present coordinating state over millions of years of evolution. Operations of the mind are sometimes misunderstood with operations of the brain. These are two separate entities.

GENOME

The genome is an extremely complex mechanism within each and every cell of our bodies. Each part of its physical existence is a determinant of a particular feature of the body and brain. In early life there were perhaps only 50 or 100 of these determinates that regulated formation of the very basic operations of life processes, such as to form enzymes that would in turn form cells with the purpose to accept outside molecules, that were a form of energy as food. The genome, for example, determined enzymes that formed cells to become a basic alimentary canal that would extract energy from the food molecules taken in, and force expulsion of the waste.

UNITY

All these elements combine to describe, "What Makes Us Human?", and different from other species of life on Earth. Environment-caused mutations that would add cells, energized by enzymes from the genome, to build a nerve and other systems, including a central control for nerve synapses of ending and starting, that would receive sensations, and originate energy flow to cells related or affected by that sensation.

Need caused mutations to build other aspects of physical life, including a system of fluid (blood and enzymes) flow, and pumping action to transfer food energy to cells, to cause energy transmission throughout the living organisms. Likewise, other cells were constructed having complex systems within the genome, adding to the length and complexity of the genome. Study of the genome would be like study of history, each addition to the genome recorded by a new construct in the genome, or comparable to steps on a seemingly endless ladder of chromosomes, that would transmit in birth to each future generation of that organism and species. These determinants, increased vastly in number and chromosomes, would identify the qualities of each and every cell in the organism, growing to the extremely complex arrangement seen today in animals, fish, plants, and humans, determinants numbering into millions within a tightly bound genome, duplicated within each cell of the living organism in the body and brain.

Need for additional mutation is fulfilled largely by environmental causes, and with advent of soul, to within the physical body, soul, brain, and mind to the physical life constructs—the total communication system. The cells of the body and brain remain, through inherited generations, like a complex chain, but are now influenced by the soul and mind since some undetermined eon, probably about at inception of Homo erectus about 600,000 B.C. The soul communicates to the genome by electrical impulses as a stimulus, and causing modifications in enzymes transferring to cells in organs of the body and brain.

Today, the body can be defensive to outside disease and stimuli. It can generate chemicals to combat poisons, offensive odors, body heat or cold, to have sensations in the five senses, and by a multitude of stimuli.

Unfortunately, there are still diseases, famine, and conditions that overcome the body and mind's combat-effectiveness, but analytical progress is being made to understand the genome and its parts that will increasingly modify the genome, body, brain, and mind elements in correctional or preventative actions.

SPACE/TIME DIMENSIONS

In recent years there have developed theories of additional dimensions in space and time, and in multiverses, both involving antimatter and void of matter. There are three visual dimensions of Earth and cosmos matter, width, depth, and height. All matter can be physically described by these three dimensions.

A fourth dimension is that of time, that may change the values of the dimensions. For example, time may bring about chemical decay, such as rot in wood, or by formation over time of prehistoric forests on Earth converted to coal by decay and pressure.

Scientist explored the existence in addition to those four dimensions. The additional dimensions have not been fully described, but are recognized to exist in the form of antimatter. Antimatter is described as the invisible component of Earth and the cosmos that possesses opposite electrical charge. Instead of possessing molecules having negative electrical charge, those molecules would possess positive electrical charge, such as ends of a magnet would have respective positive charge or negative charge.

By attaining positive charge, particles change their characteristics, including visibility and substance. In our common view of life the particles are not recognized to exist because we cannot see, touch, hear, smell, or feel them. They are antimatter.

But antimatter is real, existing in another dimension of space/time, a fifth dimension. Scientists have postulated up to 11 different dimensions, and some postulate an unlimited number of dimensions.

The description of soul and mind are of an additional dimension within our own bodies. Soul and mind work cooperatively with our physical brain and genome. The genome can instigate change in enzymes that in turn cause change in body organs and brain action. An electrical impulse from the soul may, for example, change the rhythm of the heart, or affect some other organ, for better or worse, or affect the mind, and consequently the brain, to think of and actuate certain action. In this way the body, brain, enzymes, genes, soul, and mind are all interrelated for operation in daily living.

The soul and mind are qualities of antimatter and additional dimensions. The mind, comparable to computer software, is the judgmental quality, or result, of our existence and cause the brain to carry out the prescribed actions.

The soul is believed to be either a small part of the Supreme Power (God) within our existence, or to have a power of direct communication to God, or to "entities of His organization" of archangels, saints, and Guardian Angels. The element of God is present within the soul of man and women.

This is an interrelationship of matter and antimatter, between our recognizable world of matter and of higher dimensions. Experimentation and research into the existence and qualities of antimatter are being currently undertaken by the CERN Large Hadron Collider, the world's largest and highest energy particle accelerator, beneath the Switzerland and France border, near Geneva, Switzerland. From this research may come a better understanding, and perhaps control, in the relationship between our different dimensions in space/time.

MIRACLES

Regarding the accomplishments of miracles, we often pray to god for knowledge, understanding, or a solution to a perplexing situation in our lives. In exploring ideas presented here, we believe the "mechanism" in accomplishing a miracle, or at least a solution or better understanding, occurs wholly within ourselves, or more correctly, by the soul and mind within ourselves.

The soul is presented as an extension of God's image and presence. The soul is like a portion of God that is within each of us, providing insight, presence, re-evaluation of our past knowledge, understanding, and thoughtful communication with other relevant souls, all combining to make the "miracle" happen. This in fact was affected entirely within ourselves, with oversight of God, his archangels, angels, and our Guardian Angels (souls).

The view is taken, largely because with nearly 7 billion people on the present world, and situations changing moment by moment, it would be near impossible even for one God to act on the needs of each person, with soul and mind, all at the same instant of time.

In the language of modern day administration, the situation requires what is determined as 1) span of control, and 2) delegation of responsibility. The president of a world-wide corporation, for example, must decentralize his operations to subordinate executives, managers, supervisors, and eventually to workmen, to fulfill the objectives and dictates of his superior office. It would be impossible for the corporation president to bid attention to each and every question and problem that might arise at each instant in a world-wide scope. The situation would have to be administered on a localized basis, utilizing application of corporate policies, procedures, and intent.

Such is the likeness in actions of God, individual "Princes of His Kingdom", individual souls, and Guardian Angels, within the soul and the mind of each of us, to give immediate and local attention to all factors pertinent to the arising questions.

The challenge may, on occasion, be insurmountable. In some situations, thoughtful prayers, and well wishes in Afterlife may need to satisfy.

In some instances, the mind and soul may be instrumental to accomplish change in body organs, brain activity, and mind-communication with relevant others, such as doctors, other medical people, family members, friends, associates, and others having spiritual or technical power to affect the well-being of the individual.

We pray to God, but the corrective action is based with the soul and mind of the individual. It has been said, "The Lord helps him who helps himself".

MULTI-VERSES AND ADDED DIMENSIONS

"Could our universe be just one of a multitude in the cosmos, each with its own reality? It may sound like fiction, but there is hard science behind this outlandish idea". The view is presented by respected scientists of a closely related view to multiverses in our existence. Multiverses, or other worlds parallel to our own universe, are existences occurring in other dimensions. Many versions of inflation theory predict an infinite number of universes (or dimensions). "It is a mistake to think of the multiverse as a theory invented by desperate physicists. The multiverse is a prediction of certain theories, most notably of inflation, plus string theory. The question is not whether we will ever be able to see other universes (in other dimensions), it is whether we will ever be able to test the theories that predict they exist."

"We find ourselves surrounded in our telescopic viewing, by an opaque barrier, past which we, so far, cannot see—The Big Bang. The distant universe might be uniform, or it might be full of different universes (dimensions) scattered throughout space. The condition of our local environment might be unique consequence of fundamental laws of Physics, or they might just be one possibility out of a staggering number. WE JUST DON'T KNOW!"

"The proper scientific approach is to take every reasonable possibility seriously, no matter how heretical it may seem."

Source: "Out There", by Sean Carroll, a theoretical physicist at Caltech, focusing on inflation and the Arrow of Time, presented in "Discover", October 2011.

More on this subject can be found by computer, clicking on Google, at Mind, Body, and Spirit.

DINOSAURS AND SOULS

Did dinosaurs have soul? How did soul emanate to current-day situations whereby elements of soul are possibly found in cats, dogs, Brahma cattle, horses, and some other domesticated animals, in addition to humans?

Dinosaurs living between 350 million years ago to 65 million years ago, so far as has now been found, did not possess evidence of soul to the degree found in today's humans, and some believe found in pets and other domesticated animals.

Glacial and interglacial periods happened at least ten times in earth history since 542,000 years ago. The last glacier being the Wisconsin glacial period ending 15,000 to 10,000 B.C. after millions of years of glaciers and interglacial periods covered much of northern and southern hemispheres of earth, after first glaciation perhaps 2.3 billion years ago. We see today evidence of this geologic activity through striking features in Michigan, among other locations on earth, generally in northern or southern areas of the globe, sometimes forming as beaches, gravel deposits, large and small hills, mountains, valleys and other geologic features.

Some researchers hold that a near-by gamma ray burst could well have caused a near extinction of existing mammal life around 350 million years ago when dinosaurs started, and changing of the genome to allow advent of the dinosaur age of mainly out-sized reptiles, with its many bizarre species and forms as are unearthed by anthropologists.

How did the age of dinosaurs ever get started? It is obvious that life forms at that time became suddenly radical in forms, from Brontosaurus, Tyrannosaurus Rex, and many other mammoth and exotic forms of dinosaur life, only to be undone at about 65 million years ago by a catastrophe believed caused by collision to Earth with a large cosmic traveller. That collision may have again set off a Gamma Ray change that affected the genome, this time we think for the better.

Continuation of the evolutionary march of smaller mammal species developed, especially fast without the competition of goliaths in the reptile and in some mammal families. With demise of many reptile species of dinosaurs, mammals expanded their presence and proliferation in many species from Sabre Toothed Tigers to Mammoths and Mastodons. Great Ape species of Gorilla, Chimpanzee, Orangutan, and intelligent monkeys of 6 million years earlier exhibited an intelligence with development of intellectual and physical tools, raising their competitive abilities above that of other primates. Australopithicus afarensis, personified by "Lucy", the first primate found to be bi-pedal, is dated to 3 million years ago.

Ardipithicus ramidus (loosely named "Ardi), at 4.4 million years ago, is the most complete female specimen known to the hominid groups. Several species of Australopithicus and Ardipithicus appeared on earth. About 2 million years ago new families of Homo species appeared as Homo rudolfensis in eastern Africa, Homo habilis in sub-Saharan Africa, Homo ergaster in eastern Africa, and Homo erectus migrating to eastern Africa about 600,000 years ago and lasting on Earth until about 250,000 years ago. Homo neanderthal in Europe and western Africa then became the dominant species before Homo sapiens, a new and more complex species in both physical and emotional traits, appeared about 120,000 years ago, outlasting the demise of Homo neanderthal at about 28,000 B.C.

Dinosaurs were "animals of the forests" whose bodies were formed either as reptiles (born from eggs) or as mammals (formed and given birth from the reproductive organs of the parents).

No symbols of soul existence are found in dinosaurs, with possible exception of "mother care" occurrence (a finding of fossils by archaeologists showing multiple young in nest, a parental event which may have been simply required for continuation of species extension in animals of the era, and without soul).

Dinosaurs became extinct about 65 million years ago.

Soul for humans would have been incarnated at start of a new species of Homo erectus, Homo Neanderthal or Homo sapiens. It is believed that soul was incarnated to man somewhere between 600,000, or 250,000, or 28,000 years ago, the span of Neanderthal man, or perhaps earlier. Homo sapiens development on earth started about 120,000 years ago as recorded in the annals of anthropology.

At about 20,000 B.C. Homo sapiens, Cro-Magnon, or Modern Man had free reign of their part of earth with gradual demise of Neanderthal man by about 28,000 years B.C.

Homo sapiens were first living a Neolithic hunter-gatherer subsistence. After about 12,000 B.C. Homo sapiens appeared in agriculture and herding-based community civilizations and cultures in north and eastern Africa, Egypt, and Mesopotamia. During this span from 35,000 years B.C., and at various times, man had populated different areas in Southeast Asia, Mesopotamia, Central Europe, British Isles, Southern Russia, India, the American continents, and other parts of the earth.

Man at the time, quite possibly including both Homo Neanderthal and Homo sapiens, by then different species by fact of each being a new species, or possibly sub-species of Homo erectus, possessing soul by incarnation, were able to survive the time of the dinosaur's demise. Thus, soul was likely a factor in the survival of man.

Incarnations of soul would have taken place about 600,000, or 250,000 B.C. when Homo Neanderthal specie evolved, or about 120,000 B.C. when Homo sapiens specie came upon earth, or even earlier than Neanderthal when elements of soul existed in modern Homo erectus around 600,000 B.C. The abilities of the new species changed, but ways of living in caves, and a hunter-gatherer subsistence culture, remained constant into the later Neolithic period. In this period men drew pictorials on their cave walls, as expression to some unseen god, for guidance in upcoming hunting expeditions, an expression of soul and recognition of a higher entity.

Even if soul were incarnated as late as with advent of Homo sapiens, about 100,000 B.C., during a time characterized by hunting and gathering, it still took tens of thousands of years before man emerged into civilized cultures.

This remained until about 12,000 B.C., near end of the Wisconsin glaciation, when agriculture and herding started to become the cultural norm. Probably little immediate outward progress in culture had been made in the interim years leading to 12,000 B.C., as soul continued to integrate with the physical body and mind of man.

In the period 12,000 B.C. to about 9,000 B.C. culture changed. Kingdoms developed with kings at their heads. Mud brick villages were built, with individuals specializing in different crafts and art. Development of an alphabet, writing, laws, and construction of religious shrines, and temples with offering tables for sacrifices that would become a part of the culture. Cemeteries and reverence for the dead became common. Clay vases were fashioned, with metal and copper developed for tools and weapons.

Successive nations and cultures would develop within Mesopotamia as Akkad, Amorite, Assyrian, Babylonian, Chaldean, Hittite, Kassite, and Sumerian cultures came into political power. In Egypt similar dynasties and Pharaohs arose with power, in an agricultural and herding economy. The culture demonstrated having respect for the dead, with mummification, burial procedures, and civic projects of building pyramids, temples and burial shrines.

Many pet owners attest that their pets display elements of soul. The Hindu religion holds the Brahma bull as sacred, with elements of soul. As revolting as it may seem, sexual intercourse between man and animals (bestiality) is reported in both past ages and currently. In this union it is possible that qualities of genetics of soul were passed through genes from man to domesticated animals, allowing current testimony that certain pets possess soul in their genome and are able to exhibit qualities of soul. Rare examples exist whereby different species have sexually intermingled and produced offspring, contrary to accepted rules of genetics.

QUANTUM MECHANICS IN ANCIENT TIMES

Soul is a product of the universe of Quantum Mechanics. Quantum mechanics is, even today, after being first recognized by Albert Einstein in the 1920's, not well understood. Unlike Newtonian physics, these are not definite, but only probabilities that certain qualities of an ethereal subject exists, and that the future qualities are totally unpredictable. A physical analogy might be like trying to catch a greased pig, and blindfolded as well. It is nigh impossible until someone in the future shows the way of how it might be accomplished, as classical mechanics was revealed, step-by-step, in our past.

Physicists are constantly seeking more definitive descriptions of matter and antimatter. This presentation does not reflect that exploration, but attempts to explore only the result of quantum mechanics, specifically as it pertains to the soul—how quantum mechanics can account for this existence and qualities of soul, its continuation, and its operations of communications and presence.

To gain insight into the effect of quantum mechanics 2,000 and more years ago, we first must determine and recognize their effects today. I believe we can assume that these effects of quantum mechanics were as evident 2,000 years ago as they are evident today.

First of all, quantum mechanics involve effect of minute elementary particles, anti-particles, or antimatter (elementary particles with opposite electrical charge). A common element of matter on Earth might be composed primarily of electrons, orbiting a nucleus of neutrons and some protons (plus a genome of organic matter and other defined characteristics).

In antimatter of positive-charge particles, the electron orbits are absent. They were replaced back at time of the Big Bang by Annihilation, a physical or chemical process that eliminated protons from an existence that was generally equal between negative-charge and positive-charge elementary particles. Split seconds after the initial "explosion" the annihilation took place. The author theorizes that negative-charge particles went on to form items of matter. Positive-charge particles formed an unseen world known today as antimatter, present but invisible.

Antimatter has been first recognized and analyzed only in the past 20th century. It will be analyzed further in coming operations at CERN laboratory in the Swiss Alps, on the border between Switzerland and France. Negative-charge particles and positive-charge particles will be collided and antimatter produced.

These articles of antimatter will produce tangible knowledge of a completely new world—that of antimatter. Antimatter will have many different qualities. They may be in dimensions of space beyond those four presently recognized of height, width, depth, and time. Just what those additional dimensions will show, how they are comprised, and what their qualities are, plus many other descriptions, will be discovered.

A world of antimatter will have unusual qualities. This might involve activities that are not normal or physically viewable, forces that were on Earth, sometimes seen, but previously unexplained, communication between souls without necessity of hearing, speaking, sight, or other worldly senses of man—a form of thought transmission, an ageless existence where all of history is available for souls to experience all at any one time, explanation of facts that have led to myths in our Earth culture, a further explanation of the Supreme Power (God, or by some other monotheistic name), an understanding of just how an Afterlife "works", the pleasure (or drawback) to experience meeting a past deceased family member, social acquaintance or someone known prior to Afterlife, or perhaps how Astrology alludes to what seemed in secular living as strangely and uncannily mystical, and other very strange happenings.

Listed here are some characteristics of quantum mechanics, antimatter, matter, existence in Afterlife, and possibly beyond.

1. Quantum mechanics originated at the Big Bang when forces of electrons annihilated the weaker forces of protons.

2. Forces of electrons went on to form matter in the universe that can be physically observed.
3. Forces of protons also went on to be the base of antimatter.
4. A world encompassing soul evolved in the world of antimatter, as all varieties of physical life evolved in our world of matter.
5. The two elements of matter and antimatter developed in parallel over life of the cosmos and earth and were joined in the mind of man as soul.
6. What is modern man?
7. How is modern man unique in existence and qualities of soul?
8. What is soul? Soul is Biblical, stating that God created Man in his own image. An image of God and His spirit therefore exists within man, present as soul.
9. Homo sapiens (and perhaps other species of hominoid on earth) developed with an "arrow of direction in plan".
10. Soul manipulated the RNA and DNA of the genome, with resulting change of body, mind and neurological organs.
11. Intelligent development determined man's evolution.
12. Man (through hominoids) has progressed from a tree-living animal to an intelligence that put men on the moon, ability to define and explore space, to talk, communicate, reason, plan, and organize.
13. The scope of quantum mechanics involves particle physics, elementary particles, parallel dimensions, existence of other universes, mystical (difficult to comprehend) occurrences such as dreams, foretelling, non-verbal communications, non-physical life, prayer, insight and insightful problem solving, guidance of an internal Guardian Soul, ghosts as implementation from another dimension, and many other aspects.
14. Quantum mechanics and particle physics is the "doorway" to understanding and study of soul and spirit.
15. The future will hold development of inter-dimensional communication and perhaps travel (mentally, emotionally, and spiritually) between worlds of matter and antimatter.

This revelation into ancient quantum mechanics in fact is as real as it exists today and will explain many events of history that have formerly been unexplained, and are today no longer subjects of myths, miracles, and wonderment.

In worldly terms, Afterlife will be somewhat like a first trip to a foreign, and extremely strange country, to view an existence we never knew before, and to be expanded in our scope in soul, to reconcile reality in two different cultures—to reconcile the culture of past life on Earth, now to the permanent culture of Afterlife.

Whatever qualities that antimatter are discovered to have are the same qualities that existed 2,000, 8,000, and more years ago. (The primary difference in the ages is that of increased knowledge.) They may explain dreams, a better description of soul, forces of nature that are now mysteries, and explanation of the operations of Afterlife, activities of soul in Afterlife, an existence of even other dimensions of space, of possibly reincarnation, and a further existence undreamed of previously. This is the expected scenario in existence, since the advent of Modern Man and from the previous generations of souls.

SEX IN EVOLUTION

There was time on earth when reproduction of the species was not by internal birth. Reproduction was sometimes by self-insemination (bi-sexual activity or self-replication), and later by laying and hatching of eggs by fish and reptiles, followed by the present biologic system of insemination and birth.

Introduction of mammals (about 250 million years ago) produced young to be formed and born by means internal to the body, supposedly preceded by sexual intercourse and insemination by the two sexes. This provided miraculously intelligent protection against stealing and eating of eggs by predators. That would have been the logical ending of species reproduction, and eventually preventing mankind as known today.

The Paleozoic Era (600 to 250 Million years ago) consisted of the Ediacaran, Cambrian, Ordovician, Silurian, Devonian, Carboniferous, and Permian Periods, with uplifted fossils to mountains, of sea-living creatures, each identical to others in the species.

First mammals appeared on the earth about 250 million years ago at end of the Paleozoic and start of the Mezozoic Eras.

The Mezozoic Era (250 to 70 million years ago) encompassed Triassic, Jurassic, and Cretaceous Period, common to the Age of Dinosaurs.

The Mezozoic Era was followed by the Cenozoic Era (70 to 5 million years ago), consisting of the Tertiary Epochs of Paleocene, Eocene, Oligocene, Miocene, Pliocene, Pleisocene, the Neolithic Epoch, and finally by our present Holocene starting about 10,000 years ago and characterized by development of first Modern Man from the Neolithic stage onward. First primates appeared in fossils originating about 68 million years ago.

Australopithicines, including "Lucy", roamed Africa's surfaces from 1 to 4 million years ago, with Ardipithecus ("Ardi") at about 4.5 million years ago.

The ancient population of living organisms was not always divided into male and female. In earliest eons, when living organisms of amoeba, bacteria, virus, and yeast were prokaryote cells, very simple cells, not having cell walls, consisted of only simple electrons (and elementary components of these), all cells of the simple species were alike. We can observe that even today, in yeast, there is no differentiation between one yeast cell and the parent cell.

Sexual differentiation dates from about 500 million years ago, when eukaryote cells, having cell walls, nucleus, RNA, DNA, organelles, and other complexities of various cells, was common.

There existed in the eukaryote cell X chromosomes, which determined male, and Y chromosomes, which determined female of the various species. Physical, as well as brain characteristics, developed in new-born animals, and the very means of reproduction changed, requiring sexual union to produce offspring of the species. At this time, the sexes were technically two species, or perhaps one being a sub-species of the other. Whether the basic species was male or female is undetermined.

The differences between male and female can be seen today in many observable, physical, medical, and emotional attributes. Even variations and strengths of these attributes can be observed in various forms of heterosexuality, features of one sex showing physical or emotional differences within the species. Rare, but existing homosexuality appears to be result of physical, DNA, or brain deviation, to varying degrees, from birth designation.

The differences of male and female of the Modern Man (Homo sapiens) species is more extreme than differences, say, between Homo ergaster and Homo sapiens. The differences are so extreme that anthropologists could well designate the male and female of Modern Man, or Homo sapiens, as separate species, or at least as sub-species.

The female of Homo sapiens seems to be better, for example, at discerning color values and color combinations. The female, dating from days of the primates, has mammary glands to function in breastfeeding their young. Females seem to have different values in their mentality for decision making, and in time of early physical maturity. Females seem to have less intestinal flatulence than do males and in various internal differences in biologic and anatomy make-up. Average weight and height of females is less than averages of males, and of course females have different sexual organs that allow childbirth and mental emotions regarding care of their young. Females seem to differ from males in nearly every category of biologic and mental concerns, as modern medical analysis can attest. In many ways it seems erroneous that male and female are even in the same species, and one or the other might well be an ancient sub-species of the other, going back in evolutionary history even to primates in mammals.

Homo ergaster, a close species to Homo erectus, lived on earth at a time generally estimated during a 200,000 year period between 1.7 and 1.5 million years ago, generally paralleling the early existence of Homo erectus. Although bi-pedal and erect in walking, the body of Homo ergaster was covered with thick hair, Homo ergaster may not have had a larynx (voice box) and had little means to communicate, other than by basic guttural sounds or body movements.

H. erectus has been determined by anthropologists to directly precede Homo neanderthal, and Homo sapiens (Cro-Magnon or Modern Man) on earth. The diaspora of H. erectus was widespread with fossils found in Africa, Europe, and Asia. Preceding species of H. erectus were Homo habilis, H. ergaster, and H. rudolfensis.

The evolution-through-need advocates, would undoubtedly claim that the larynx and loss of body hair "just happened", which is a very unscientific approach to the facts. They might argue that the rudiments of a larynx always existed in the hominid, but just wasn't developed. The loss of body hair, it would be argued, was from lack of need for hair in a temperate climate. If this is so, why do Eskimo, or Inuit, people of the far north not have continuation of a hairy body?

Advocates of Intelligent Evolution, on the other hand, would argue and visualize change in the body through the genome, which would designate a new existence of a larynx as specified in the genome. Homo erectus has been determined by anthropologists to directly precede Homo Neanderthal (Spain, France, and Europe), Homo heidelbergensis (Germany), and Homo sapiens (Cro-Magnon or Modern Man) on earth. The diaspora of Homo erectus was widespread with fossils found in Africa, Europe, central Asia, and Southeast Asia. Preceding Homo species of Homo erectus were Homo habilis and Homo rudolfensis. Would an inactive larynx have existed in each of these preceding species?

In the first years of life on earth (about 3.8 billion years ago) there existed various algae, virus, yeast, and bacteria. As far as is known there was no differentiation by sex. All generations were clones of their preceding or contemporary generation.

Sex eventually developed about a billion years ago, evolved within ancestral single-cell prokaryotes, first in plant life with a myriad of genome forms in production of pollen, different with various plants, that would germinate and cause growth in the respective plants. Millions of the pollen developed for the chance that one might succeed to make a successful sex match. At the female ovule of a plant, which contains an egg cell, the pollen grain attempts to travel a tube to connect sperm and egg. If the pollen lands on the wrong species of plant, or is too weak, or too old, the tube does not form, and no successful product of sex occurs at that particular time or place. Fertilized blossoms developed into fruit having seeds to start a new generation of the plant and species.

A similar process occurred in all forms of animal life of various species. Sperm joined ovaries to cause reproduction, if it was a fertile match, which in turn developed to birth and maturity, thereby causing further and multiple reproduction of species.

At some later time (about 440 million years ago) fishes developed on Earth, and later amphibians and reptiles. Reptiles, like fish, replicated themselves by a birth process from hatched eggs. Whether or not a prior fertilization process took place, male to female, or a self-fertilization process, is not known. We can visualize an insemination process between two Tyrannosaurus Rex, or like beast, as quite a stressful event.

By definition, mammals birth their young from internal body functioning (with some exceptions such as the Platypus), resulting from sexual union of male and female. When and if this happened during early evolution of mammals has so far been undetermined.

At the molecular level, cells reproduce themselves by division, at that time without outward consideration of sex determination. This happens on either a systematic, heredity process, or by fortuitous mutations, appearing at random, that provides biologic characteristics, directed by DNA. Sex is determined and manipulated in the Mitochondria DNA for female, and the X chromosome of DNA of the male.

It seems obvious that at some stage of hominid development there was an essential process of male and female union, which of course happens today. Again, it has not been explained when this sexual process started, even as determined by archaeological studies. It has been shown that "Lucy", of Australopithecines specie, was a female, determined largely from anatomy studies of the fossils. This is dated at about 3.75 million years ago. Primates and hominoids of apes, chimpanzees, monkeys and Bonobo, in both zoos and the wild, have been observed in sexual union activity.

Reptilian dinosaurs (starting about 315 million years ago) produced from eggs, were extinct after about 65 million years ago. Some lesser size reptiles, birds, fish, and some sea life continued in life even after this extinction event, and related species continue today.

ADDITIONAL DIMENSIONS IN SPACE/TIME

There are many theories regarding various stages of birth and development in cosmology. These include concepts of String Theory, extra dimensions in space/time, vacuum Energy, limits of the universe, and more, not to mention Dark Energy and Dark Matter. Each has potential to change our views on the birth, development, and the laws governing the universe. Any one of these concepts could change time-honored theories of the universe.

String Theory holds that energy in space is not located in precise points in space, but instead chains or strings in the cosmos that affect qualities and actions of the cosmos. Under string Theory many additional dimensions besides the four we presently recognize of width, depth, height, and timer, must exist with a brane, or membrane-type structure, for each. It has been proposed by cosmologists that as many as eleven different dimensions exist and are necessary to validate String Theory. Others have ventured to state that the number of dimensions is unlimited.

Vacuum Energy is energy that existed in space and continues to reside in the vacuum of preexistence of an early concise space. It is not yet clearly defined, but Vacuum Energy is believed to be associated with Dark Energy and Dark Matter.

Vacuum Energy also could be source of the Big Bang, reflected in creation of not only the cosmos and our Milky Way Galaxy, but repeating also in creation of other galaxies in space.

The author ventures to propose that Vacuum Energy is the source for creation and of constantly expanding space in our universe, defining the changing limits of the universe, currently 13.7 billion years old. The phase of expanding space seems to be attracted to the Great Wall of Galaxies which has been observed at extreme known limits of space, which conceivably could account, by gravitational attraction, for the increasing velocity of space growth. This also leads one to wonder where and how this Great Wall of Galaxies originated.

Perhaps developments in geology and evolution of the physical Earth in early characteristics of Earth were products in science, whereas creation and existence of soul, mind, and body systems were planned by God, to be made actual through natural means in birth and growth.

Fundamentalist religious individuals, especially, may not agree with my views that challenge traditional religious Biblical dogma of long-standing, but hopefully individuals of these groups will at least understand there are arguments that alternative possibilities exist. Bringing realism to replace folklore and myth should actually strengthen beliefs in God through truths in the concepts of spirituality. After all, it was once believe, before evidence to the contrary, that the stars and sun revolve around the Earth. The weight of evidence has changed with increase of accumulated knowledge at present, with all respect given to the necessary preliminary thinking of the past by Giants in development of philosophic thought.

There were sub-species of each primary species that reflected variations existing and developing due to environmental conditions. Such variations in sub-species continued slowly but steadily into Homo sapiens, with radical changes between sub-species of male and female of the specie, and sub-species developing due to environmental conditions.

Evidence of sub-species can be observed today, not only in sex, but also in biology, anatomy and mental or emotional make-up, fomented largely by environmental conditions,

Much research is needed in this area of study to show, as precisely as possible, now recognized in the broad range of 500 million to 250 million years ago, just when sexual union was required for propagation that took place at levels of primate, hominid, hominoid, bird, sea life, reptiles, mammals, and Homo sapiens for reproduction. Bees and flowers were essential for plant reproduction.

It is possible that reproduction continued from external eggs for non-primates until relatively recent time in the chain of evolution, perhaps as recently as 200 or 300 million years ago, when the Great Ape family of Gorilla, Chimpanzee, Monkey, and Bonobo, the Families of Ardipithecus, Australopithecus, Paranthropus, and Homo Family of Homo habilis, Homo ergaster, Homo erectus, Homo Neanderthal, and Homo sapiens, continued mammalian internal birth characteristics.

In a recent discovery it was shown that the male factor was not required in certain circumstances for reproduction by the female of the specie. (Source: Smithsonian magazine, November 2009, pg. 12). Is this a new development in biology, or merely reinstitution of an ancient procedure in nature and development? Will a day come when two females of the human species will, one possessing the near-male biologic qualities of male sperm, produce a fetus?

SCIENCE VERSUS RELIGIONS

Is the act of living a product of the mystical or a product of science? Mankind has pondered for many centuries the case of many, if not all, situations in living. The general premise in religion history followed a vector that God created everything and controls everything. Men of science have questioned this view.

I believe this premise can be divided into factors that are natural and those that are mystical.

Natural effects are those that can be explained in a scientific way from generally accepted knowledge having an effect. Mountains are caused by shifting action of geologic plates. Clouds are caused by moisture accumulation following evaporation from Earth's surface and moved by winds and rotation of the earth. Old people die often because of a simple "wearing out" of some biological component of their body or brain.

At one time each of these, and more, were believed to be mystical, but as knowledge in science increased, the cause could often be shifted from mystical to natural or rational causes.

Nevertheless, some actions in life remain unexplained and remain on a mystical side for explanation.

How the genome system got started, how a human embryo grows within the mother from a sometimes required union of male and female eggs to a fully functioning animal or human fetus, and what is the origin and function of soul? All were held to be miraculous and supported the occurrence of mystical happenings.

With some analytic thought, it can be reasoned that happenings on Earth can be categorized to be either of natural origin and occurrence, and others still retained as mystical. My list here is suggestive and many more events could be added and placed in natural vs. mystical categories.

Who knows? Perhaps someday babies can be produced and assembled in a machine and grow, like bacteria or in formulation of beer or medicines, hopefully not in some grotesque, irrational form.

Perhaps some genome-equivalent system can be evolved to plan and control biologic growth, much like an industrial production planner or engineer can devise machines and to convert raw material to a finished automobile, brand spanking new, and ready to roll off the production line.

Presented here is a table of Life Design And Operation—a division of events into natural and mystical.

DOMAIN OF THE MYSTICAL (GOD)

- Plan and design of the living organism and its integral parts
- Growth and specific evolution of the living, in many species
- Creation of the genome system
- Installation of the soul to the living body (reincarnation)
- Movement of the soul at death into Afterlife
- Response to prayer
- Direction and stimulus of a culture of great persons in history
  Mind-controlled actions of humans (Ethical, reasonable, moral)
- Activity in Afterlife (Guidance to present Earth-living man-persons.
- Actions of conscience or guardian angel
- Location of extra dimensions (e.g. on Earth, NOT in space)
- Cellular system of living cell specialization in forming various and specialized organs, considering need for specific organs and assignment of T-cells)
- Design of brain for specialization and operation
- Directing vectors for specie evolution and changes.
- Intelligent design direction (Origin and vectors for fur, spots, coloration, anatomy, etc.)
- Non-physical communication, thought transmission
- Re-incarnation with mostly extermination of memory of Afterlife experiences.

- Cellular system of living cell specialization in forming various organs (Needs of organs and assignment)
- Extension of existence of soul in Afterlife
- Actions along improbable, irrational phenomena (Guardian Angel?)

DOMAIN OF THE NATURAL AND EXPLAINABLE

- Scenic mountains and plains (Result of water, erosion, geology, and geography)
- Air (Result of atmosphere composition and changes)
- Working of the pre-designed genome system (Systematic once installed into life organisms)
- Living time: Aging of body and brain (Biologic and medical concerns)
- Culture development (Natural evolution of ideas)
- Technology development (Ideas of the brain and mind)
- Warfare (Acts of man)
- Weather: Daily, seasonal, and cyclical (Nature generated)
- Catastrophes of nature with causes explained (nature generated)
- Physical catastrophes, earthquakes, volcanoes, floods, extreme temperatures, etc. (nature-generated and explainable)
- Clouds (Atmospheric conditions and laws of nature)
- Actions supported by explainable reason (Logical events)
- Evolution by natural selection (Physical needs for biologic changes—Darwinism)
- First and subsequent structure of living cells (Evolution)
- Secular death
- DNA system of gene inheritance
- Winds: (Result of atmospheric conditions of relative air pressure lows and highs and rotation of Earth.
- Long range temperature and climate fluctuations of Earth: (glacial actions) and Earth travel in cosmos.
- Living time: Aging of body and brain (Biologic and Medical causes)
- Secular death: (A natural event)

- Logical events: (Actions supported by explainable reasons)
- Species changes and sub-species (Evolution by Need—Darwin)
- Energy Conversion of elements and forces into body (Biologic operations of lungs, intestines, kidneys, etc.)
- Directed changes in new species and sub-species (Directed by DNA)
- System and operation of gene inheritance to a next generation (natural biology)
- Determination of cells through bi-sexual union (Genetic modification, cloning)
- Temperature/climate fluctuations of Earth
- Operation of brain, after initial design
- First life on earth (First, and natural, "spark" with a cause and effect for start of life)
- Operational specie changes and sub-species (Evolution by need)
- Operation of RNA and DNA (Genetics)
- Energy conversion of elements and forces to body (Biology)
- Operation changes in new species and sub-species
- Gene/protein/enzyme/cell system association
- Population growth and changes
- Ability to comprehend short or long-range future, using existing brain mechanisms

---

In the age of primates hominids and hominoids there existed a pre-history of cannibalism, that was changed as man (Homo) realized the worth of fellow hominoids, with realization that destroying and eating a fellow hominoids is spiritually wrong.

The Big Bang is held here as the start of the essence of soul, evolving to its present form, from annihilation of positive-charge elementary particles into antimatter and the base soul.

Perhaps someday the secular equivalent of soul can be developed by scientific training and by shaping the ethics and morality of humans as they develop from infancy to maturity. But today these all seem improbable and remain as mystical events, perhaps the doing of God.

With this observed division, much of the conflict of society can be resolved between what is function of science and what is function of a mystical God.

It is NOT an either/or question.

WONDERFUL WORLDS

PART IV

AN AGRARIAN CULTURE

THE GREAT LEAP FORWARD—
INTELLECTUAL DEVELOPMENT

It seems to be a popular debate in philosophy and theology whether mankind developed through planned sequential acts, each building upon a previous status, to define a path towards Modern Man, scientifically known as Homo sapiens.

Pursuit of the actual pathway need not be a choice of the traditional theological dogma, nor solely on principles of science. There is mid-ground for reconciliation of philosophies. The method of appreciating intelligent design lies in the merger, or symbiotic relationship of theological philosophy, and that of science. The "arrow of life", so to speak, of evolution, points to a breakthrough in understanding the merger of science and theology in body, brain, mind, science, and soul. This breakthrough most assuredly will happen in the future. Science is coming near in its explorations into antimatter. In addition, theology is gradually recognizing the truisms of science, sometimes at the expense of age-old religious dogma embellished with stories of myth, miracles, gods, and demons.

Mankind today must continue to be open-minded in screening possibilities. World history is well-known for its once misunderstood, so-called impossibilities, that later, through developed knowledge, fell into oblivion (such as Earth is the center of the universe, the errors of the Inquisition, belief that Heaven is in the sky, the Earth is flat, etc.

"In a nutshell," the author recognizes that the soul exists within the psyche and body of each of us. The existence of soul can be identified by its defined qualities. The soul is dynamic within the body and manipulates the RNA and DNA system, or genome, of the biological body, spurred by electrical energy directed by the soul, thereby causing changes in the body's biological functioning.

This concept may be difficult to fathom in the present state of accepted world knowledge. The author holds that the soul is an extension or microcosm of the Supreme Being (God).

Manipulation of the RNA and DNA by soul has usually been for the good and health of the body or mind, but bad health, or bad development direction, can also result from manipulation, mutation, or from overview by the soul to its responsibilities, or by some intended action or inaction.

The soul therefor is the prime mover, under the direction of the Supreme Being, that causes changes of the body, as well as in the mentality, and therefore is the controlling element to the individual body's existence and direction.

Some persons of Faith express that the power of God was with them, or some such expression, when they experienced the power of prayer. Such statements, the author believes, are only partly correct. Operationally, the power that is expressed is action of the person's own soul that provided physical ability or mental clarity, allowing change to take place within that individual. It seems that in earthly terms there is a "chain of command" that takes place, with a "span of control " at any instant, by God's system, over a world population of 6 billion or more people, needs changing moment by moment, plus other entities that possess a soul, would be a bit much to expect, even of God.

Intelligent design is real and plausible. Through series of planned changes, or mutations, throughout the ages of evolution, forerunners of human mankind have developed in a direction away from that of other animals. A variety of life, including sea life, mammals, reptiles, amphibians, amoeba, virus, bacteria, and both microscopic and normal size plants, have undergone change over the eons in their component features.

Various forms of man-kind species have existed, some leading all the way to Homo sapiens, today known as Modern Man. Other species have expired in their experience toward continued existence (extinction) other than Homo sapiens, Homo neanderthal, Homo erectus, Homo ergaster, and others which have been archaeologically exposed for over the past thousands of years.

Examination of this physical evidence of past man-types is proof enough that mankind developed progressively over eons in a steady progression toward humanity that we experience today. The author, for one, finds it inconceivable that progression of mankind happened by an unplanned series of mutations and biological accidents.

Without planning and implementation complete chaos in anatomy and biologic design would have been a logical result. With existence of recognized benchmarks in life, evolution would have happened, developed, and have been evidenced over eons through intelligence and planning. Creation and change would be proposed, but disposed by implementation of natural processes.

When it comes to choosing which theory of evolution development is true, strong arguments have been advanced by both sides of the question in ideas of whether evolution occurred strictly by natural, random selection, or by a planning force (intelligence).

For those favoring the Biblical representation, it is stated in the Book of Genesis that "God created man in his own image". Image of God refers to an intangible existence, different than having physical body and brain as do mortal beings. Image of God conjures up an existence of non-physical soul or spirit, existing within man, and beyond death in an Afterlife existence. "In God's own image" certainly does not imply that God had a human body, brain, and emotions of man's secular life. Until insight is gained into higher dimensions of space/time, secular life will remain unknowing of the true circumstances of evolution and creation. We should maintain an open mind and understanding of all arguments in support of varying ideas.

Professor Sean Carroll, Biologist at the University of Wisconsin, has written that 600 million years ago (incidentally coinciding with presence of the first member of the Homo family, Homo erectus), certain patterns in growth of multicellular organisms appeared. Those patterns proved so useful that versions of the genes governing them are carried by nearly every species that has arisen since. These several hundred "tool kit genes" are molecular evidence of natural selective ability to hold on to very useful functions that arise.

Research on how and when "tool kit genes" are turned on and off has helped explain how evolutionary change in DNA gave rise to Earth's vast diversity of species. Determination of an organism's form during an embryo's development is largely result of a small number of genes that are turned on or off in various combinations and order.

There is no doubt that evolution, the change and development of a species, happened: it is only a question of how it happened. Did it come about from natural, self-selecting devices, or did it happen from some overall planned activity?

Changes result from direction in the inner cell activity of RNA and DNA, going back nearly to year one in life history of earth. Reading the genome of a particular life form, whether it be plant, animal, fish, bird, or reptile, is like reading a book of history concerning development of that life form.

The genome of mankind, now over 1.5 billion genes long, was once much simpler, perhaps only 100 genes long, that provided for only the basic functions of life for intake of food for energy, digestion, and a waste expulsion system, but grew as life forms became more complex, not only in mankind, but also in primates of Chimpanzees, and apes, which microbiologists have shown. DNA for Orangutans are within about 98 percent of the same for modern humans.

What caused this change in similarity? It seems true that some developments occurred through natural sources. Green algae, that somewhat slimy substance found today on sedentary water, is undoubtedly much the same today as in the "year one".

Bacteria too, although greater in variety today, were once extremely simple in biology, as were microbes and other single-cell prokaryote organisms, such as sponges and yeast. Some development occurred simply to support first activities of life, such as circulatory systems to distribute and use the energy evolved from food intake and digestion in first animals on earth, with other systems to support systems of life.

But what prompted change in the "mechanics" of bio-life? Our culture would still, for example, be back in the first development stage before our bio-life machines if intelligence had not been applied to convert energy of steam in industrial systems, acting upon a piston, to push a rod, that converted lineal motion to circular motion, such as in the steam engine of railroad locomotives allowing the Industrial Revolution, and later the steam powered ocean ships and railroad transportation, progressing to the modern electrical, electronics, and computer chip technology of today.

Intelligence is the key. Analogy is made of today's junk yard. Items in a junk yard would remain inert but for the intelligence of some mechanical or design genius person who could combine parts to perform some planned or derived system. This motivation, like intelligence, was also the moment of inertia in providing change in biology of life in humans, and continues through additional complexities in evolution in life of animals, sea life, plants, bacteria, and all forms of complex life.

The source of intelligence that instigated change in biology of life will be discussed as "Probability". However it seems inconceivable that changes occurred strictly by accident or by advantageous mutation of change in RNA and DNA of genomes. Again, an analogy in industry, perhaps parallel in plant growth, of some worker, if acting alone and performing perhaps purely with even less thought deviation or judgment than elementary robots, would not provide intelligence to effect change. Bodies of coordinators, planners, administrators, managers, and supervisors, as well as engineers, product managers, and even marketers, will affect first and subsequent happenings to bring about intelligent and planned changes of commercial products.

Homo erectus and Homo neanderthal produced great strides toward modern humanism. Before Home erectus, predecessors of Homo neanderthal, there was relatively little that resembled todays humans.

Unlike the commonly depicted features of Neanderthal Man as a heavy-boned, hairy, robust, animal-like creature, with large skull, the cave man of 250,000 years ago, Homo neanderthal, possessed definite gradation toward humanism. Unmeasured is the workings of his mind.

Modern human anatomy, as far as studies in anthropology can reveal, appeared in the fossil record at least 1.8 million years ago with later advent of Homo sapiens. There are three sites in Africa that provide good evidence in support. Two are cave sites in South Africa. One is called Klasies River Mouth, and one is called Border Cave. The third site is called Omo Kibish. All three have skeletons that are without a doubt modern.

What does tend to be disputed is the date. The best date is around 125,000 B.C. That is not precise, but definitely after the first appearance of Homo neanderthal at about 250,000 year ago.

Klaisies River Mouth is a site showing that early modern humans lived on the west South African Atlantic seacoast during a glacial period affecting northern climes. They set up their living area between the actual cave and a large dune that would have sheltered them from the winds coming in off the ocean water. Studies showed that diet indicated efficient extraction of seafood from the area. These early Homo sapiens ate clams, mussels, and larger creatures including penguins and seals.

Two additional sites are in the Middle East, specifically in Israel, that suggests these might be as old as the prior sites mentioned. These two sites in Israel, Skuhl and Qafzeh, have been dated to around 100,000 years ago, with overlapping range of dates, leaving question whether the Israel sites are a bit younger or older than the other sites.

Homo sapiens are a relatively young species. At 65 million years ago we had an ancestral population of mammals that began to evolve in the direction of primates. In the year 2001 anthropologists discovered a form of 7 or 6 million years ago that appears to be the first hominoid, the first example on the hominoid lineage with a bipedal feature. Its location is at Lake Chad, in what is now a desert, over a thousand miles away from the Olduvai Gorge in east Africa. The Lake Chad fossil, Sahelanthropus tchadensis, is now considered the beginning of the human lineage. It had an ape-like brain size and skull shape, combined with more human-like face and teeth. It also sported a remarkably large brow-ridge.

This finding suggests several locations on Earth from which different species originated: 1) Oduvai Gorge in east Africa, origin of most species prior to Homo erectus, 2) A Caucasian race, and later Caucasian "sub-races", originating from the Georgia area of southern Russia between the Caspian and Black Seas and the Caucuses Mountains, believed to be the beginning of the Caucasian race human lineage and origin of Homo erectus populations, migrating to Mesopotamia, Eastern Africa, Delta Egypt, north Africa, India, central and northern China, Siberia and the Alaska entryway,

Indochina, Sumatra, possibly Australia, Pacific Polynesian islands of Fiji and Hawaii, to the western shore of South America and later to Central America as early fossils of 12,000 B.C. and earlier in central North America, 3) Lake Chad area of Sahelanthropus tchadensis species mentioned above, and 4) other areas of southwest Africa, who may have migrated to the southern tip of Africa, then to Antarctica, and to the tip of South America, as forerunners of fossils in South America.

At 2.4 million years ago hominoids were found in our own Homo genus, Homo habilis, and later Homo erectus, emerged or re-emerged perhaps continuous from 600,000 B.C. to 120,000 years ago, and later to Modern Man.

So much for human and pre-human anatomy. What we are really interested in is the origins of modern behavior. We want to understand what types of behavioral advances were tied in with the change in evolutionary time. It is possible that even with that first manifestation of man's soul, characteristics began to manifest in the creatures of the Homo species. These individual manifestations appeared gradually, one by one, or as clusters of innovations as the Homo species developed.

Technology would be included (e.g. stone tools, projectiles, etc.), art, trading (over some long distances), and a strong interest in reconstructing symbolic representations of images such as cave wall paintings and spiritual dolls, including spiritual thoughts of intangible events emanating in the minds of these early hominids. For quite some time anthropologists traced modern human behavior to a particular place and a particular time. That place was Western Europe (cave life) and the time was around 35,000 years ago and before. This period has been named the "behavioral revolution" or "the great leap forward" that happened about 35,000 years ago. Everything was supposed to have made a relatively sudden change, so we have to look at one part at a time of man's existence to understand how we really became human.

Evidence for pinpointing modern human behavior in time and space can start with technology when the Paleolithic tools appear in Western Europe, a time and place that Homo sapiens succeeded Homo neanderthal in the progressing lineage to Modern Man. Cave art dates to various times, some as early as 65,000 years ago. The Lascaux Cave in France dates back to 17,000 years ago, with other caves in Italy and Spain, and other sites that are still being discovered. Other art forms such as clay spiritual dolls were portable in size and nature, different than anything that transitional hominoids or Neanderthals have produced. There seemed to be a new level of symbolic representation with Homo sapiens. New excavations have been made at Katanga in the African Congo, and at Blombos Cave in South Africa, originating at 90 to 80,000 years ago.

Rock Art and painting at Apollo 11 cave (named following the 1969 space fight landing on the moon) in Namibia in Southwest Africa, dated at 27,000 years ago, 10,000 years before Homo sapiens were painting Lascaux Cave art and living in Namibia. It is thus shown that ability in art existed with behavioral changes and symbolic representation, and before future speech and language developed.

WHAT MAKES HUMANS DIFFERENT?

David Christian, Ph.D., Professor of History, San Diego University proposes in Great Courses, The Teaching Company, that what makes humans different from other species is our energy and control of that energy. He states, "What is really weird about our species is that over about 250,000 years we have slowly gotten our hands on more and more of that energy. And I am going to argue that this is a process that is hard to see in the early stages of our history, but it actively began 250,000 years ago".

"As a species we are controlling a staggering amount of energy. What is it about us, that enables us to do this?" He states further that, "human control of energy and resources increased very slowly in the Paleolithic Era before about 10,000 years ago. We can see it start to build up as agriculture appears, and then in modern human societies. More energy, more control of energy, allowed our species to multiply. More people with more energy, helps explain why human society, particularly in the past 10,000 years or so, have become more and more complex in their existence.

With this said, your author postulates that bio-electricity and energy provided the source for soul to later be activated as part of

Modern Man specified in the DNA. It has been advanced that our DNA possesses a "God Gene" that enables modern humans the many qualities commonly associated with soul. Humans are adapting. For Darwin's finches, each adaptation was genetic. Humans seem to be able to adapt in other ways, such as intellectually. No other species we know of in the history of Earth has advanced like humans.

Regarding the availability and depletion of energy, the reader is directed to the concept of an energy storage sink within glacial ice, detailed by the author in another chapter.

DID HUNTING MAKE US HUMAN?

Homo erectus were, according to archaeological studies, greater meat eaters (and hunters) than were earlier Homo habilis. Successful hunting accounted for large brain size and increased intelligence and cunning, by necessary communication, and cooperation. Hunting requires development of plans, conceptual vision, problem solving, communication, perhaps belief in a guiding force or system, knowledge of total life and relationships, and with knowledge of comparative body and anatomy of man and animals.

As males hunted the females were more active in their gathering, gardening, cooking, and familial roles. Meat was, in effect, currency to entice females to perform as males desired and may have developed societal factors that help shape how gender gets incorporated into models of human evolution. There are scenarios that depict females in control. Females became active in cooking as fire was controlled.

DO HUMAN RACES EXIST?

A group in a body with knowledge of anthropology, contend that there is only one race in humans. However medical science, for one, acknowledges that certain ethnic groups of the human population are more susceptible, such as persons with sickle cell anemia present a frequency in Negroid groups, and certainly differences in frequency of medical conditions occur between male and female.

If we hold that there is only one race in humanity, that races are merely subjective divisions, these differences may arise historically from past diets, environments, living customs, and the like. Perhaps designation of race is merely a sub-specie of Homo sapiens. We must not forget that groups are not now, and never have been, distinct from each other in the first place. This is argued for the non-existence of different races, opposed to common thought as obvious today.

The natural color and complexion of original primates in Africa was dark or black, as members of the Ape family are today as Gorilla, Chimpanzee, monkey, or Bonobo. If we hold that there was only one race in humanity, that races are merely subjective divisions, these additional physical differences may have arisen historically from local diets, environments, skin sensitivity, living customs, and the like. Perhaps designation of race is merely sub-specie of Homo sapiens or Homo eretus. We must not forget that Homo groups are not now, and never have been, radically distinct from each other of that Homo group.

EMOTIONS IN EARLY MAN

At the beginning of the time of primates, it would have been a fine line between emotions peculiar to primates, and the basic emotions of animals of the forests. Emotions of primates would be primarily oriented to satisfying basic survival needs, in food hunting and in battle of survival. Emotions involving even nurturing their offspring of primates would likely have been only minimal. More refined emotions would develop over hundreds and thousands of years, leading to complex emotions that humans experience today.

MODERN, HUMAN EMOTIONS

Are the mind and emotional genes of yourself nearly identical to the mind and emotional genes of your ancestors? This would be difficult to prove since there was no testing of the genome 50, 100 and 200 years ago. We can only study results of mind and emotion genes for comparison. Certain qualities seem to repeat from one generation to the next, or may even skip a generation. Both good and undesirable-behavior genes seem to run within families. Many times families comment that a certain child is "just like his father" (or grandfather), or a daughter is like her mother or grandmother. Repetition of physical attributes, such as facial or height, is often more immediately obvious.

Memory transfer between generations is discussed elsewhere in this book as that pertains to Monarch Butterflies in their multi generation flights from northern climes over hundreds of miles to wintering locations in the warm Sierra Madre Mountains near Mexico City, and of subsequent generations that return to spots of "ancestral" origin three and four generations later within the same year. Miraculously, this is accomplished with the minute size brain of an insect, and through multi-metamorphosis. Memory is also discussed of Salmon fish in return to their river of origin at their maturity.

Professor Robert C. Solomon, Univ. of Texas at Austin, teaches that emotions incur intelligence and strategies that are rational judgments for survival. He proposes that emotions are our doing. An emotion is not just a product of evolution, but a product of cultivation, and sometime of personal choice. For example, is an emotion concerning family, or religion, say, a learned reaction, or an ingrained quality? Also, we sometime learn of persons who have committed heinous crimes, demonstrating either no, or absence, of society-accepted emotions, or misdirected emotions. Such emotions may be acceptable in particular cultures, or even taught and encouraged in circumstances of warfare, or by crime circles. Example: acts of the Taliban, such as killing and wounding of multiple numbers of people by public bombings supposedly performed in the name of a religion. (Source: Great courses).

Dr. Daniel N. Robinson is a member of the philosophy faculty at Oxford University. He writes that consciousness is essential to human existence, but points out that a computer (a machine void of consciousness) has defeated a human being (having consciousness) in a match of chess. (Ibid).

WERE ALL WORLDLY HOMINIDS THE SAME?

It seems to be popular thought among most anthropologists that human life on earth as Homo sapiens or Modern Man started in one place and migrated elsewhere. From the Rift Valley of Africa, it was believed, pre-historic people migrated northward and eastward to what is now Abyssinia, Egypt, Mesopotamia, the Indus Valley and Indian continent, perhaps central Asia, to southwest Asia, and even southward and westward in Africa.

In contrast, not all persons seeking the trail of development of humans on earth agree with a one-source theory. Homo erectus species as Caucasians is believed to have emanated from the modern-day Georgia region of Russia, between the Black and Caspian seas at the Caucasus Mountains. They migrated to Mesopotamia as the Aryan people, and eastward to India, to central and northern China, Siberia (the Denisova hominines that may be predecessors of the North American and Aleut Indians.), to Indochina, Sumatra, possibly Australia, to South Pacific Polynesian islands, and to South America at about 20,000 or more years ago.

One can observe tribal differences in physical qualities of stature, complexion, height, robustness, facial features, biomedical features, and perhaps mental acuteness in people emanating from various locations on earth and their racial sub-species.

These characteristics can be described, but of necessity are general stereotypes of the localized populations with common variations from the described average.

1. Ethiopians—Located near the recognized birthplace of human development in Eastern Africa, a few hundred miles north of the Rift Valley that produced fossil evidence of mankind several million years old. Skin was presumably black, and height was average on Earth. Early kings and civilizations of Upper Egypt in present Abyssinia were predominately black in race.
2. Egyptian of the lower Nile Delta—Skin color and complexion are of a noticeable difference, being of a lighter or ruddy color as opposed to Ethiopians.
3. West Africans—Archaeologists have unearthed the most ancient of human presence in Chad, several hundred miles west from the Rift Valley.

4. Mesopotamia—Lighter skin color, even ruddiness in complexion. Some anthropologists hold that early migration to Mesopotamia was from the EAST, perhaps describing Aryan sources from the Indus Valley of present-day India, or from central Asia. They may also have come from the north, about of the present area of Georgia in Russia, a reported origin of Homo erectus species.
5. Indus Valley—Caucasian persons of light but ruddy complexion, with somewhat delicate skeletal features.
6. Central Asia—People of lands generally north of the Himalaya Mountains and on plains of Asia. Persons have ruddy complexions and heavier skeletal presence.
7. Oriental Chinese—Caucasian, but having appearance of mild yellowish skin coloration and oriental features.
8. Oriental Japanese—Descendants of mainland China's Korea Peninsula. Much like Oriental Chinese but noticeably different in facial features.
9. Oriental Southeast Asia—Features parallel to other Oriental groups, but sometimes smaller stature with less yellow-skin coloration (e.g. Denisovan hominid).
10. Pacific Islanders—Common black in skin coloration, and of average to heavy skeletal features of "solid Negroid race". Ancestors probably migrated from east Africa.
11. Likely Polynesian descendants from Samoa, Fiji, or Tonga Islands, an extension of races from the Southeast Asia archipelago.
12. Northern Russian—possibly descendants of Inuit people of arctic region, or first people from central Asia as North American Indians (or Denisovan hominids)
13. North American Indians—Descendants of people from west of the Bering Strait. Skin coloration having tinge of redness and unique facial features. Average skeletal features.
14. Baltic—Northwest Europe feature of light skin coloration, light color in cranial hair, modern-day Scandinavian stereotype from Sweden, Norway, Denmark, Northern Germany, and Vikings.

15. Balkan—Southeast Europe features of Caucasian with heavy skeletal features.
16. North African—Generally parallel features of Egyptian Delta civilization, ruddy complexion.
17. Central European—Stereotype of Roman, Anglo-Saxon-European and early American, with light color complexion and average height, average skeletal features.
18. Inuit or Eskimo—Like the American Indians, this civilization is believed to have migrated from west of the Bering Strait, probably the arctic Russia area, to migrate across the arctic ice and lands, but unlike the American Indians, the Inuit did not possess the reddish skin coloration of his brother in near geographic origin.
19. American Indians—Predecessors crossed the Bering Strait land formation. Probably originated from Russian Arctic area, and before from northern Scandinavian locations (e.g. Denisovan hominids).
20. Species having mental capacity to comprehend complex ideas, and to organize and separate related ideas.

It is noted that wherever in the world that primates, hominids, or hominoids were discovered archaeological study reveals that there was evidence of similarity. This commonality existed regardless of geographic differences, whether in Africa, Europe, Asia, Southeast Asia, or original natives in North and South America. It is granted that most of this similarity can be related to worldwide migrations, although it is yet undetermined, or proved, the origin of diminutive-size, brownish skin humans in South America, or northern Asia (Russia) people before migration across the Bering Strait of present day Alaska, or of black or almond-color skin natives of Polynesian south sea islands of the Pacific Ocean of, Java, Fiji, or Hawaii. How did these persons, descendants of Homo sapiens, happen to be located in islands of the Pacific Ocean, and of a different culture than otherwise observed in pre-history man? It is possible that local conditions of weather and environment caused each to acquire unique physical and perhaps mental traits. Skin coloration could result from intense (or lesser) exposure from the sun and cosmic rays over many generations, either darkening or lightening the skin and hair color. Susceptibility to various medical conditions, unique to different civilizations, may account for certain weaknesses or abnormalities. But each group is 99 percent identical to each other, with genome action controlling their unique features. In some, variations are so radical, such as Negroid from Caucasian, and variations in between, that it is very difficult to explain how the differences would come about, even by mutations that may have evolved by some force or power.

There appears to be some differences with variation of physical, neural, intelligence, or medical propensities. The question and debate among anthropologists continues, with both groups presenting valid arguments of a single source, "out of Africa" theory, or a "multiple source" theory evidenced here (e.g. Caucasian people from the Black and Caspian Seas and Caucuses Mountains of the present Russian Georgia state).

Present day descendants of different areas, because of multiple generations involving the change, each have characteristics unique to themselves.

MATHEMATICAL SUPPORT FOR INTELLIGENT DESIGN AND PLANNED EVOLUTION.

We can call upon the science of mathematics to prove the validity of Intelligent Design and Planned Evolution. Evolution, if planned, allows near perfect change, or 100 percent probability (designated in probability calculation as 1.0) of development steps in the seemingly never ending series of steps in evolution actually occurring successfully as if planned. In a series of successive steps, deriving probability can be calculated. The probability that a series of events in evolution would happen favorably or unfavorably to form a chain of successive events leading to Modern Man, relying on pure chance, is unbelievably low.

The evolutionary steps in development of the first animal, perhaps the worm, is used as an example. If the reader is unfamiliar in calculating probability, I will lead you through the various points in reasoning and calculating probability of success when in a series of related events. You will see that the chance of success becomes lower and lower when chance is based on random choice, but remains high in success when the events are planned.

Choice #1

The probability of the first event that a cell would remain flat, or would curl up, is 0.5, a 50 percent chance that the cell would curl up to form a tube, and 50 percent change that it would not. The reciprocal of the chance of failure is the probability of success.

Choice #2

A next probability that food particles would enter the tube and provide energy to the animal-tube is 0.5 again, a 50 percent chance Yes and 50 percent chance No. The probability that both the two consecutive events would turn out favorably is 0.25 (0.5 X 0.5 = 0.25), or 1 chance in 4 that the event would happen favorably.

Choice #3

Probability that the tube would successfully process food as energy would, again be 0.5 percent, Yes or No. The probability that three events in a row would happen favorably is 0.125 (0.5 X 0.5 X 0.5 = 0.125, leaving 1 chance in 8 (1.00 divided by 0.125= 8) that the three events in the series would happen favorably.

Choice #4

Each development event of evolution would be 50 percent chance if happening or not happening. A fourth event in a series would have probability of 0.0625 (0.125 X 0.5 =0.0625) or $0.5^4 = 0.0625$, or 1 chance in 16 (1 divided by 0.0625) that the four events in a row would happen favorably.

Extension

Each development event would subsequently be 0.5, or 50 percent chance of happening advantageously or not. A 5th event in a series would have probability of 0.03125 (0.0625 X 0.5 = 0.03125), (or 0.5^5 = 0.03125), or 1 chance in 32 times (1.0 divided by 0.03125 = 32), and greater to 0.015625 (1 chance in 64), and so forth, higher each time with lower probability that an evolutionary event required a Yes or No choice of success in evolution.

It is easy to see that after only a few events in evolution, the probability of an event happening successfully, based on pure chance, becomes a million and more to one.

On the other hand, if a series of events are PLANNED, the probability of success each time retains at 1.0, or near sureness, that the event will be successful. The next event, IF PLANNED, would also have a probability of 1.0 (near sureness), maintaining a 1.0 probability, and so forth through the series of planned events.

It will be seen that the only way to assure success in a series of events (or near it depending on outside factors) is TO PLAN THE SUCCESS OF EACH EVENT. This is what Planned Evolution is all about. It is extremely disingenuous to hold that evolution happened successfully in a series of events based strictly on random chance in nature.

Perhaps this natural choice in evolution is what happened in the Age of Dinosaurs, and deep sea life, resulting in much weirdness in those populations. Weirdness in dinosaur population was wiped out at the 65 million years event of extinction, but weirdness in deep sea life has continued after destruction of surface life. Basic plants, sea life, and small mammals would have continued unaffected by the disaster.

In our mind's eye we can visualize the number of individual steps in a series occurring during the evolutionary process, with each step decreasing the probability. The end result of progressive steps is unbelievably low, with comparable low probability of success still continuing today. Evolution was not left to random chance where probability would remain at 0.5, a 50-50 chance but decreased in chance with each step of the series in evolution. The approaching end result of progressing steps, still continuing today, is unbelievably low.

The only way to successively operate a probability series such as this is to PLAN the steps so that each step has a probability of 1.0 (certainty). This is mathematical proof that evolution was planned. Planned, intelligent evolution was the way of life, and the planning was directed by some force. The force is assumed to be that of a universal God that has the intelligence and power to influence success at each evolutionary decisive point.

(The essence of this argument has been expressed in a book, "Chance and Necessity", written in 1970 by Jacques Monod, a French biologist and Nobel Prize winner.)

In the Charles Darwin observations this principal of probability versus natural mutation, changed. It must be reasoned that the finches' change in formation of their beaks were due to high probability of immediate need for change at the very next generation, else to hinder their continued survival, rather than by time-consuming changes.

With first one-cell animals there was eventual need for change, as opposed to no immediate motivational need for a two-dimensional cell to curl up into a round formation of a worm. Change in beak would likely have been a planned change in genome, based upon immediate need. There was definite design developed, without any haphazard, experimental, or multi-choice changes. This would suggest that both Natural selection and Planned Evolution were at work, or that tools of Natural Selection are available to implement Planned Evolution.

INTELLIGENT EVOLUTION

There seems to be two diametrically opposed theories regarding intelligent planning in evolution. The first is that there was no overriding intelligence in evolution, that animals merely existed and randomly changed by environmental needs and mutations, with changes in amoeba, yeast, bacteria, and whatever were first life on earth. They simply performed what was necessary in order to continue life.

For the amoeba and yeast there was really no action required. The existence of yeast continued from the Sun whether there was food energy or not. With amoeba the energy required was totally from our sun and components of the prokaryote cells were maintained by the sun's rays and energy therein, even to our modern day.

Bacteria had some incentive to eat whatever was available, to transfer its food intake to energy, and to expel the residue. The food for bacteria consisted of decaying matter, and perhaps for neighboring amoeba and yeast.

The very wonderment of the developed human body, brain, and mind include phenomenon that might have been determined over the ages by chance, but has too much intelligent organization than to have been developed by a series of haphazard mutations, and were difficult even by subsequent trial and error of mutations. Complexities giving multiple probabilities were:

1. The initial energy conversion system of elementary animals with a food intake organ, alimentary canal to convert food to energy, and an excrement system.
2. A blood or fluid circulation system and pumping heart to distribute energy to various organs.
3. An oxygen conversion system to take elements from water or air by lungs or other body devices.
4. A kidney functioning system to purify the body and expel contaminants.
5. A growth system to allow increase in size and abilities.
6. A skin system to contain and protect internal parts of the body.
7. A skeletal system that allows structural flexibility.
8. A muscular system that provides power for movement.
9. Pancreatic and other internal body systems that allow more complex life systems.
10. A brain and nervous system, a central control mechanism that coordinates living processes.
11. A non-physical mind system that allows more flexibility in life, greater than simple subsistence.

12. Semblance of an intangible, and so far unexplained, non-verbal communication system, characterized by ESP, mind reading, thought projection, recognition of higher psychic belief, and existence of faith
13. Realization of existence of soul qualities as an element that may transcend secular life to a future existence.
14. A bipedal and erect walking and running system and ability, although some anthropologists hold that this action was evolved to allow sight over tall grasses for hunting and protection.
15. Super hearing and eyesight organs and systems, unique in each animal world.
16. Other unique complexities known to the medical and psychology worlds.

It is illogical to believe that these complexities happened in evolution without planning of some force, and not of solely natural choice by demands of the environment.

First prokaryote cells, flat as they were, supposedly curled themselves to a tubular form, essentially a flat worm, taking in food at one end of their tube, absorbing whatever energy they found into their own energy, and expelling the waste material from the reverse end. It is granted that the two-dimensional cell may have somehow naturally formed into a tube-like structure.

What happened after the tube formed was miraculous, and I believe was directed by some power outside that of the cell. The worm developed what today is called an alimentary canal to extract energy from food intake. Next, a pulsating system was developed to better extract expended food, and the forerunner of a heart, with a fluid circulating system to follow that would aid in the pulsating. A kidney system was needed to aid in energy extraction, and fluids were pumped through the worm. A nerve system was developed to control operations of the energy-gaining process, and a means of locomotion or movement was developed, with other organs added as needed. Each development required intelligence.

Prehistoric eyes first detected only a glow of light. A hearing system received vibrations indicating presence of food or danger. Movement was devised through undulating expansion and contractions allowing change in position. Other organs and appendages were added over millions of years that formed first sea plants, and later fishes and a variety of sea life, both soft plants and calcified shell sea animals.

These growth events did not occur by themselves. There was no mathematical way that the internal system and organs were added and made operational as a natural event. There was some force other than natural means that caused the advancing complexity of first and early life. It has been said that "God is the greatest scientist of all".

To illustrate, if an unlimited number of unknowing monkeys were set forth in a junkyard, it would be improbable that they could, even eventually, devise and assemble a fixture to accomplish a desired outcome. Likewise, it would be impossible for a body system, even after numerous trials and failures, to produce by chaos a functioning organ within itself. An intelligent guiding force would be the only explanation of how this operation could be accomplished.

Other more complex plants, fishes, sea life, reptiles, amphibians, and eventually land mammals evolved, the latter possessing legs and feet, lungs to exist in an oxygen atmosphere, protective covering such as seen on crocodiles, or fur covering of animals to provide protection and warmth. From this stage more complex animals were devised, each occupying a certain niche in life's existence and variations, as subspecies and new species developed.

Again, these added creatures were instigated, not merely to fill some unclaimed void in nature's environment, but to produce increasingly complex animals that would lead to more complex animals, primates, and later hominids and hominoids, and later to result in the Homo series of man-like creatures of Homo ergaster (which had some evolutionary deficiencies). This species was followed quickly by Homo erectus having 600,000 years of successive development, later Homo neanderthal, and at about 120,000 years ago, Homo sapiens, who became refined at about 30,000 to 25,000 years ago to Modern Man, with anatomy structure similar to today's humans of hairless (or nearly so) body, ability to think and reason as no previous specie had been able to do before, to further develop his mind, emotions, and qualities of soul, and which had evolved, to be a "full member" of mankind in brain, mind, and body. Among the existences of mind and soul, man realized an existence of a life-guiding force, attempting to describe that force in paintings on walls of caves and in basic art forms.

THE GODS

Man demanded that the guiding force be tangible and devised a system and hierarchy of gods, some of human form, who controlled all conceivable aspects of life and environment. Several god-persons evolved, each to explain varieties of philosophies for life's existence, most recognizing an existence after one's secular death, as in an Egyptian "Underworld", or Christian period belief in Heaven, or an "Afterlife", with possible reincarnation to a later life. Some ancient kings and leaders devised a dogma that they themselves were gods.

Upon the existence and death of Jesus Christ, a new philosophy of Christianity developed, which has continued for over 2,000 year: In Christianity, one God is recognized who is in control of all events to occur on earth and in Afterlife. Dogma has been developed over those 2,000 years regarding proper worship and adoration of God, His community, and His abilities, including the philosophical mind of man.

Science has progressed on a path that both supports Christianity, and also clarifies various aspects in ancient and Biblical history, the genome, and a realism of multi-dimensional communication, to be hopefully developed in future years.

With a popular belief and conviction in a Supreme God, that was believed to control all aspects of life on earth, it follows that evolution has not only happened, but that each step in evolution was planned and executed by a Supreme Force, placed in action by natural means, many within the description by science, and some by means today that are recognized as supernatural (or unexplained), or of higher space/time dimensions to be explored and explained in future years.

Charles Darwin's "Natural Selection" is not opposed to intelligent Design. What Darwin calls Natural Selection was actually an intelligent "engineered change" in biology, caused by a change in the DNA and RNA of birds and mammals that he studied. Specific engineered changes in DNA and RNA sent messages through the natural genome system, sending enzymes and proteins to biologic organs of the studied animals and birds. For example, certain changes in DNA and RNA caused change in beaks of certain birds, caused by needs of some specific birds for cracking of hard plant seeds. The same process would have pertained to elongated beaks of Hummingbirds' tubular beaks for drawing of nectar from flowers, and by other species or varieties having peculiar organs and needs.

SUCCESSION OF INCARNATIONS

It could be argued that spirituality and reincarnation seem, somehow, to follow family bloodlines. If a soul is to be reincarnated, any peculiarities or coincidences appear within a family, as opposed to reincarnation into a far-away person such as in another or foreign country. It is difficult to prove, as spirituality itself is difficult to prove, but emotions and characteristics of a person, say, in a modern generation, often relate back to those of a previous generation within the family bloodlines.

A person of modern generation with high abilities in, say, mechanical aptitudes, understanding of people, or extreme spirituality, as example, may be able to trace back into ancestry of previous generations for like aptitudes.

> It is often said that a child is "just like" his, or her, father, mother, grandfather, or grandmother in certain features of emotions or talents. For example, I never knew my grandfather. He was born about 1840 and died in 1929, 2 years after I was born. He "grew up" as a farmer's son. At age 21 he enlisted as a Private in the Union army of the American Civil War, and was discharged in June of 1865.

The military record shows that his army unit participated in battles at Antietam, Gettysburg, Kennesaw Mountain, Atlanta, Savannah, and other battles. To come through 3 ½ years of the Civil War, still alive and without battle injury, I believe indicates he was a unique person. I believe he must have been resourceful, devoted to a cause, perhaps being individually spiritual, emotional in a general sense, and driven to complete a project. His father, my great grandfather, born probably about 1820 in the Boston, Massachusetts area, of parents working in the business of the early shipping industry, moved to Burlington Vermont, to New York state, and again to Michigan. He was a farmer, and raised sheep in Vermont, New York state and in Michigan, before moving to northern Michigan where he opened a meat market. I believe my grandfather was much like my father, a self-employed businessman, and I believe I, also self-employed and self-driven, share much of the pioneer traits that each of those ancestors had. Perhaps a person's soul is the reincarnated soul of, in this case, my great grandfather, with the genome acting in association with the soul.

A spirit or soul, of some former person living outside of a person's own bloodlines will probably not be reincarnated to a person outside of the bloodline. A soul of, say, Adolf Hitler, Genghis Kahn, a south African tribal leader, or even Mother Theresa, are all outside of, say, my own bloodline for example, and will not be forerunner of my own soul. Any forerunner of my own soul must be a blood ancestor. The blood ancestor would have died, many years, decades, or centuries before my own birth, assuming that incarnation of soul took place at time of my birth.

It is not wrong to question existing dogma or doctrine. This process has existed virtually from the very start of technology, generating new and improved understanding of how and why things exist. A series of guttural sounds and hand motions composed first communication. Connection of those sounds, and in form of hieroglyphics and symbols, formed our first oral and written sentences of communication. Communication allowed different ideas to be discussed, and changes for the better resulted. A system of counting resulted in arithmetic, with extension resulting in subtraction, division, and multiplication in mathematics. Mathematics grew in complexity to algebra, geometry, trigonometry, squares, and square roots, with untold advances in number theory, all because question was directed to status quo, with new ideas developing.

This questioning extended to all aspects of knowledge. It is only through questioning of accepted concepts in knowledge that new knowledge can transpire. Philosophers of today can, and need, to be questioned in order to improve on existing concepts.

Through this procedure, new ideas and concepts have risen in the development of not only philosophy, but in all aspects of knowledge. Questioning in religion, spiritualism, and philosophy is no exception, nor in politics, science, art, education, and all fields of study.

A written medical article advances a thought, "Medical myths, lies, and half-truths: What we think we know may be hurting us". This thought applies also in philosophy, spiritualism, antimatter, alternative dimensions, and other fields of knowledge.

An existing species can develop its own timely improvements during its species lifetime. Our modern-day species has "lifted itself up by its own bootstraps" by self-improvement in all fields of our culture, in science, mathematics, art, language, writing, knowledge expansion, computers, and many other aspects of culture.

TO AN AGRARIAN CULTURE

Our biologic, anatomy, mental, psychological, and spiritual progression had advanced greatly when man made transition from the Neolithic Age in a cultural breakthrough to the agrarian mode of civilization.

Life had advanced from single-cell amoeba to multi-cell Eukaryotes, having developed a cell wall, enclosing a nucleus of protons, RNA, DNA, and one or more electrons orbiting an atom's center. The cells had joined to form molecules, then formed life-allowing organs, other life-enhancing organs and appendages like arms, leg, or fins, with various coverings of fur, hair or shell. Two goals in life were 1) continued existence, and 2) reproduction of its species.

Form of man developed from primate animals to hominids having man-resembling skeletons, appendages, protective skin covering, and five senses of sight, hearing, smell, taste, and touch.

From this hominid world a branch of life developed as hominoid, specifically of the ape family of Gorilla, Chimpanzee, Monkey, and Bonobo.

Further change produced a variety of species that had some anatomy features of man, but the miraculous breakthrough was emergence of Homo, or man-like species, which developed, with some changes occurring purely from environmental needs, but more importantly there appeared to be a direction in evolution, as opposed to chaotic development, even if useless-developed features disappeared with succeeding generations. Spiritual development was an inherent part of this progress.

ANCIENT LAKE TURKANA (KENYA) PREHISTORY:

Australopithicus anamensis (4.4 to 3.9 million years ago B.C.)— Ape-like in some regards but walked upright. Earliest bipedal hominid. Possibly our oldest known ancestor.

Australopithicus afarensis (3.5 to 2.5 million years ago B.C.)—"Lucy" is 3.2 million year old fossil discovered near Afar, Ethiopia.

Homo habilis (2.6 to 2.5 million years ago B.C.)—Successor to A. robustus and A. boisei. H. habilis used stone tools. Some anthropologists believe H. habilis evolved into H. erectus and would thus be a direct ancestor of modern man. Possible "missing link" between primates and man.

Australopithicus aethiopicus (2.5 million years ago B.C.)—Evidence that evolution leads not only to divergence but possibly also to convergence.

Australopithicus boisei (2.1 to 1.2 million years ago B.C.)—Believed to not be our ancestor and species eventually died out.

Homo erectus (1.8 to 1.6 million years ago B.C.)—Anthropologists agree that Homo erectus was definitely our ancestor. First hominid to emigrate from Africa, spreading all the way to China and Indonesia, with a group deviating to Russian Georgia and Caucuses Mountains/Black and Caspian Sea area. Branches leading to H. neanderthal and H. sapiens, or these may be sub-species of H. erectus. Ancient H. erectus was a tool maker, which provided the ability to obtain meat from relatively large grazing animals. Demonstrated basic emotions, thoughts, analysis, and other complex reasoning, with belief in spirituality and Afterlife. May have possessed incarnated soul. Less complex in mind and soul than later Homo erectus of 600,000 B.C. The hominoids increased their mental abilities of reasoning, analysis, conceptualizing, and belief of an external controlling spirit to which he should recognize and honor.

Modern Homo erectus, at about 600,000 years ago, was a species that advanced its abilities and migrated to various and far away locations of the ancient world. His RNA and DNA produced greater complexity in the mentality, analytic, spiritual and technological ability, and with advancements in developing tools, some of purposefully broken cutting stone, and later from metals found in certain locations which could be formed from fire-heat and later mixing or smelting of copper, tin, and other elements to form bronze, a harder and more malleable metal, useful for weapons in primitive warfare and in fashioning more complex tools and instruments to aid in his living.

There had been a long span of time after the Phanerozoic, the eon of multi-celled organisms, and during the Pliocene leading to man-like hominids, later hominoids at about 6 million years ago, and when ancient Homo erectus migrated at about 600 thousand years B.C. a humanistic culture developed.

During the Pleistocene early man had lived in bands for mutual protection and cooperation in hunting activities. Control of fire and basic cooking processes developed. Forms of communication, rudimentary at best, developed to exchange ideas. Ancient man used natural caves as shelter against bad weather, cold, and wild animals. It was a hunter-gatherer culture.

At about 12,000 to 8,000 years ago, an idea had erupted that caused change from hunter-gatherer culture to an Agrarian culture, of using land to plant, cultivate, and grow food in a controlled system, and to herd groups of animals to aid efforts of man, to group bands of sheep together to harvest their wool for clothes and material and to provide food on a regular basis, all in an agricultural culture.

The agrarian culture led to radical change in man's way of living. In having a more social civilization, various groups of men became specialized in different trades, to improve their own lives, and to gain by trading products of their making for their different products. Trade avenues, by land and by sea. developed with men of different settlements. A form of wealth developed based on ownership of number of animals and by amount of land under an individual's control.

Kingdoms developed of land ownership and services provided by kings to groups within the kingdom, for mutual protection and ease of living. Rivalries, jealousies, and want of another king's possessions developed, with resulting warfare between kingdoms.

This social culture developed, as shown by archaeological evidence, in the Egyptian Old Kingdom of about 8,000 B.C., and in Mesopotamia at about the same time. Agrarian culture developed at various times in different geographic locations.

With eventual agrarian refinement, the first substantial kingdom in Mesopotamia was the Akkad kingdom, led by Sargon the Great (2270 to 2215 B.C.). His empire extended from Elam, through present day Iraq and Syria, to the Mediterranean Sea. His was first to form a multiethnic, centrally ruled empire, his dynasty controlling Mesopotamia for 1½ centuries and was conqueror of Sumerian city states of the 22nd century. Successive cultures, kingdoms, and kings followed. Pyramidal structures had been erected as early as 8,000 B.C., i.e. Step Pyramids of Aquaba on the Nile River during the Old Kingdom in Egypt were burial places for honored kings and leaders

Structures, monuments, and places to worship the multitude of gods that were believed to control nearly all aspects of living, and that would control a person's fate in different aspects of life. Kingdoms of kings and warriors fought to gain dominance of land, people, religion, or ideas.

Cultures and kingdoms developed in various locations of Greece, Macedonia, Persia, the Italian Peninsula, Crete, Carthage, Sicily, Turkey, and various cultures of Mesopotamia, including Canaan, Israel, Judea, Phoenicia, and other locations, as well as on the Greece and Turkey peninsulas. Each developed their own culture and beliefs, with desire to expand their geographic areas, culture, or religion that would meet their changing needs in a growing kingdom.

Many notable kings came into power over the various cultures. In Macedonia Cyrus I gained power, followed by his grandson, Alexander the Great, who warred and developed an extensive kingdom of Persia, lasting to his death from battle, and consequent division of the empire land holdings among four of his surviving military generals.

Further refinements in culture, science, mathematics, and religions developed. Hebrews first recorded in the Old Testament their beliefs and history starting about 600 B.C., the forerunner of Christianity years later, with recordings of the New Testament at the time of Christ and after. The abilities of Homo erectus species merged, with "design improvements", to a new species, (or perhaps sub-species) of Homo neanderthal and Homo sapiens, and a later contemporary development, either within or separately, (anthropologist are not fully agreed on this evolutionary development of 600,000 years age, evolving to the Homo sapiens species and to Modern Man, starting about 120,000 years ago, later questionably being the sole specie of man on Earth from about 28,000 years ago with extinction of Homo neanderthal. The recent finding of evidence for consideration of Denisovan hominids added considerable complexity to the anthropology trail.

Homo sapiens man was still Paleolithic Man however, living in caves, participating in familial relationships, recognizing an external spirit having powers greater than himself, but physically untouchable. He had ability to reason, to plan, to analyze, to hunt, to pick berries and plants, to differentiate between nature's animals for hunting and in fishing, and to develop a social existence in family and community.

The Neolithic Period could be generally defined, in part, as the time on earth in development of Homo sapiens prior to the agricultural mode of living. We commonly picture this as days of the caveman, void of housing, and subsisting in families by day-to-day hunting for food and cooking by open fire. Clothes consisted of animal skins. In some locations art drawings pictured on rock cave walls had existed from their ancient time. Tools of axes and tools for cutting were fashioned from stone, with some fashioned from chance findings of iron or copper.

Such was the mode of living in Mesopotamia until about year 10,000 B.C. At that approximate time the mode of living gradually changed to an agrarian culture whereby seeds were planted to provide a planned source of food. Meat from domesticated animals was preserved by rubbing with salt found in dried-up salt water sea beds.

Starting about 12,000 to 10,000 B.C. (the timing varies with different geographic areas), man discovered another, easier, more sociologic, and perhaps a more sensible way of living from his previous hunting, fishing, and live-by-the-moment way of existence. Nature's plants that he had encountered in the wild could be replanted in a localized area, and seeds of plants could produce growing plants near where he lived. Animals previously encountered in the wild, some in the canine family like wolves, coyotes, and dogs, and milk-producing animals, could be tamed and controlled to a domestic state.

With this new stability in his life, man could abandon his cave dwelling and construct a form of house of mud or timber to accommodate himself and his family in a permanent, or at least a semi-permanent, location tending his new form of agriculture and a form of controlled animal husbandry.

Change from a culture of hunting and gathering of daily food, to an agrarian culture of land development and utilization spreading to different geographic areas. Over many years, guided by economic decisions, greater time allocation was given to agrarian activities.

A society developed with association of people, both of near and far distances. Production of food and technical goods in excess of his immediate needs allowed trading to acquire additional and different goods to enrich his life.

Success would often breed further success, with individual wealth and power developing. Communities developed with impressive wealth and sophistication of the time, important trading centers, and major players developed in the early game of this new civilization. Communities developed with various persons making and assembling tools and features that others of near and far communities would desire, the rudiments of trade. Housing provided permanence to their location and building of communities. Some men became land holders while others were servants upon the land, beholden to his betters.

Communities of people would find it advantageous to band together for purposes of trade, exchanging skills in various specialties, and protection from outsiders. Communities also banded together into kingdoms with designated leaders. Many kings and kingdoms arose and many were warring kingdoms, prompting construction of walled cities.

Other contemporary features of the Agrarian period included Temples, known as ziggurats that contrasted with Egyptian ziggurats of about 8,000 B.C. built to contain the deceased remains of rulers. Populations continually feared before dreaded invaders. Priests made sacrifices to gods believed to rule over and protect the community.

In Babylon Hanging Gardens were built. Social classes consisted of emperors and staff, merchants, and farmers or serfs. By the 5th century there were cities and urban classes, invention of writing, and other technologies, with an urban style of living, becoming increasingly sophisticated. Cultures expanded and cities like Uruk and Ur grew in population that required development of new ways to sustain themselves.

Various geographic and political cultures developed, each rising and subsequently falling, often by defeat by a neighboring military force. It seemed that urban civilization and organized warfare emerged hand-in-hand.

The following is an account of the Hadza culture in Africa, a living history of pre-agriculture existence, (from National Geographic, December 2009).

> A group living today, the Hadza, of about 1,000 people, live in their traditional homeland, a broad plain encompassing the shallow, salty Lake Eyasi, and sheltered by the ramparts of the Great Rift Valley in Tanzania, Africa. Some Hadza have moved close to villages and taken jobs as farmhands or tour guides.
>
> But approximately one-quarter of all Hadza remain today as true hunter-gatherers. They have no crops, no livestock, no permanent shelters. They live just south of the same section of the valley in which some of the oldest fossil evidence of early humans are found. Genetic testing indicates that they may represent one of the primary roots of the human family tree—perhaps more than 100,000 years old.
>
> The Hadza appear to offer a glimpse of what life may have been like before the birth of agriculture 10,000 or so years ago. Time has not stood still for them, but they have maintained their foraging lifestyle in spite of long exposure to surrounding agriculturalist groups, and it is probable their lives have been changed very little over the ages.

For 99 percent of the time, since the genus Homo arose nearly 2 million years ago, all persons lived as hunter-gatherers. Then, once plants and animals were domesticated, the discovery sparked a complete reorganization of the world. Food production marched in lockstep with greater population densities, which allowed farm-based societies to displace or destroy hunter-gatherer groups. Villages were formed, then cities, then nations, and in a relatively brief period, the hunter-gatherer lifestyle was all but extinguished. Today only a handful of scattered peoples, some in the Amazon, the Arctic, Papua New Guinea, and Africa, maintain a primarily hunter-gather existence.

Agriculture's sudden rise, however, came with a price. It introduced infectious disease, epidemics, social stratification, intermittent famines, and large scale wars. Today, traditional Hadza live almost entirely free of possessions. The things they own, perhaps a cooking pot, a water container, and axe can be wrapped in a blanket and carried over the shoulder. Hadza women gather berries, tree fruit, and dig edible tubers. Men collect honey and hunt. Nighttime baboon-stalking is a group affair. Typically, hunting is a solo pursuit. They will eat almost anything they can kill. They dine on warthog, bush pigs, and baboon. The poison that men smear on their arrowheads, made of boiled sap of the desert rose, is powerful enough to bring down a giraffe. If hunters come across a dead elephant, they will crawl inside and cut out meat, organs, and fat, and cook this over a fire. Hadza camps are loose affiliations of relatives, in-laws, and friends. Each camp has a few core members that come and go as they please. The Hadza recognize no official leaders.

Source: National Geographic, December 2009.

There were many city-states in Mesopotamia and the Levant (ancient Israel) as different areas transitioned from a Neolithic to an Agrarian culture. An agricultural economy of animal herding, grain and crop planting with harvesting, and trade of goods and food developed with neighboring city-states of near and far. Groups of city-states joined together for mutual benefit and for protection against raids by other marauding city states or bands, and local kingdoms that arose.

A series of empires developed from local kingdoms. A first empire was that of Akkad (2300-2080 B.C.) with Sargon I (or Sargon the Great) as king. Sargon I set a pattern of empire government organization, military strength, an agricultural and urban society, and trading.

There were many qualities of ancient Mesopotamia. Contemporary features of the agrarian change were religious temples, known as ziggurats (that contrasted with ziggurats of Egypt of about 8,000 B.C., built to contain deceased rulers). Populations constantly feared destruction before a dreaded invader. Priests made sacrifices to gods believed to rule over and protect the city. Hanging gardens were built in Babylon. From Neolithic times there were classes of emperors, merchants, farmers, and servants available for warfare to attack neighbors or to protect the kingdom. Vast empires existed of Assyria and Persia, with cities and urban areas of the 5th century B.C. expanded such as Uruk and Ur. There was invention, writing, and other technologies. An urban style of living became increasingly sophisticated, and increasing populations was required to develop new ways to sustain themselves. Royal graves contained silver, gold, human sacrifices, and crafted artifacts. In Egypt "Amarna letters" (Amarna was an Egyptian city) between Akhenaten and contemporary rulers in the Near East, written during peaceful times with cooperation pledged among neighboring rulers, with niceties of gifts, royal marriages, dispatches of royal physicians if needed. "Kanesh tablets" were outposts for Assyrian trade in Turkey early in the 2nd century.

There were long shadows of the ancient Greek world. King Leonidas and 300 Spartans defended a key pass at Thermopylae against an overwhelming Persian invasion. Highest minds of ancient Greece, Socrates, Plato, and Aristotle, evoked ideas about morality, justice, government, and ethics. Political battles resulted in the Persian Wars, Peloponnesian Wars, civil wars between nobles and lower classes, and first stirrings of democracy, with complex structures and trial by jury. Draconic rules called for death to all who break the law.

Wealthy landholders, land workers, and merchants associated with agriculture activities under sponsorship of the king were obliged to form armies as needed by the king, and also for aggression led by their own king to expand the empire, a duty owed to the king in exchange for the king's protection and leadership. Other empires and local kingdoms developed to extend or reconstitute the dynastic and familial empires. Sargon II, Sargon III, and other kings continued or restored dynasties in kingdoms and empires.

Numerous kings of Sumer, detailed in literature of pre-Akkad rulers, and based on archaeological data, are listed in exaggerated time periods of "Sars" (based on periods of 3600 "years") and ranging upwards to 28,000 "years"). It is believed that originally a year was essentially one day, from perhaps sun-up to sun-up. Later years were defined as the period of the moon phase, from full moon to full moon. Still later, years were defined as the period from one annual planting time to the next annual planting time.

Written records relate to various pre-dynasty kingdoms of early Bronze Age (which ended in Mesopotamia about 2900 B.C), and which included the legend of King Gilgamesh and a great flood reported occurring about 2900 B.C. Periods also covered the several kingdoms of Kish (after 2900 B.C.), Uruk (2700-2600 B.C.), Ur (2600 B.C.), Awan (2600 B.C.), Lagash (2500-2271 B.C.), Hamazi (2500 B.C.), Second Dynasty of Uruk 2500 B.C.), Second Dynasty of Ur (2500 B.C.), Adab (2500 B.C.), Mari (2500 B.C.), third Dynasty of Kish (2500 B.C.), Akshak (2500-2400 B.C.), Fourth Dynasty of Kish (2400-2300 B.C.), Third Dynasty of Uruk (2296-2271 B.C.), to the Dynasty of Akkad 2270-2083 B.C.) which included Sargon I and the First Empire, Fourth Kingdom of Uruk (2091-2061 B.C.), Second Kingdom of Lagash (2147-2050 B.C.), and the Fifth Kingdom of Uruk (2055-2048 B.C.). Other kingdoms arose in Greece, Italy, Macedonia, Egypt, Mycenae, Cyprus, and other geographic areas where organized civilizations developed and were extended.

Our story progressed towards the period leading to the time of Christ and the end of our exploration from pre-cosmic and pre-man existence, through the various species of hominid, hominoid, and man, cultural and technology developments, to where time dating turns from Before Christ to Anno Domini.

We have discussed many new, and some original with the author, ideas, some not in agreement with those expressed by experts in their fields, but which seemed quite logical, reasonable, rational, and believable, and stood well in evidence for those presentations to balance against nearly unbelievable events presented heretofore. It is suggested that the reader consider these thoughts with an open mind to gain a greater truth to the questions posed.

FROM AN AGRARIAN CULTURE
PART I

FROM NEOLITHIC TO A CULTURE OF KINGDOMS

The Agrarian revolution in mankind marked a radical change of man and his culture on the face of Earth. There was, for the first time, in the many glacial cycles that Earth had experienced, a form of man and culture unlike anything before.

It seemed that all the past species of cellular development, four-footed creatures, upright walking mammals of the Great ape family, and various forms of pre-man, had now converged to persons not radically unlike we appear, exist, and think like modern man today.

Climate had changed, with the ice glaciers withdrawing from the central and temperate latitudes of the earth, exposing and allowing culture of the land for new purposes. The cold-loving Mastodon, Wooly Mammoth, Sabre Tooth Tiger, and other mammals from the Neolithic age were hunted nearly to extinction. Grains could be grown and harvested in a yearly growth cycle, and animals that previously ran wild could be domesticated to provide milk, meat, and hides for a new source of unending nourishment and clothing, to help man in his daily pursuit of a new form of living, and provide protein for growth of brain and body. Man would emerge from Neolithic cave-living to construct crude and simple houses for himself and his family unit in a community of people, traveling daily outward to personal or community plats of land devoted to agriculture or pasturage, and reaping the autumn harvest and summer berries to nourish his family during the cold winter, and trading some excesses in food products for other things or services he would need. He might also gain a form of money from his efforts to exchange in gain for a better way of life, or even to produce a form of relative personal wealth.

Certain men grew to be leaders in their community, and some developed in a culture where they would become a king, exercising powers over lesser men, extending their land holdings and wealth, sometimes by force of an army of persons devoted or bound to that king, against some neighboring king and persons servile to him. Kings were made and honored in ancient cultures: 1) There was ancient observation that all things have their superior types such as in trees, animals, and food locations, 2) Self-appointed kings likened themselves to gods or that they either were gods or had a special communication ability with gods. This theory of kingship developed with god relationships through the Roman Empire, and even today with political theory in certain countries maintaining a kingship.

Extended kingdoms of increased land developed, and greater wealth was created. Government, or a basic form of group organization developed, and there was development of technology in all facets of living that had never existed before—in agriculture and animal husbandry methods, in weapons and systems to persuade animals and other men, women, and children to do to his advantage, and the offering of new forms of culture requiring a form of counting and language with which to communicate and compare, and development from ideas in new technology, such as the plow to turn up soil and grassy land, all greatly expanding a community culture.

Man's mind and reasoning produced a belief that either unseen or self-appointed leaders, or even personal gods, existed and were influential for his and his family's welfare in all facets of living. He imagined a hierarchy of gods and produced systems that gods would be worshipped and pleased, and man would benefit to his advantage in exchange for making various forms of sacrifices, some of humans, their children, or animals, to the gods—a first form of formal religion, religious dogma, and practices.

A society had developed in which man, and groups of people, not radically different in physical appearance and existence from today, would associate and work together for their common advantage—and this was only 12,000 and 8,000 years ago, a relatively short time in the evolution of man. This is a time, much shorter in years and in evolved bodily changes, than was experienced by Homo erectus, Homo Neanderthal, or even Homo sapiens, and their preceding primates, that had previously lived on Earth.

A procession of events and notable persons is detailed in the following Timeline.

TIMELINE OF HISTORY

TIME	EVENT
One million years ago to before 13,000 B.C.	Pleistocene Age before the Neolithic Period. This was a period over thousands of years causing periodic swings of frigidness and extreme heat (causing extremes of glaciers and arid deserts). Homo erectus had existed since 1.8 million years ago. Modern humans (or near so) existed in southwest Asia. Agrarian cultures developed in various locations at various times: in the Euphrates-Tigris rivers basin, the Nile valley in Sudan and Egypt, Modern Pakistan and India, China, and the Americas. There was clear evidence of humans in the Americas by findings of "Clovis points" of shaped stone at present Clovis, New Mexico at 13,500 B.C., and stone cutting tools found along the Atlantic coast near present-day Maryland of 22 to 20,000 years ago, similar to the Solutrean tools found in southwest Europe. It is possible that ancient fossils exist in areas other than African Olduvai Gorge. Most original efforts of anthropologist have been performed in various sections of Africa. This is where most concentration and preservation of fossils has occurred. Different anthropological results could develop by extensive exploration in the Georgia-Caucuses Mountain region of Russia, where the author believes was origin of Caucasian Homo erectus, who after 1.8 million years ago seeded many sub-species in Mesopotamia, Europe, and eastward through India, Indochina, Papua New Guinea, and beyond. Olduvi Gorge is a steep-sided ravine in the Rift Valley that stretches through eastern Africa. It is in the eastern Serengeti Plains in northern Tanzania, and is about 30 miles long. The gorge

	is an important prehistoric site, sometimes referred to as "the Cradle of Mankind". Millions of years ago the site was covered by a large lake. Around 500,000 years ago seismic activity diverted a nearby stream, which cut down into sediments, revealing seven main layers in the walls of the gorge. The earliest archaeological deposit, known as Bed 1, has produced evidence of campsites and living floors, with stone tools made of flakes from local basalt and quartz. Since this is the site where these types of tools were first discovered, these tools are called "Oldowan". It is now thought that the Oldowan tool-making tradition started about 2.6 million years ago. Bones at these sites in Africa are not of modern humans, but of primitive hominid forms of Paranthropus boisei species, and the first discovered specimens of Homo habilis. The site displays the oldest known evidence of Elephant consumption, attributed to Homo ergaster around 1.8 million years ago. Another layer displays more sophisticated hand axes of the "Acheulean" industry, made by H. ergaster at about 1.75 to 1.2 million years ago. (Source: Google)
Undetermined	A Negroid race migrated outward from Olduvi Gorge, east coast of Africa, and Abyssinia north, south, east, and southwest, to form various ethnic groups. Much later some ethnic Negroid travelled eastward (author's theory), by ocean on rafts or other floating vehicle, and colonized northern Australia, Borneo, Papua New Guinea, and various islands in the Pacific Ocean.
40,000 to 30,000 B.C	Homo ergaster and Homo neanderthal, as separate species disappear. Humans in Ice Age Siberia and southeast Asia (Denisovan hominids), Caucasians, Orientals, and people of India, in addition to historic Negroid, as separate races, with many sub-species within races. (Examples are relatively short stature

of Peruvians in South America, short height of Pygmies of Africa, and diminutive "Hobbits" (Homo floresiensis) of Papua New Guinea, robust features of Slavic peoples, and fairness in skin pigmentation of the Nordic. Caucasian, and various other races, some with ruddy skin pigment variations developed, such as European, Asian, Egyptian, Mesopotamian, Indian, and Oriental.

Evidence of humans at Terra del Fuego in South America (spear point found). This may indicate migration from southern tip of Africa to southern tip of South America, traveling by water and land along face of southern glaciers at the Antarctica continent, to the land of South America, and perhaps northward to eventually form an ancient civilization in the Americas. (Author's Theory).

Homo neanderthal (or Homo erectus) species were living in Croatia and Yugoslavia areas from the Georgia, Russia location.

Skeleton fossils of slain animal, at age of about 13,500 B.C. have been found at present-day Clovis, New Mexico.

A new theory places Stone Age Europeans in North America at least 20,000 years ago, forerunners of migrants from Siberia who crossed the Aleutian Archipelago about 15,000 B.C.

Early hunters may have been Solutrean, with their distinctive stone cutting blades, anvils, and other tools, found in eastern North America Atlantic sites of Chesapeake Bay islands. The blades strongly resembled those found at dozens of Stone Age Solutrean sites in Spain, Portugal, and France where they lived about 25,000 years ago.

A Caucasian race, supposedly originating in southern Russia at the present Georgia region of the Black Sea, Caspian Sea, and Caucuses mountains developed from an unknown process. These persons

	could well have migrated eastward through India, to China, Indochina, fording low-level waters or a land bridge of the Malay Peninsula to Java, Papua New Guinea, and rafted outward into the Pacific Ocean to settle Polynesian islands, and perhaps to South America. It has also been proposed that original people of south and central America emanated from Japan. Evidence of the flow of migration, leaving successive fossils is still incomplete, but a logical trail of migration can be assembled, based on fossils found. Direct migration from east Africa would account for unseemly presence of Negroid race in northern Australia, Borneo, and islands of the Pacific where Caucasian race from migration would be more logical. Caucasians had migrated from southwest Asia (Russian Georgia area) to Mesopotamia, India, Indochina, Malaysia, Java, Papua New Guinea, and westward to Polynesian islands. Hominoids of the Negroid race seemed to have avoided areas where species of Caucasian race had settled, but instead continued their own migration pattern, constantly eastward to islands, using floating vehicles such as rafts or crude boats, riding streams and trade winds to miraculously cross ocean waters from Africa. (Reminiscent of Christopher Columbus, hundreds or thousands of years later, crossing the Atlantic Ocean and finding land.) Migration and exploration seem to be a basic trait of all Earth's people, observed throughout history.
20,000 to 8,000 B.C.	Late Neolithic times. Palo Indians in eastern, western, southwest, and other regions of North America.

15 to 12,000 B.C.	Future North American Indians cross land bridge from northeastern Asia across Bering Strait to Aleutian Islands and mainland Alaska. Migration southeast by land and southward by land and watercraft along the coastal ocean, pursuing hunting and fishing grounds. Migration was eastward in far North by future Eskimos (Inuit). 130 animal species become extinct in span of 400 years, including Woolly Mammoth, Mastodon, Sabre Tooth Tiger, and giant rodents.
15,000 to 10,000 B.C.	The Straits of Hormuz partially enclose the Red Sea at the eastern end. The author believes waters now cover what was once either dry or swampy land at outlet of the Tigris and Euphrates rivers delta. The melting at this time of earth's extensive glaciers that covered the earth down to 35 or 40 degrees North and South Latitude, caused levels of ocean waters to rise significantly. At the time, the present mountain chain at the Straits of Hormuz formed a barrier against the Arabian Sea and Indian Ocean, except perhaps for a drainage river through a low valley in the chain into the lower-elevation Arabian Sea. Additional evidence for the continuity of land at the present Strait is that reportedly one migration route to the Mideast by early species of man passed over a mountain chain at the present Straits of Hormuz.

Weight of water over the formerly low swampy area could have caused shift in the tectonic plates, causing earthquakes at the Straits of Hormuz.

Two or three factors may have caused flooding of the Persian Gulf lowland. First is the rise of inrushing water of the Arabian Sea, causing a backwash through the mountain chain and river, flooding the low-land beyond. Second, there may have been earthquakes, in that earthquake-prone part of the earth, caused by shifting geologic plates and increased weight on |

	geologic plates from expanses of additional sea water. Erosion would have widened the Straits. Third, increased flow of river water from the mountains in Turkey that are sources of the two rivers, could add to the volume of water in the Persian Gulf. This theory varies from other published events that river waters and rain water alone caused flooding of the Persian Gulf. It seems illogical that high river levels alone, even floods, would have had this effect, even in a year or period of catastrophic weather, or extensive glacial run-off. Flooding purely from mountain water run-off would normally have existed for hundreds of years and been common.
13,000 B.C.	Clear evidence of humans in the Americas. Clovis points found in present Clovis, New Mexico, and ancient people existence in several southwestern states. Start of change from hunter/gatherer to agrarian culture in Asia, Europe, and Africa.
12,000 B.C. (Approx.)	Neolithic period in Egypt and Mesopotamia following glacial withdrawal. Various waves of migration, following hunted animals, into Alaska from Asia, such as Aleut, (Eskimo), Algonquian, and tribes of Pacific Northwest.
10,500 B.C.	Palo-Indians in Kentucky.
10,000 B.C.	Holocene Epoch: Period of human development and intervention. Civilization starts and evolves to various kingdoms. Extinction of the Sabre Tooth Cat (Tiger) of North and South America. Neolithic culture using ground and finished tools.
Undetermined- Possibly 10,000 B.C. or before.	Possibly migration from Southeast Asia, across Pacific Ocean, from Polynesian or Aborigines cultures to western South America, for first Andean civilization.

10,000 to 8,000 B.C.	Hairy Mammoths and Mastodons become extinct as combination of environment and being hunted by man. End of last (Wisconsin Glacier) Ice Age. First signs of agriculture and animal husbandry.
9,600 to 8,200 B.C.	Dawn of civilization as product of the Human mind: Sites in western Turkey (Gobekli Tepe and Nevali Cori) built by hunter-gatherers at time of Neolithic Revolution, and locations near the Fertile Crescent at Turkish and Mesopotamia mountain origins of Tigris and Euphrates rivers. This culture was slightly reminiscent of stone and wood structures at the Stonehenge area of Great Britain, but 7,000 years earlier. Stonehenge and Woodhenge are believed to be religious and season-determination sites. Nabulian villages, dating 13,000 to 10,800 B.C. were found in the Levant (which encompasses Israel, Palestine, Lebanon, Jordan, and western Syria), suggests that developments in needs of a shared religion, and communal living that preceded and caused the rise in agriculture and agrarian culture.
9,500 B.C.	Withdrawal in area of last (Wisconsin) ice age. First signs of agriculture and animal domestication. Major events following turn from Neolithic to Agrarian culture: Change in climate, biology of man, mind, spiritual, and emotional outlook affected agricultural and cultural orientation. Lobe of Mediterranean Sea into upper Sahara Desert provides salt deposits upon entrapment and evaporation of sea water.
9,000 to 6,000 B.C.	More centralized and settled agricultural society along Nile River. Sahara Desert green with flora and abundant with fauna. Only a few archaeological records for Egypt of this time have been found.

8,000 to 6,000 B.C.	Transition period in Mesopotamia and Upper Nile of Egypt to agrarian farming and animal herding culture.
7,000 B.C.	Evidence of pastoralism, cattle herding, and cultivation of cereals in eastern Sahara and northern Africa. Lighter skin color in racial features of Egyptian and north Africa people derived from genetic blending of Caucasians from Mideast and Negroid from Ethiopia and Lower Nile River Sources (author's theory).
7,000 to 1,000 B.C.	Archaic Period of North and South American Indians.
6,000 B.C.	Sumerian and Amorite groups living near area of the joining of Tigris and Euphrates rivers.
6,000 to 4,000 B.C.	First pyramids in Egypt; some of pictured royalty in tombs were black Nubians. Oldest of stepped pyramids is Saqqara Pyramid of Djoser built by King Zoser. (Great Pyramid of Giza is estimated to have been built between 3,200 and 2,200 B.C.)
6,000 to 2,000 B.C.	Sahara (now desert), was green and fertile: Hunters at 6,000 to 4,000 B.C., herders at 4,000 to 1,500 B.C., horses at 1,500 to 600 B.C., camels at 600 B.C. to present. Sahara was a nursery for cultures that spread throughout the African continent and ancient world. Present salt deposits from dehydrated ancient saltwater bodies of lobe from Mediterranean Sea.
5,500 to 3,100 B.C.	Pre-dynastic period in Egypt. Upper and lower Egypt are separate (dividing line is about where Cairo is located). Ubaid period in Mesopotamia (start of emergence of urban society) consisted of farming, animal husbandry, pottery, metallurgy, the wheel, circular ditches, tell (mounds), megaliths, and Neolithic religions.

5,000 to 3,000 B.C.	Uruk period 4,100 to 2,900 B.C.: Region settled in area of southern to northern Mesopotamia. Ancient communities were Eridu, Uruk, Ur, and others, all along Tigris and Euphrates rivers. Peopled groups were Elam, Akkadian, and empires of Amorites, Babylonia, Assyria, Hittites, and Kassites.
4,800 to 3,100 B.C.	Egyptian pre-dynastic period.
4,200 to 1,200 B.C.	Egyptian Bronze Age.
4004 B.C.	Date of biblical Creation, and of Adam and Eve, as calculated by Bishop James Ussher (1581 to 1656 A.D.) from Biblical generations cited in Bible. (Other Biblical theories range from years 5,872 to 4,658 B.C.). There are many accounts of Creation in Pagan and Babylonian cultures. Stories may be allegory for then-present life existence.
4,000 B.C.	Sumer immigrants in Mesopotamia settled swampy delta land near Tigris and Euphrates rivers. Construction of Babel, Akkad, Ur, Uruk, other communities. Mortar is used in building construction. Cave art found at north Sahara Desert location 900 miles southeast of Algiers, showing oxen and hares in a green environment. Possible domestication and hunting of animals in Americas. Egyptian hieroglyphics present.
3,500 B.C.	Copper Age. First evidence of man in England. Mesopotamian civilization in Tigris-Euphrates valley. Oldest city of Ur in Samaria. First Sumerians in Mesopotamia, followed by Semites (2,500 to 2,300 B.C.), Sumerian New Empire (2,300 to 2,000), Elamites (2,000), Amorites (2,000 to 1,750), and Kassites (1,750 to 1,300 B.C.).

3,500 tp 3,100 B.C.	Neolithic farmers: Communities of Akkad in north, Sumer in south. Start of extensive trading centers.
3,400 B.C.	Upper Kingdom and Lower Kingdom of Egypt are separate; Nubians are from Ethiopia.
3,200 B.C.	Upper Kingdom and Lower Kingdom of Egypt are united under King Menes, first Dynasty of 6 in Old Kingdom.
3.246-2,288 B.C.	Earthquake opens mountain range forming Straits of Hormuz and allowing waters of Indian Ocean to flow into lowlands, forming Persian Gulf. Flooding and wash could cause Biblical "Great Flood" up Tigris-Euphrates valley. Possible flooding and destruction of "Garden of Eden" near city of Ur. (Theory of author).
3,200 B.C.	Oldest found evidence of Egyptian king, Scorpion I
3rd Millennium B.C.	Leaders in Mesopotamia: Nimrod, in Assyria, appoints self as first king, was mentioned in Old Testament, son of Kush, was a black Ethiopian, contemptuous, rebellious toward God, and considered an anti-God, warred with Abraham. Tower of Babel built, just south of Baghdad, "201 years after biblical Great Flood". Abraham, nephew Lot, family, and personal group migrate from Ur to Baghdad area, and at age 75, is deeply involved as obedient Messenger of God, has four wives (Sarah, Hagar, who is Sarah's handmaiden, Rebecca, a "patrilineal parallel " cousin Keturah), and sons (Ishmael, Isaac, Zimran, Jokshan, Medan, Midian, Ishbak, and Shuah) who each founded new communities. Abraham's birth is stated in the Hebrew Bible at 1,948 years after the biblical Creation, or year 1,948 A.M. (Anno Mundi, Hebrew "Year of the World", or about year 2,056

	B.C. Another source places birth of Abraham at year 3,312 A.M., and the Samaritan version at year 2,247 Anno Mundi. The Old Testament states Abraham died at age 175. Sarah, his first wife, died at age 127. The Bible places Abraham within a generation of Noah, and is a contemporary of Shem, Noah's son, and is an acquaintance of King Nimrod. Christians, Jews, Islam, and Baha'i all recognize Abraham and his deeds. Abraham's descendants include Jesus (through Isaac) and Mohammad (through Ishmael).

In a Biblically reported discussion of God and Abraham in the latter's tent, a possible explanation is that the dual-universe soul of Abraham acted as a medium between an inanimate God and the mind of Abraham. (Author) |
Near above events.	Sodom and Gomorrah, two of the "five cities of the plain" on River Jordan are destroyed.
Prior to 3,100 B.C.	Egypt in pre-dynastic Neolithic Period.
3,100 B.C.	King Minos, possibly known also as King Narmer, founder of 1st Dynasty and first king of Egypt, United Upper and Lower Egypt. Egypt will be closely tied to religion for next 3,000 years.
3,100 B.C	Beginning of written history.
3,000 B.C	Presence of Indians at Frankenmuth, Michigan, USA; Copper extracted by Indians in Michigan Upper Peninsula.
3,000 B.C.	Cuneiform writing developed in Mesopotamia.
3,000 to 1,200 B.C.	Aegean civilization (Crete).

3.000 to 700 B.C.	Troy I through IX (was built, destroyed, and rebuilt several times). Homer writes books, " Odyssey" and " Iliad", of period during and after Troy battle.
3,000 B.C.	First cities and states in northwest Africa and southern Mesopotamia. Minoan civilization in Crete. Chickens first bred in India from jungle fowl (3.000 B.C.), in Egypt by 1,400, and in British Isles by Romans.
3,000 to 626 A.D.	Assyrian Empire.
3,000 B.C.	World's first written script in Sumer and in Egypt (in Cuneiform and Hieroglyphics).
3,000 B.C.	Ancestors of Celts (Kelts) began migration westward from Russian Steppes at Black Sea near Caucasus Mountains, through Danube River valley, Switzerland, northern Italy, Spain, Gaul, Normandy, England, and Ireland. Kurds of Turkey and northern Iran (ancient Hittites) are probably of Celtic derivation.
3,000 to 2,900 B.C.	Reign of legendary King Gilgamesh. Report of a great flood.
3,000 to 1,000 B.C.	Worship in Mideast of Baal and other gods.
2,950 to 1,600 B.C.	Stonehenge in England built by Druids; influence parallels Mycenaean and Minoan architecture and bronze artifacts; Druids (who are believed to have migrated to England) showed high degree of competence in astronomy.
2,890 to 2,686 B.C.	Second dynasty in Egypt Old Kingdom.

2800 to 2300 B.C.	Bronze Age in Greece. Religious beliefs with icons possibly started when a ritual activity, such as human burial, turned into a religious activity relating to beliefs about living fortunes, life, death, and perhaps a hereafter.
2.700 to 2,200 B.C.	Old Kingdom in Egypt. First pyramids.
2,700 B.C.	Chinese invented process for weaving silk.
2,686 to 2,134 B.C.	Old Kingdom of Egypt comprised 3rd to 6th Dynasty. Capital was Memphis. In 4th Dynasty son of Snefru (Khufu) and grandson commissioned and built the Great Pyramid.
2,649 to 2,268 B.C.	Intermediate period in Egypt, characterized by disunity and period of relative decline. Many pyramids built. Egyptian pyramid at Giza completed at 2560 B.C.
2,630 B.C.	Saqqara Step pyramid in Egypt built by Pharaoh Snefru for burial and spiritual purposes. Imhotep was architect. Snefru was assumed to be a god who assured annual flooding of the Nile River. (Perhaps a Pharaoh's identifying himself as a god was really a first experience in recognizing his soul as a part or reflection of a god. He would feel his thoughts and actions as "godly", not recognizing that his subordinate, common people, also had these emotions of consciousness, love, empathy, and the like. Research and survey was not yet present to any stage of public study. It would be illogical that Pharaoh would consciously exercise an audacity to set himself up as competing with his god that he recognized as supreme.)—Author's theory).

2,630 to 2,250 B.C.	Several major pyramids built, of 138 total (years are approximate):
	
	Pyramid of Saqqara in 2,630 by Pharaoh Djoser.
	    " Maidum in 2,600 by " Snefru.
	Bent Pyramid in 2,600 by " Snefru
	Red " in 2,600 by " Snefru
	Great " in 2,566 to 2,558 by " Khufu (or Cheops)
	Pyramid of Khafre in 2,520 by " Khafre
	    " of Menkaure in 2,490 by " Menkaure
	    " of Pepy II in 2,250 by " Pepy II
	
	Most pyramids were built as tombs for the country's pharaohs and their consorts during the Old and Middle Kingdom periods. The step pyramid (Saqqara) was designed to serve as a giant stairway by which the soul of the deceased pharaoh could ascend to the heavens. The most prolific phase of building was during the greatest degree of absolute rule: Over time, authority became less, the ability and willingness to harness the resources required for construction on a massive scale decreased,
	
	and later pyramids were smaller, less well-built, and often hastily constructed. In the Sudan, in the 600 B.C. to 300 A.D. period, there was a pyramid building revival which saw over 200 pyramid tombs built.

2,613 to 2,589 B.C.	Pharaoh Snefru built Great Pyramid of Giza (built by 4th Egyptian Dynasty), and Pyramid of Cheops are oldest and largest of the 3 pyramids.
2,600 B.C.	Rise of Indus civilization in India.

2,589 to 2,566 B.C.	Pharaoh Khufu/or Cheops (same person).
2,566 to 2,558 B.C.	Great Sphinx built. Believed to be face of Khufu/Cheops
2,558 to 2,532	Egyptian capital city at Alexandria.
2,550 to 2,000 B.C.	Start of Semite civilization: Elamites, Amorites from Babylon.
2,500 to 2,000	Hatti empire in Mesopotamia. Also a Hittite, Old Kingdom empire. Indo-European race and culture in India (Aryans).
2,500 B.C.	Climate change creates dry Sahara with deposits of salt (residue of salt water from previous lobes of Mediterranean Sea) in certain locations of northern Sahara.
2,349 to 2,278 B.C,	Reign of Sargon I of Akkadian empire, creator of history's first empire in Mesopotamia.
2,348 B.C. per Bishop James Ussher's Timeline	Great Flood, categorized in status of myth (per Google computer service). Myth is believed symbolic to avenge sins and rebirth. Existence is questionable from archaeological studies.
2,300 B.C. or older.	Tower of Babel in Babylonia. Dispersion of population to Canaan, Egypt, Chaldea, (Ur), Greece, China.
2,300 to 2,080 B.C.	Akkad Empire: precedence for later empires.

2,300 to 1,750 B.C.	First dynasty (Hiah) in China. Original Japanese people reportedly migrated from the Korean Peninsula, and those people originated from Mongolia. Times of migration are not known. It is possible that Koreans, formerly from Mongolia, were original Inuit Eskimo who migrated from Asia, Alaska, and northern Canada, to Lapland Europe, arctic Russia, far northern Asia, Siberia and Mongolia. Eskimo migrated to Alaska from Siberia at a time after migration of the future North American Indians and chose an Arctic way of life. Facial features and skin coloration, average height, environmental living ways, and subsistence modifications of arctic Asians and Eskimo are historically similar. (Author's theory)
2,288 B.C.	Great flood and Noah family epic reported by Hebrews; Other reports range from 3,246 to 2,348 B.C). Legend of King Gilgamesh predates and reports story of a great flood, possibly model for Biblical tale of a great flood in time of Noah. (King Gilgamesh was probably a real person who lived about 2700 or 2500 B.C., the 5th king in 1st dynasty of Uruk.—(Source: Google)
2,249 to 2,150 B.C.	High point of Akkadian empire in Mesopotamia. Walled cities in Old World for protection and containment purposes.
2,181 to 2,055 B.C.	Egyptian First Intermediate period.
2,150 to 2,193 B;C.	Akkadian empire in Mesopotamia disintegrating.
2,112 to 2,047 B.C.	Third dynasty of Ur (Ur III) and Uruk founded. Amorites. Abraham (Abram) leaves city of Ur, travelling through the "Fertile Crescent" to Babylon and Jerusalem,.
2,100 B.C.	Old Babylonia Empire.

2,100 B.C.	First ziggurats built in Mesopotamia.
2,055 to 1,650 B.C.	Egyptian Middle Kingdom period (Dynasties XI to XIV). Noted for expansion of trade outside the kingdom.
2,028 to 2,004 B.C.	Ur III empire collapses.
2,000 B.C.	12th Dynasty in Egypt: First democratic kingdom in history. (Contrast Greece) Empire Period in Egypt starts.
2,000 to 1,770 B.C.	Mesopotamia reverts to patchwork of city states. Introduction of Potter's Wheel.
2,000 B.C.	First multi-city empires as cities and states in India and China.
2,000 B.C.	Minoan civilization develops in Crete.
2,000 to 1,200 B.C.	In India, Hinduism, Aryans, Rig-Veda are oldest major religions of those recognized today; zenith of Hittites who were considered Indo-Europeans, or were Aryans with Hindu religion of Rig-Veda.
1,994 to 206 B.C.	Chinese dynasties: XIA dynasty: about 1,994 to 1,766 B.C,; Shang: 1,766 to 1,027 B.C.; Zhou: 1,122 to 256 B.C.; Qin: 221 to 206 B.C.
1,950 to 1,750 B.C.	Assyrian trading colonies.
1,900 to 1,200 B.C.	Hittite Empire (Anatolia, or modern Turkey), Troy built.
1,921 B.C.	Abraham (Abram) begins journey to Haran and Canaan from city of Ur.
1,900 B.C.	Thebes made capital of Egypt Old Empire.
1,894 to 1,595 B.C.	After fall of Assyria: Time of Medes, Scythians, Kassites, Mitanni, Chaldeans, Amorites, a god named Marduke, King Nebuchadnezzar, King Hammurabi.

1,840 B.C.	Famine in Canaan. Joseph, son of Jacob. Brothers attempt to abandon and leave Joseph to die, travel without Joseph to Egypt, and later meet Joseph who, fortuitously, is then in service of Egyptian royalty.
1,800 B.C.	Ancient Indian city uncovered in southwestern Michigan, USA on floodplain of St. Joseph River.
1,800 to 1,450 B.C.	Period of Minoan naval empire.
1,792 to 1,750 B.C,	Reign of King Hammurabi. Development of "Laws of Hammurabi". Hammurabi was 6th king of Babylon city and first king of Babylon Empire from 1,792 to 1,750 B.C.
1,766 B.C.	Foundation of Chinese first dynasty, the Shang, in the north of the country.
1,750 B.C.	Kassites and Hittites overthrow Amorites (Babylonians).
1,750 to 1,000 B.C.	2nd Dynasty (Shang) in China.
1,680—1200 B.C.	Hittite, Old, Middle, New Kingdom.
1,674 to 1,549 B.C.	Egyptian second Intermediate period (Dynasties XV to XVII. Hyksos capture Memphis capital, but Egyptians drive Hyksos back to Asia.
1,630 to 1,600 B.C.	Founding of Mycenae and their Aegean Sea communities for trade.
1,600 to 1,100 B.C.	The Mycanae civilization flourished during the period roughly between 1,600 B.C., when the Hellenistic culture in mainland Greece was transformed under influence from Minoan Crete, and 1,100 B,C. when it perished with the collapse of the Bronze Age civilization in the eastern Mediterranean. The collapse is commonly attributed to the Dorian invasion, natural disaster, and climate change.

16th Century	Kassite kingdom in Babylonia.
1,590 to 1,550 B.C.	Hurrian kingdom of Mitanni in Mesopotamia.
1,570 to 1,080 B.C.	Egyptian New Kingdom (Dynasties XVIII to XX) saw Egypt's greatest territorial extent into Nubia and Near East, fought Hittite armies for control of Syria. Egyptian New Kingdom was Egypt's most prosperous time and peak of its power.
1,530 to 1,500B.C.	Dynasty XVIII in Egypt. Hyksos driven from Lower Egypt. Egyptian frontier advances more into Nubia (now Ethiopia).

FROM AN AGRARIAN CULTURE
PART II

1500 YEARS B.C. TO THE TIME OF CHRIST

The timeline of history continues. The date of 1,500 B.C. nicely divides our Timeline when Queen Hatshepsut, stepmother of "King Tut" served as Egypt's female Pharaoh, and when King Hammurabi devised a code of Laws that was model for nearly all Semites of Babylon, Assyrians, Chaldeans, and Hebrews.

This also divides when Egypt's New Kingdom was at its apex as a regional power, from when various religious Books were assembled for inclusion into the Old Testament of the Hebrews.

The author concludes with his view, based on study of over 13.7 billion years of world events, what he feels lies in the future for humans in the next 50 to 100 years.

Date	Event
1,500 B.C.	Early Babylonians: Code of Hammurabi is code of nearly all Semites (Babylonians, Assyrians, Chaldeans, and Hebrews). Earliest glass vessels.
1,500 B.C.	Sumerian polytheistic religion: "Existence of man is for this world only, with no offer of hope for an afterlife."
1,495 to 1,425 B.C,	Pharaoh Thutmose I leads Egyptian army to Euphrates River and expands Egyptian empire into central and northern Palestine and southern Syria.

1,479 to 1.425 B.C.	King "Tut" (Tutankhamen), son of Pharaoh Akhenaten and possible step-son of Queen Hatshepsut. Tut is grandson of Amenhotep III.
1,450 B.C, to 1,350 B.C.	Wave of destruction sweeps Crete. Minoan civilization ends. Mycanae dynasty rules from palace. Mainland Greeks take over the islands and colonize the Aegean. Mycanae believed remnant of Hittite culture of Mesopotamia and following religion of Greece.
1,440 B.C.	Exodus from Egypt to land of Canaan. Many dates have been proposed. Actual existence of Exodus has been questioned by some anthropologists and Egyptologists. (Per Google).
1,430 to 1,330 B.C.	Hittite New Kingdom expands into Anatolia (present Turkey, Macedonia, and northern Syria).
1,427 to 1,400 B.C,	Reign of Amenhotep II. Battle with Mitanni for control of Syria.
14th-4th Century	Scholars generally agree that the 39 Books of the Old Testament were written over a period of 1,000 years from the 14th to the 4th century B.C. The Old Testament is divided into three major divisions: Historical, Poetical, and Prophetic. The Historical division begins with the first five Books of Moses, called the Pentateuch, also called The Torah. Job, Psalms, Proverbs, Ecclesiastes, Songs of Solomon are the Poetical Books. There are 17 Prophetic books with 4 major books of Isaiah, Jeremiah, Ezekiel, and Daniel.

1,400 B.C.	Hebrews move from their home near city of Ur. One group, under leadership of Abraham, lived for time in city of Ur; another group or tribe taken to Egypt and was there held in captivity. Physical presence of God, Yahweh, in tent of Abraham (per Bible). Hittite and Hattie groups.
13th Century	Final collapse of the Mycenaean trade.
1,386 to 1.349 B.C.	Amenhotep III ruled Egypt at peak of its glory
1,367 to 1,350 B.C.	Amenhotep IV, son of Amenhotep III and his chief queen Tiy rule Egypt
1,360 B.C.	3rd dynasty in China.
1,354 to 934 B.C.	Middle Assyrian Empire. Assyria begins rise to power after defeat of Mitanni in Syria.
1,350 to 1,300 B.C.	Signs of rising culture in Greece.
1,341 to 1,323 B.C.	Tutankhamen, or "King Tut", the boy pharaoh. Reigned for 9 years. Died at age 19. Son of Akhenaten and Nefertiti (an incestuous relationship of parents caused disastrous physical health of son). Tutankhamen ended worship of god Aten and restored worship of Amun to supremacy. Built temples at Thebes and Karnack.
1,320 to 1,295 B.C,	Rebellion among Syrian and Palestinians nearly destroys Egyptian empire in Mesopotamia.
1,303 to 1,213 B.C.	Pharaoh Ramses (Ramses The Great) led Egyptian army to recover Israel and Palestine.
1,300 to 700 B.C.	Ionia Collusion in Greece. Assyrians arise.
1,292 to 1,290 B.C.	Ramesses I (First of 7 named "Ramses")

1,290 to 1,279 B.C.	Sety I is son of Ramses I and Queen Sidre. Is father of Ramses II.	
1,290 to 1,224 B.C.	Ramses II and Queen Nefertiti rule Egypt.	
About 1,285 B.C.	Egypt is attacked from the west by Libya.	
1,279 to 1,213 B.C.	Reign of Ramses II. (Eleven different pharaohs are named Ramses.)	
1,265 B.C.	Hittite power weakened by civil war.	
1,250 to 1,230 B.C.	Greek Trojan War. Final collapse of Mycenaean trade empire.	
Prior to Exodus	Moses in Egypt debates with Pharaoh.	
1,250-1,100 B.C.	Moses and Hebrew exodus from Egypt, migrating to Promised Land east and west of Jordan River, then occupied by Canaanites of Semitic speech who had lived there for centuries. Conquest was slow and difficult. Combat with Philistines from Asia Minor gains Palestine.	
1,230 B.C.	Elamites devastate Kassite Babylonians.	
1,220 to 1,210 B.C.	Hebrew exodus from Egypt and settle in land of Canaan, one of several calculated dates of Exodus, (if this existed at all—per Google).	esopotamia reverts to patchwork of city states.
1,200 B.C.	First Olmec ceremonial center built.	
1,200 to 1,180 B.C.	Waves of destruction sweep over coastal zone of Levant. "Judes" (of Judea) rule autonomous Israelite clans in Canaanite hill country.	
1,184 B.C.	Generally agreed date of the sack of Troy.	

1,200 to 1,100 B.C.	Arabian Peninsula, not completely devoid of flora and fauna, would allow Exodus travellers and livestock to possibly subsist for legendary 40 years on route from Egypt to Canaan. Most of desert, before present arid condition, was abundant in plants and nature life. (Author's Theory)	
1,200 to 800 B.C	Neo-Hittite kingdom.	
1,200 or 800 B.C.	"The Exodus is the biblical story of departure of the Israelites from ancient Egypt as described in the Hebrew Bible. Some recognize the reports as theology told in the form of history. There are many naysayers of the time, or even existence of the Exodus, based on various conflicting references and traditions (in Books of Exodus. Leviticus, Numbers, and Deuteronomy). It is also numerically extraordinary that the reported 600,000 Israelites and others, their families, and animal herds would logistically be able to complete the journey. It is calculated that 600,000 people (Exodus 12:1:37), traveling 10 abreast, not counting livestock, would be a procession over 50 miles long. There is no evidence that Egypt suffered loss of this many people, that would suffer a demographic and economic catastrophe, nor that the Sinai Desert ever hosted (or was able to host) this extent of people and their herds, nor of a massive population increase in Canaan, estimated to have population of only 50,000 to 100,000 at the time." "View of the mainstream modern biblical scholarships is that the Exodus story was written, not as history, but to demonstrate "God's purpose and deeds with his chosen	

people—Israel". The archaeological evidence concerning the origins of Israel is "overwhelming" and leaves no room for an exodus from Egypt, or a 40 year pilgrimage through the Sinai wilderness. For this reason, most archaeologist and Egyptologists have abandoned the archaeological investigation of Moses and the Exodus as "fruitless pursuit". It has become increasingly clear that the kingdom of Israel and Judah has its origin in Canaan, not Egypt."

Source: (Google, click "Exodus", Wikipedia)

In addition, exact details of Moses, God, and The Ten Commandments at Mt. Sinai are questioned. (Moses may have etched rules on animal skins, based on the similar Egyptian, "Book Of The Dead", of which he was familiar from his upbringing as a youth in an Egyptian palace), as was the account, possibly inaccurate, of the biblical portrayal in debate of Moses and Pharaoh (Ramses III).

1,200 to 800 B.C.	Homeric Age (named for time and writings of author Homer) of Greece: Iliad and Odyssey written about 900 B.C.
1,200 to 600 B.C.	Italian Peninsula occupied by Etruscans and Greeks.
1,200 B.C. to 200 A.D.	Cultures and cults of India develop.
1,200 B.C.	End of Hittite New Kingdom empire. A Neo-Hittite kingdom from 1.200 to 800 B.C. Iron Age in Mesopotamia.
1,179 to 1,173 B.C.	Ramesses III rules in Egypt and settles the Philistines on the southern coast of Canaan.

1,150 to 1,140 B.C.	Egypt has last remnants of its empire in the Levant.
1,125 to 1.103 B.C.	Elam kingdom destroyed
Late 11th Century	Egypt disintegrates into feuding principalities.
1,100 to 400 B.C.	Egypt Late Empire (21st to 30th Dynasties).
1,080 B.C.	Egypt loses control of Nubia. Egyptian empire is finished.
1,030 B.C.	Philistines begin to expand into the interior of Canaan.
Between 1,079 and 1,007	Young David, a future king, kills Goliath of the Philistines during reign of Saul.
1,069 to 653 B.C.	Egyptian third Intermediate period.
1,025 B.C.	Start of Hebrew "judges" in Old Testament records: Kings, Saul (first king of Israel), Samuel, David, Solomon. Union of 12 tribes and dispersal—later 10 northern tribes form Israel, 2 southern tribes form Judah.
1,020 to 914 B.C.	Some leading Kings of Israel and Judah: Saul 1,020 to 1,000; David 1,000 to 993; Solomon 960 to 931; Jeroboam (in Israel) 931 to 910; Rehoboam (in Judah) 931 to 914.
1,004 to 588 B.C.	Original Temple to worship God started by Solomon, and home of worship.
1,000 B.C. (Approx.)	Joshua conquers Jericho.
1,000 B.C.	David makes Jerusalem capital of the Israelite kingdom.

Undetermined-Possibly 10,000 B.C. or before.	Possibly migration from Southeast Asia, across Pacific Ocean, from Polynesian or Aborigines cultures to western South America, for first Andean civilization.
1,000 B.C. to 1,500 A.D.	Mesoamerican cultures: Olmec, Toltec, Inca, Maya, and later Aztec.
1,000 B.C. to 700 A.D.	Adena and Hopewell Mound Builders in North America. La Venta or Vera Cruz, Mexico are possible origins of Hopewell culture. Adena and Hopewell people are pre-Mississippians. Woodland Indians will conquer Mound Builders, who become extinct, absorbed, or migrate to elsewhere (to Mexico or Mid-America?—possibly Aztec).
1,000 to 500 B.C.	Mexico Middle Classic Period.
1,000 to 753 B.C.	Estimated settling of Italian Peninsula by Etruscans and Greeks. Founding of Rome in 753 B.C.
960 to 931 B.C.	Solomon reigns as king of Israel and begins construction of the Hebrew Temple.
934 to 612 B.C.	Neo-Assyrian empire, capital city of Nineveh.
931 B.C.	Northern clans of Israel reject Solomon's son, Rehoboam, and anoint Jeroboam as their king. Israel disintegrates into Northern (Israel) and southern (Judah) kingdoms.
928 B.C.	Death of King Solomon.
883 to 824 B.C.	Assyria becomes great power once more.
850 to 600 B.C.	Phoenicians (Carthaginians) colonize the western Mediterranean area.
800 B.C.	Carthage in Africa is founded by king and merchants in Tyre at eastern Mediterranean Sea.

800 B.C.	Homer composes the "Iliad" and the "Odyssey".	
8th Century B.C.	Pharaohs in 22nd and 23rd Dynasties	
776 B.C.	Greeks hold first Olympic games.	
753 B.C,	Rome founded: Romulus said to have plowed furrow around Palatine hill to mark boundary of new city.	
744 to 630 B.C.	Zenith of the Assyrian empire.	
732 to 730 B.C.	Late period: When Egypt became province of Rome.	
721 to 705 B.C.	Sargon II expands Assyrian power into southeastern Anatolia.	
685 to547 B.C.	Lydian kingdoms at location of modern Turkish provinces.	
673 B.C.	Egyptians fight Assyrian invasion. Although Assyrians capture Egyptian capital of Memphis and Egypt Delta, Egypt later captures Thebes and Upper Egypt.	
672 to 343 B.C.	Egyptian Late Period—Dynasty XXXI. (Last 2 pharaohs are from Libya.)	
645 B.C.	Kingdom of Lydia submits to Assyria.	
638 to 558 B.C.	Solon, lawmaker at Athens	
About 628	Zoroaster (Iranian Persian prophet and philosopher; founder of Zoroastrianism. (Belief in one universal and transcendent God.)	
626 B.C.	Neo-Babylonian empire formed.	
600 to 576 B.C.	Cyrus returned Jews to Jerusalem in 586 B.C. after fall of Babylon.	
600 to 530 B.C.	Achaemenid empire in Anatolia, with Cyrus The Great as king, followed by Darius I, and Alexander The Great.	

600 to 100 B.C.	Many Books of Old Testament recorded in writing of Holy Books.
586 B.C.	Jerusalem falls to Egyptian king Nebuchadnezzar. The Temple is destroyed and thousands are deported to Babylonia. Time of Isaiah. Writing/deposit of historic accounts in cave of Dead Sea Scrolls.
586 to 539 B.C.	Period of Hebrew captivity at Babylon. Jerusalem plundered and burned, Temple destroyed.
580 to 500 B.C.	Pythagoras (Mathematician) in Greece.
About 563	Buddah (Siddartha Gautama in India), philosopher and religious figure. "Nothing has a soul, and any grasping at permanence ends in suffering and failure." Confucius, in China (551-479 B.C.) was contemporary of Buddah. Developed program for lifelong moral growth. "Ignorance is the night of the mind, but a night without moon or star."
560 to 559 B.C.	Xenophanes (Greek philosopher). Cyrus becomes king of Persia, destroys and annexes the Mede kingdom.
551 to 479 B.C.	Time of Confucius (Philosopher).
573 B.C.	Carthage establishes control over the Phoenician settlements of the western Mediterranean Sea.
539 to 150 B.C.	Alexander, grandson of Cyrus, starts from eastern shore of Persian Gulf (now Iran). Persians conquer Chaldeans and start Empire. Capture Babylon.
538 to 515 B.C.	Second Temple built by Jews in Jerusalem.

535 B.C.	Carthage defeats the western Greeks off Corsica, establishing itself as defender of the Phoenician communities against Greek aggression.
525 to 404 B.C.	Darius I conquers Egypt, which is annexed to Persia. Many religious traits are parallel to those of Judaism, later Christianity, and Islam. Jews freed from Babylon for return to Jerusalem.
522 B.C.	Darius I is king of Persia and launches an expedition into the Ukraine against the Scythians.
519 B.C. to 485 B.C.	Xerxes (4th king of Achaemenid Empire of Persia)
510 B.C.	Romans overthrow the monarchy and found a republic.
510 to 440 B.C (approx.)	Parmenides (Greek philosopher)
509 B.C.	Carthage makes a trade agreement with Rome after Rome expels Etruscan kings.
500 B.C.	Sparta, in Greece, flourishes. First coins (Persian and Greek).
500 to 100 B.C.	Sumerians, living between Tigris and Euphrates rivers, invent writing and making of beer.
500 to 200 B.C.	Zapotec culture in southern Mexican highlands. Mexico's late pro-classic period.
499 B.C.	War breaks out between Greece and Persia—Persians are defeated by Athenians at Marathon in 490 B.C.
490 to 439 B.C.	Zeno (Greek philosopher).
486 to 465 B.C.	Xerxes The Great of Persia.

480 B.C.	Series of Wars: Xerxes, successor to Darius I of Persia, battles Greeks at Thermopylae. War between Carthage and Sicilian Greeks, but Carthage is defeated at Syracuse.
478 to 465 B.C.	Athenian-led Delian League captures remaining Persian outpost in Europe
469 to 399 B.C.	Socrates (Greek philosopher). Socrates, Plato and Aristotle, all of similar philosophy, span 469 to 399 B.C.
461 to 429 B.C.	Age of Pericles in Athens.
460 to 453 B.C.	Egypt rebels against Persia.
460 to 370 B.C.	Hippocrates (Greek medicine)
447 B.C.	Work begins on the Parthenon at Athens.
431 to 404 B.C.	Peloponnesian Wars between Athens, Sparta and Persia.
428 to 348 B.C.	Plato (Greek philosopher)
424 to 405 B.C.	Reign of Darius II of Persia.
412 to 323. C.	Diogenes (Greek philosopher)
404 B.C.	Egypt rebels and remains independent of Persia, control for 60 years.
400 to 300 B.C.	Olmec civilization in Meso-America. Adena culture living in central and eastern Kentucky.
400 to 25 B.C.	Egypt Ptolemaic Era.
395 B.C.	Persia alternately aides both Athens and Sparta.

384 to 322 B.C.	Aristotle (Greek philosopher). Tutors future Alexander The Great.
380 B.C.	Egypt again under control of Persia.
375 to 225 B.C.	Greek Hellenistic Period following death of Alexander the Great and high development of Greek culture.
358 B.C.	Egyptians defeat second Persian effort to recover Egypt.
351 to 343 B.	Persia finally recognizes Egypt as autonomous.
350 to 100 B.C.	Several Books of Old Testament collected, revised, chronicled. Some ascribed to an ancient author and recorded as such.
343 to 332 B.C.	Egypt Persian period.
336 B.C.	Phillip II of Macedonia invades Anatolia.
334 to 301 B.C.	Kingdom of Alexander The Great of Persia.
332 B.C.	Alexander The Great takes Egypt. Alexander The Great killed in 323 B.C. Empire is split among his generals.
332 to 30 B.C.	Egypt Ptolemaic period: Cleopatra VII of Egypt offers throne to Marc Anthony of Rome.
305 to 364 B.C.	Seleucid Empire.
300 B.C.	Aggressive cultures in Gaul (Gallic tribes) and Scandinavia (Vikings).
290 B.C.	Colossus of Rhodes destroyed by earthquake.
280 B.C. to 1,480 A.D.	Lighthouse at Alexandria, Egypt.
265 to 220 B.C.	4th Dynasty in China.

264 B.C. to 146 B.C.	First Punic Wars between Carthage and Rome. Romans defeat Carthage. Later, Carthage defeats Rome, but after, Roman Fleet defeats Carthage.	
260 to 256 B.C.	Carthaginians use Sardinia and Corsica as bases from which to raid Italian coastal towns.	
237 B.C.	Rome seizes Sardinia and Corsica.	
220 B.C to 400 A.D.	5th Dynasty in China.	
218 B.C.	Second Punic War starts with Hannibal crossing Alps to invade Italy and inflicts defeat on Roman forces. Scipo (Roman general who defeats Carthaginians and later is Africanus of Rome) invades North Africa. Carthage empire is dissolved, with city of Carthage destroyed by Romans at end of Third Punic War.	
200 B.C.	China first unified by Emperor Shi Huang di. This was a traditional agrarian tribute to king. Tributes came from land. Elite groups despised commerce.	
200 to 70 B.C.	Dead Sea Scrolls written.	
200 B.C to 200 A.D.	Mexico Pre-Classic Period.	
196 B.C.	Rosetta Stone carved.	
190 B.C.	Kingdom of Armenia.	
164 B.C.	Hebrew Temple re-dedicated—celebrated today at Hanukkah.	
146 B.C.	Greece becomes a Roman province.	
133 to 27 B.C.	Roman Republic	
106 to 43 B.C.	Cicero of Rome. Stoics believe "virtue is sufficient for happiness and tranquility of mind is the highest good".	

100 to 44 B.C.	Julius Caesar (17 years older than his military rival, Mark Antony).	
87 to 44 B.C.	Sulla, Pompey, and Julius Caesar: Role of Roman Senate in democracy is defined. Caesar is assassinated in 44 B.C. by senators for fear that Caesar would make himself Dictator. End of Roman Republic.	
83 to 30 B.C.	Mark Antony, and love affair with Cleopatra VII.	
74 to 4 B.C.	Herodias (Roman historian).	
69 to 30 B.C.	Queen Cleopatra VII is last Ptolemy of Egypt.	
58 to 51 B.C.	Caesar conquers Gaul.	
50 B.C.	London, England founded.	
46 B.C. to 120 A.D.	Plutarch I (Roman historian)	
44 B.C.	Caesar assassinated.	
37 B.C. to 14 A.D.	Octavious, nephew of Julius Caesar, becomes Augustus Caesar and over Roman Empire. Reforms, expands, strengthens, and centralizes administration of government. Purifies Senate. Outlying proconsuls made departments of central Rome. Rome's "Golden Years" are in progress in all areas.	
31 B.C.	Naval Battle of Actium (at Alexandria, Egypt) between Augustus Caesar and Mark Antony, both of Rome.	
30 B.C. to 285 A.D.	Egypt-Roman Period.	
27 B.C to 330 A.D.	Roman empire: Roman period of principles. Much progress is made in science, art, and literary works under many Caesars.	
5 B.C to 33 A.D.	John The Baptist (Known as the man who "prepared the way" for Jesus Christ). beheaded by acts of Salome.	

7 or 2 B.C to 30 or 36 A.D.	Jesus Christ. Little is known about his youth, aside from his rebellion in the temple, until days of instruction and crucifixion. Essenes group, of which Jesus lived and was associated, is described in the Dead Sea Scrolls.	
Time of Christ	Gnostics believe "spiritual knowledge is revealed directly by God".	
January 1, 0000	Roman calendar, based on Sun, is instituted from Jewish, which is based on lunar months. January 1, 0000 of Roman calendar equals years 4,758 in Jewish calendar. Jewish year 5,743 is year 1983 in Roman calendar. Pope Gregory will decree further revision of Roman calendar in year 1,582 A.D. to Gregorian calendar.	

ANOTHER VIEW—600 B.C. TO TIME OF CHRIST

An important time in development of society, culture, religions, and philosophy existed from 600 B.C. to the time of Christ. Cyrus the Great (595-529 B.C.) and his grandson Alexander the Great (356-323 B.C.) from Macedonia, expanded and ruled the Persian Empire from Mesopotamia to the Arabian Peninsula, Egypt, and to the borders of India. Alexander died at the palace of Nebuchadnezzar II in Babylon and his empire was divided among his generals.

Citizens of the city-states of archaic and classic Greek Troy and Athens defeated the Persians at The Battle of Marathon in 490 B.C., preserving the independence of their land and island culture. Advances in Greek culture were made with lives and philosophies of many in poetry, sculpting, writing, philosophy, oratory, painting, medicine, astronomy, music, pottery, architecture, mathematics, geography, theology, grammatics, navigation, comedy, politics, athletics, and military actions.

Great thinkers included Archimedes (mathematics), Hippocrates (medicine), Homer (writing), Sophocles (military), Ulysses,

Xenophanes, Socrates, Plato, and Aristotle in philosophy, and many others.

The Peloponnesian Wars (431-404 B.C.) between city-states of Troy and Athens ended with an uneasy unity in Greece. The Trojan War at the city of Troy in present-day Turkey was battle with Athenians against Trojans upon the pursuit and return of the Grecian Queen, Helen (of Troy). The Greeks ended the Trojan War with the secret scheme and entry of the Trojan horse containing Greek warriors for breaching the gates, invasion, and defeat of Troy.

Recognized periods in Greek anthology are Archaic (750-500 B.C)., Classical (500-323 B.C.), Hellenistic (323-146 B.C.), Roman-Greece (146 B.C. to 330 B.C. A.D.), with a final phase in Antiquity of the 4th and 6th centuries A.D. Greek culture advanced in many fields of philosophy, mathematics, medicine, religions, recording of history, and government, with representation, as a first democracy.

Mycenaean people are believed to be descendants of the Hittites of Mesopotamia (now located in northern Iraq as the Kurd people), but who followed beliefs of Classical Greece. (The Hittites, now Kurds, would have migrated southward a short distance, 200 miles, from the Caucuses Mountains area of Russian Georgia, the larger group migrating westward, up the Danube River valley, through Switzerland, a branch to Germany, to Etruscan Italy, Gaul, the Iberian Peninsula, and eventually to the British Isles and Ireland—Source: "The Celts", St. Martin's Press). An island empire of Mycenae developed in the eastern Mediterranean Sea, having an advanced culture, but was reported largely destroyed by earthquake. By 1200 B.C. the power of Mycenae was declining. Mycenae dominance collapsed within a short time after 1200 B.C. All the palaces of southern Greece were burned, including that of the Mycenae, for unknown reasons. Displaced populations escaped to former colonies of the Mycenaens in Anatolia and elsewhere. A further theory is that the destruction of the palaces is related to the attack by the mysterious Sea Peoples, who destroyed the Hittite empire and then attacked the 19th and 20th dynasties of Egypt.

The people of Mycenae were strong in their hierarchy of gods. The construction of palaces had similar architecture to that of southern Greece, with interior presence relating to Egyptian pharaoh life. By 1200 B.C. the power of Mycenae was declining and palaces of southern Greece were burned. Much of the Mycenae religion survived into Classical Greece in their pantheon of Greek deities. They worshipped deities of Zeus, Jupiter, Hera, Athena, Poseidon, Agamemnon, Perseus, Andromeda, Artemis, and nymphs. Later, gods revealed in human forms with an animal as a companion or symbol, surviving as Satyrs or the goat-god, Pan. They believed in a "priest-king" system and the belief of a ruling deity in the hands of a theocratic society. Mycenae developed in the idea that each man was a servant of the gods, and sought a "moral purpose". They also believed in a "cult of the heroes", whereby great men of the past were exalted after death because of what they had done, and would, after death, travel to an island called "Elysion" for a happier existence. They believed that only the heroes could live in human form, and the souls of the rest would drift unconsciously in the gloomy space of Hades. The Mycenae probably believed in a future existence.

A theory argues that earthquakes played a major role in the destruction of Mycenae and many other cities at end of the Bronze Age. However, no conclusive evidence exists to confirm any theory of why the Mycenae citadel and others around it fell at time Mycenae was no longer a major power. In 468 B.C. troops captured Mycenae, expelled the inhabitants, and destroyed the fortifications.

First settlement of the present Apennines or Italian Peninsula was by the Etruscan people at the Tiber River at 758 or 728 B.C. Consolidation of the Roman city-states, influenced by Greek traders and Greek neighbors in the Hellenic civilizations at south of the peninsula, was completed. Rome became the most powerful empire of the world. Greek culture was a powerful influence in the Roman Empire. The civilization of the ancient Greeks was an immeasurable influence on the language, politics, educational system, philosophy, science, art, and architecture of the modern world, fueling the Renaissance in western Europe, and again resurgent during various Neo-classical revivals in the 18th and 19th century Europe and the Americas.

Carthage in Africa, at the southern shore of the Mediterranean Sea, was founded by Phoenician traders in Mesopotamia in the 1st millennium. Carthage was a major urban center, a strong military and trading force for nearly 3,000 years, conflicting with the Roman Empire, but was defeated by Rome after a series of wars.

Both Rome and Greece held polytheistic religious beliefs with a hierarchy of gods, with many gods of Greece presented in parallel with like gods of Rome. Most of Hebrew religions of Mesopotamia held belief in one god, with construction of a temple devoted to the believed—one God, Yahweh, in Jerusalem. Among sects of Hebrew thought were the Essenes near Jerusalem, of which Jesus was a member.

Egypt, as a separate military and cultural power with pharaohs, was replaced about 300 B.C. with outside powers from Libya and Greece. Many rulers of Greece replaced the pharaohs, including Cleopatra VII, who enchanted the Roman Mark Antony in a seemingly political love affair, to unite Egypt with the Roman Empire, thus ending the 10,000 year existence of Egypt as a separate power and pharaoh culture.

Vikings of the Norwegian north made periodic voyages and raids into Mediterranean Sea locations, and later extended and settled communities in the Normandy area of present-day France (and perhaps Gauls or Visigoths of the 5th century A.D. that invaded Rome in 476 A.D.

Homo erectus species and sub-species continued their migration from the Caucuses Mountains area into Mesopotamia, eastward to India, China, Mongolia, Indochina, Papua New Guinea, Australia, New Zealand, Java, and eastward to islands of the South Pacific Ocean. The author believes they migrated further to South America, migrating north from Peru to Central America, building a civilization paralleling their former beliefs, formed and modified over hundreds of years in spiritualism, culture, pyramid-building architecture, and a hierarchy of gods.

Two other figures are large in ancient philosophy and religion. Confucius (551-479 B.C.) was a Chinese thinker and social philosopher. His philosophy emphasized personal and governmental morality, correctness of social relationships, justice, and sincerity.

Gautama Buddha (Siddhartha Gautama) (563-483 or 400 B.C.) was a spiritual teacher from India and is considered as the Supreme Buddha. In Hinduism he is considered as one of the avatars of the

God Vishnu. He is also regarded in the Ahmadiyya Islamic religious movement, and the Baha'i faith. Baha'i is a religion and philosophy. Buddha is recognized as an awakened or enlightened teacher, to escape a cycle of suffering from rebirth.

THE ANTHROPOLOGIC VOID IN SOUTH AMERICA

There currently is a void in anthropology evidence in South America and Central America between 13,500 B.C. and 1,500 B.C., a 12,000 year period of believed occupation, with little currant evidence of a people existence.

It is believed that Peru was first settled at a very early time, perhaps as early as 15,000 B.C. or earlier, by migration from Southeast Asia, Java, perhaps New Zealand, Australia, Fiji, and the Hawaiian Islands. Anthropologists theorize early settlers to South America were from Japan, a population who had originally migrated from the Korean Peninsula prior to the modern era. A Korean theory proposes that their heritage was from Mongolia and diaspora also to the Philippines.

There is evidence in anthropology of probable sub-species of Homo erectus progressing their way from one Southeastern Asia Island to another, northward to the Hawaiian Islands, with voyages eastward, eventually to the Peru area on the Pacific Ocean coast. Many South Pacific islands would have been settled by one or more people groups of Negroid race from eastern Africa.

A civilization of Peru migrated along mountain highlands and valleys northward to present Ecuador, Panama, Costa Rica, and the Central America mountains, valleys, and coastal areas. Ancient civilizations in Peru date back to 1,500 B.C., with ancient cultures of Olmec, Maya, and later Aztec. A Norte Chico civilization (dated to 9,210 B.C.) lived at coastal central Peru. (Please explore: Archaeological Project Caral at Google for further information). There is evidence of civilization in Ecuador at 50,000 B.C. and cultures of Valdiva (3,500-1800 B.C.), Chorrera (1300-300 B.C), and other cultures of Tolito, Carchi, Chavin, Paracus, and Moche.

Asian culture repeatedly crossed a land bridge across the Bering Strait from Siberia, probably about 20,000 to 15,000 B.C., migrating southward and eastward by land in pursuit of migrating animals, and also by sea on rafts of hand-hewn logs or canoes along the shores and islands of present Alaska, British Columbia, Washington state, Oregon, California, Mexico, Central America, and also by inland routes. North American Indians populated mountains and plains of the western United States and Canada, and Eskimos (Inuit) migrated eastward along the frozen Arctic waters and lands to Greenland, Iceland, and perhaps Labrador, Newfoundland, Prince Edward Island, and other North Atlantic locations, perhaps in the Arctic Circle to Scandinavia and Russian Siberia. Indians populated northern Canada, the Laurentian Shield area, Great Lakes, and southward into the present southern, northern, and western United States. Indians migrated south into the American West, west from the Great Lakes area, and south to the present Carolinas, Kentucky, Tennessee, and Georgia regions. The Iroquois of New York state migrated first south and then back northward from present Georgia and Carolina regions of southern United States.

Indians were numerous in the western prairie states from Wyoming to Mexico. Arrow points have been found near Clovis, New Mexico, but Indians further populated the west extensively. It is undetermined whether the Clovis Points of 13,500 B.C. were historically from south-migrating North American Indians, or from a northbound civilization from Peru and Central America. It is noted that pyramids of Central America have a commonality with pyramids of ancient Egypt. The later Aztec Indians of Central America, about 500 A. D., are believed to have migrated south from southern United States and northern Mexico.

Therein lies the problem in tracing origins of the Clovis and western states Indians and the recorded ancestry of the Peruvian Indians. There is a void of anthropology finds of 12,000 years between the Clovis Points found in Clovis, New Mexico at 13,500 B.C., and the recorded Olmec civilization in Peru of about 1,500 B.C. It is logical that a chain of civilizations during this period is eventually to be found.

VARIOUS ORIGINS OF RACES IN THE WORLD

It is now proposed that origins of the major races in the world can be determined. The Caucasian race is believed to have originated in southern Asia at the Caucuses Mountains, in the present state of Georgia of southern Russia. These were Homo erectus species. They originally travelled a short distance of about 200 miles south to populate various groups as Hittites in Mesopotamia. Homo erectus in the Neolithic period had migrated briefly to east Africa and then eastward through India to China, Indochina, Papua New Guinea, Sumatra, Java, Fiji, Hawaiian islands, and to South America. Scientist have now compared the Denisovan genome with an additional 33 populations from mainland Asia, Indonesia, the Philippines, Polynesia, Australia, and Papua New Guinea. They found Denisovan genes in east Indonesia, Australia, Papua New Guinea, Fiji, and Polynesia.

A sub-species of Homo erectus were the widespread Denisovans of north China, Siberian Russia, and oddly enough to a branch in Java 50,000 to 30,000 million years ago. The Denisovans of north China, Mongolia, and Siberia are believed to have spawned the first North American Indians species in their crossing of a land bridge through The Bering Straits. Different sub-species merged including the Eskimos (Inuit) of the Arctic Circle.

The Japanese Islands were populated by migrating people from the Korean Peninsula. Some Korean lore maintains that, in history, Koreans evolved from Mongolia. Koreans, with Oriental features, also were in diaspora to the Philippine Islands and Indochina.

It is believed the Negroid race originated in eastern Africa as Homo ergaster species and were later absorbed as a sub-species of Homo erectus.

Ancient people of the Mideast, or Mesopotamian cultures, and natives of northern Africa typically have ruddy complexions, among other unique body features, which had developed in mating of Negroid and Caucasian races. A unique race developed in India from genetic crossing of Homo erectus Caucasians or Denisovans and Negro people from the nearby east Africa coast.

The races in Japan, Philippines, Korea, north China, Siberia, as well as Indochina, java, Fiji, Hawaii, and South America people may tie together with the recently discovered Denisovans as a linchpin to all. The whole arena of Oriental races appear to be related, with certain facial, body, or anatomy features, intellect abilities, culture, and perhaps other feature or qualities.

With this explanation there is association for origin in races of Caucasian, Negroid, Oriental, North American Indian, subcontinent India, with various other sub-species, minor racial differences, and cultures throughout the world as outgrowths of the parent species and as required by local environmental conditions.

THE AUTHOR'S OUTLOOK

Within, say 50 to 100 years, our world will experience momentous changes. The changes will be more eventful even than when Homo sapiens replaced Homo neanderthal species. Changes in scope may involve a new species, spirituality, psychology, the mind, our psyche, medicine and surgery, geology and land use, the weather, our diets, and hopefully relief for worldwide famine and world understanding.

Within the past 50 years human life has witnessed what could be called origin of a new species on earth in our worldly development. Robotic machines, and controlling electronics, have been developed by engineers to perform seemingly endless applications of routine, and now, judgmental operations.

Robotics are employed at industrial plants to weld metal, spray paint, perform quality tests, do mechanical manipulation, and perform even more processes.

Robotics have been developed to vacuum clean our household or business carpet, and to mow our lawn while we leisurely observe the process from our shaded patio and lounge.

Androids assist in human operations to relieve or aid physically impaired persons and to ease others in our household, business, and entertainment processes.

The computer is only a few steps away from independent self-control, with sensing and analysis of our human mental situations and our emotions, solving problems, with application of appropriate reflection of our "feelings" or states. These applications can be imagined and therefor soon be put into actual operation.

Robotic action can be foreseen whereby robots will independently produce other robots in ordering necessary components, handling materials and supplies, performing assembly operations, directing logistics, and more.

Robotics is employed even today in performing exacting surgery in repair of human hearts and in other medical procedures.

Parallel intelligence is witnessed today in cellular telephones, using seemingly endless memory and analysis abilities, and at increasing speed and size of instrument configurations.

Intelligent electronics is being developed further in warfare, which historically seems to be a major source for invention and development of new ideas in our society.

Like a new species, this new phase in our culture could well be named "Homo Roboticus", or some such name, to change and become our way of life over the next 100 years. It may well be that "Homo Roboticus" and Homo sapiens will exist together, like Homo neanderthal and Homo sapiens existed from 128,000 to 20,000 years B.C., one eventually destroying the other by its overwhelming technology.

Continuing continental drift will unceasingly change our Earth, with partial break-up of present continents in division at western North America (California), southeastern Africa, continuing uplifting at the Himalaya and other mountain ranges, and other land changes resulting from volcanic action, rift separation, subduction, and changes in ocean shore outlines.

Other significant changes will develop upon greater understanding of science and the nether world of added space/time dimensions in the world of antimatter, mind, and soul.

The birth and life of Christ marks a division in history, leading us through the Romans, the Christian, Jewish, Islam, and other spiritual beliefs and practices. Wars, philosophical thought, technical advancements, and cultural changes lead us to our present day. These events of the past 2,000 years are not within the scope of this portrayal from pre-cosmos existence and the development of man and his progression up to the time of Christ.

Sources:

1. The Great Courses (several courses and professors), Google—Wikipedia, The Free Encyclopedia (primarily for dates)
2. Western Civilizations, college textbook
3. King James Bible
4. Harper Study Bible
5. Archaeology magazine
6. National Geographic magazine
7. Scientific American magazine
8. Discover magazine
9. Nature magazine
10. Encyclopedia Britannica
11. "The Celts", St. Martin's Press, 1975

PART V

MYTHS, QUESTIONS, REFLECTIONS

THE MYTHS OF GENESIS

We realize that the faith of some readers will be so strong in their teachings that they will disagree with some presentations given here. Nevertheless we ask that all who read this book do so with an open mind to alternatives and how events might well have happened in reality.

It was understood in our childhood training that "God created Man in His own image". Logic would follow that we, as living, physical, humans, are not outwardly the same as God. Humans are physical and God, as far as we know, is non-physical, existent in the world of antimatter and in additional time/space dimensions.

When it is reported that God created man in his own image, it logically means that the soul, with mind of man, is like the being of God, inanimate and non-human, independent of physical body, with a existence oriented to morality, ethics, justice, humane, care of the disadvantaged, familial, omnipotent in secular life, reverent, worthy of giving spiritual praise, benevolent, and with endless life, all qualities lacking in animals of the forests.

An intelligent newspaper writer I enjoy has answered the question, "Why do we cling to beliefs, even after seeing facts that contradict them?" Marilyn von Savant, writing in many weekly newspapers, answers, "Because people get freaked out by the notion of being wrong about anything. It makes them feel insecure. If you can be wrong about this or that, what about all the other stuff you think you know? The more important the subject, the more unnerving the emotion".

"After we leave school, we tend to head down one of two roads: (1) We close our minds to new or different information while becoming more and more sure of ourselves, or (2) we watch, listen, and learn as we get older. The second road has way more bumps and curves, but it is also the path to wisdom."

It appears to the author that the writers of the Biblical stories in The Old Testament are not unlike that of storytellers, or authors of books, that tell a certain story. Take the situation of an oral history story teller. The story typically commences at a point in time. Let us say the scene is arbitrarily established at some particular era. The story develops from that point. No mention is made, typically, of the previous time or eras since the development of life, nor of the previous generations. The story teller commences his story at a certain point in time, and reports the story from that time forward.

In general, Books of the Old Testament were transcriptions of the oral histories perceived at the time of telling. This applies especially to the early chapters in the Book of Genesis.

Such, it seems, is the chronology reported in the Old Testament. Let us disregard the given story of mystical creation, reported by the Irish Bishop, James Ussher, in the 19th century, to be in year 4004 B.C. Let us instead relate that moment in time to merely when the Biblical authors chose to begin the story and for which previous actual history is not revealed. The events prior to start of his story, which were thousands of years before, incorporate many phases in mankind's historic physical, mental, and spiritual development, through various species of man leading up to that moment of time are largely disregarded. At a later time the actual previous events are ignored at start of his story.

The pre-history prior to start of the Biblical story is real, evidenced by archaeological findings of man type fossils laid out in pre-history as facts of existence. This pre-history is the story of life's start on earth, by still an undetermined event, with development and growth of animals, plants, microbes, bacteria, yeast, and like organisms. Consequently the various species were followed in mankind, to that time Biblical authors chose to commence telling the story.

The stories were conceived with certain later concepts that reflected the times of man, from start of an Agrarian culture in the 10th to 8th century B.C. to perhaps the 6th century, when many events in history were written and codified.

1. Hebrew lore reflecting thoughts and practices existing in Mesopotamia, at and before this time:

a) A concept of Heaven, or the Egyptian Underworld, and a concept of the undesirable "Hell".
b) Reflection of the Egyptian "Book of The Dead" in articles detailing the "Ten Commandments".
c) Multiple gods, although some were considered "senior" in the god hierarchy.
d) Concept and description of time, like 40 years and 40 nights, was not exact but merely intended to mean an "extended period".
e) Preparation of the dead with artifacts supposedly needed for entry into an afterlife.
f) In Egypt, flowers presented at death (a custom continuing to today).
g) In Egypt, nametags attached to toes of the deceased, to identify the residing place of the dead and soul, are possibly forerunners of present gravestones.
h) Perceived continued life in an afterlife or Heaven.
i) Existence of pagan worship of gods and animals.
j) A close association of man and gods.
k) A hierarchy in history of Priest, Saints, and Holy Men, supposedly having soul communication with gods.
l) Male superiority exercised over females.
m) Group worship, reflected at homes and gatherings later in churches.
n) Idols and artifacts in religion.
o) Preservation of the deceased body, by addition of red ocher substance to grave of the deceased, perhaps a forerunner in thought of today's casket for burial for preservation.
p) Construction of pyramids, temples, or mausoleums for religious practices.
q) Symbolism for gods (i.e. the Burning Bush at Mount Sinai).
r) Construction of mythology regarding existence and actions of their gods.
s) Organizational leadership reflected in that of Moses over the Hebrews of the Exile.

2. Many adventures, lore, and myths, such parallel as told in Homer's "Odyssey" of the 1400 to 900 B.C. period, were carried forward in the first telling of tales portrayed in the early passages of the Book of Genesis. Though exaggerations, they were perceived by people of that time as facts. Homer's "Odyssey" is a tale of hard-to-believe details applied to underlying basic events of adventure, morality, and justice, told in a mythical manner. (The story of "The Iliad" precedes the "Odyssey" in time by telling details of war in Troy and heroic persons involved.)
3. Presentation of original thoughts and extension of other mystical and religious beliefs, showing that Hebrew writers could likewise create stories of events.
4. The Creation story, told in the Old Testament Book of Genesis, was demonstration that the Hebrews too, in addition to Homer, could create epic tales.
5. The Old Testament indicates that Eve, and supposedly Adam, were created from clay of the Earth. This does not provide for the intricacies of the body, with all features of skeleton, heart, muscles, digestive and circulatory systems, eyesight, mouth, taste, and numerous other systems that support life. Logic and reason will force discard of this theory, even though it may be salvaged as "the work of God".

God works through nature to manifest his ways. In observing evolution, for example, the creations of God have worked through reasonable and logical progressions to formulate new hominoids, cultures, societal developments, and progression of life and living in our world. Commands of soul and mind were activated by bioelectrical energy to the genome, which in turn activated enzymes, changing physical cells or mental thought. There really are few miracles that cannot be explained with adequate time, effort, and knowledge. The same physical laws have existed since the creation of Earth and man.

An Engineer, for example, may have desire to build "his dream automobile", but he must work through the many and varied individual steps of constructing parts of heavy steel, stampings of sheet metal, purchasing certain parts, metal finishing, and painting before the automobile is completed and working. It is doubtful whether even God could "Abracadabra" and "zap" to make the automobile complete and working instantly. To integrate religion and science we need be reasonable and logical in our assumptions.

Caution must be observed in reading the Book of Genesis from a tendency for the writers to exaggerate in the tales, lore, and myths. We must separate reporting of myth and magic from actuality and realism.

The author believes the portrayal of some events described in the Book of Genesis is illogical and have not been substantiated in archaeology findings. It is mythology: Stories were fabricated by ancient individuals who developed, or in all honesty, repeated a story, as we are sometimes inclined to do when facts are not known. These were stories told and "handed down" through many generations as oral history, with some errors and exaggeration in the multiple retelling by pre-history mankind to describe their origins, but without direct knowledge of the events that preceded them over the prior 14.7 billion years.

These stories were adopted in ancient writings of the Old Testament Book of Genesis, and told to successive ages of Biblical storytellers as facts, supported only by intense belief in faith. In a way, those believers of the oral history were unknowing of actual facts, but were perceived by some as the true story, as were the original story tellers, again, driven by faith in believing the accepted stories.

It is illogical to believe, what the author might call a "Zap Theory", that the geography and atmosphere of the world were created quickly in seven days, or that the bodies, brain, and minds of male and female were immediately composed and constructed of clay.

It is believed the physical, geologic, and atmospheric qualities of Earth were generally, as rational, logical, and believable during Earth's and modern man's evolution, as they are today. Intercourse, as the cause of birth and biologic development of physical persons, is normal and dates back to at least 350 million years ago, occurring first in ancient sea life and mammals.

This concept is not a charge that God did not instigate, plan, and develop life on earth as a process of intelligent design and directed evolution. The author believes evolution was too organized and too logical, and not chaotic from accidental mutations and acts of biological development, to have not been controlled, and was developed in subsequent biologic actions of the genome, from the first instance of life existing on earth. Creation came about, but many thousands of years in time before being interpreted by Biblical writers.

Geologic history is much older than the 4004 B.C. time that an Irish Bishop, James Ussher proposed, but religious myth, based on hearsay and legend or spiritual history reports, may have been derived at that time.

The story of Adam and Eve may be indicative in recognition of man and women and a chain of events of a Biblical nature. With changes in culture, theology, spiritual insight, and even refinement of details and concepts in the mind, there is just too much evidence in fossils and anthropology that proves mankind is much older than 6,000 years ago. In fact, humanoid creatures and man-like species can be evidenced 6 million years ago. Presence of brain and mind changes, incorporating soul, and traceable at least to 600,000 years ago to Homo erectus, Homo neanderthal, to Homo sapiens, and to Modern Man.

To hold and state the firmament, daylight, geology, plant, and biologic life of man started only 6,000 years ago is not rational. In fact, anthropologist and historians report and cite findings and cultures in Egypt, Mesopotamia, and other world areas that evidence man much before, to Pharaohs and kings of the Old Kingdoms of Egypt, a first king in Mesopotamia, Emperor Sargon of Akkad in Assyria, and life in the agrarian culture of man as early as 10,000 and 8,000 years ago, without any reference to a Garden of Eden or related events, even in the ancient fabled story of King Gilgamesh, whose life parallels a fabled great flood in his time, a periodic event that may have been common, predictable, and caused by heavy rainfall in the mountains and headwaters near Turkey. Investigations in geology prove events of prior Earth, and in anthropology events of previous life and cultures.

The Biblical Creation story, with incarnation of soul to the body of man, might generally be true, but biologic, geographic, species development, mind creation, and refinement is much, much older. The Biblical Creation story is acknowledgement of the existence of Earth and its features, of man and women, and that the soul of man existed, and was a direct link to God. In thinking of Creation, one must discern between physical, biologic, geologic, anatomy, cultural, and mind Creation, as opposed to reporting actual history of the Biblical world creation.

The findings of science and the true realities of religion can be reconciled. The findings of particle science and quantum mechanics during only the past 100 years are bringing this dichotomy of origins closer and closer together.

1. Many elements of the Biblical creation story are symbolic. The story of original man and women, constructed per the Creation story of a universal clay material of the earth, are devised. The reptile snake is symbolic of the constant temptation to which man and female are subject. The apple is symbolic of knowledge gained outside the realm of God, that is provided by God, that has the potential of being ruinous to people on earth.
2. The Biblical story of Moses and Pharaoh are also largely symbolic. The exact determination of which Pharaoh existed at the time has not been determined. (As many as 11 pharaohs named Ramses are recorded in ancient Egyptian history.) The catastrophic events related in the debate are illogical and portrayed as if God and Moses were "like stage magicians in a road show, pulling rabbits out of a hat" so to speak. This seems ludicrous in theology.

3. The Biblical miracle of multitudes being fed by a limited number of fish, or by Manna growing from stone in the desert, is possible but greatly exaggerated.
4. Jesus may not have been brain-dead at the Crucifixion, as defined in modern medical or biologic knowledge. Soul to soul communication still existed with God, even after a state of stress, unknown in medical science of that time.
5. Lazarus, who was resurrected from a supposed state of death, may also not have been brain-dead, and therefore able to have body restored in life through soul-to-soul communication with Jesus.
6. Moses, receiving the Ten Commandments, could well have devised (or later Biblical Septuagint translator writers devised about 270 B.C.) and exercised his own interpretation of events occurring on Mount Sinai, paralleling his own deep and somewhat dramatic manner, and his knowledge of the Egyptian "Book of the Dead", known through his own residency as a child and as a youth in an Egyptian house of royalty in which he was raised, supposedly from infancy. The Ten Commandments are amazingly parallel to the "Egyptian Book of the Dead", which were developed to tell the "Keeper of the Underworld" of good things done by the Pharaoh in his life and bad things not done. It is questionable whether the Ten Commandments were written in stone, making the two tablets very heavy for transporting in the wooden ark, and permanent (the tablets have never been found). They may well have been written on animal skin, possibly even by Moses himself, and in the Hebrew language, based on his supposition of what God would have said and provisions of "The Book of the Dead". Writers may have interpreted the Moses meeting with God. The Septuagint Translators interpreted the words attributed to Moses. These were Pre-Christian writings, a Greek version of The Old Testament, so-called from the legend that the translation was made by 70 emissaries from Jerusalem for Ptolemy II about 270 B.C.
7. It is suggested that Moses may have had, to some extent, what modern psychology calls "Delusions of Grandeur", a feeling

of ability to aggrandize himself, perhaps before God, of his exceptional abilities, as performed by Egyptian pharaohs. This may have had root in the enigma of having been raised in the daily presence of an Egyptian pharaoh and the royal family and have aspired to that royal stature.

Motivation of Moses in devising an account of the event, would have been to convince and encourage his followers in the exile from Egypt, who were trending after extended time in the desert, even after "God given miracles" (such as Manna as food), to not abandon belief in God in favor of return to belief in Baal.

8. The epic of the Great Flood could be explained by earthquake action at the present Straits of Hormuz. Mountains exist on both sides of the straits, suggesting that a chain of mountains existed at one time between the two shores. This chain of mountains would have held back waters of the Arabian Sea. Earthquakes, common in the area, could have opened the mountain chain allowing waters of the Arabian Sea to flood the low land of the present Persian Gulf.

Flood waters would have washed upstream into both the Tigris and Euphrates river system causing water to overflow in a flooding action. Flooding and wash would have occurred extensively up the river plain as far as present-day Turkey, (the reported resting place of the Ark of Noah) and have been raised to mountain-level by subsequent earthquake and earth-movement action.

Flood waters would have evaporated into the atmosphere, into a weather cycle causing further rain, adding to and prolonging the flooding condition. This condition happened in the state of Iowa (United States) with continued flooding by the Mississippi River in the early 21st century, and at previous years, with subsequent evaporation, and more rainfall.

Extensive flooding in this area is told in a previous historic, and perhaps fictitious, Tale of King Gilgamesh, even before time of the reported Great Flood and Noah. In fact, the epic of the Great Flood may have been recalled and told as a happening after, and resembling the flood in time of King Gilgamesh.

These interpretations are challenged, not to dismiss the general accuracy of the events, or of the Holy persons themselves. Jesus Christ was certainly an amazing and wonderful person, with essence of God, and with His Spirit, a great teacher, with an uncommon ability in correcting man-development situations, proposing actions, and communicating to his followers. Christians believe that Jesus Christ was God, Man, and the Holy Spirit. Some Jews, and later Moslems, believed instead that Christ was merely a Prophet.

Moses, even if the event did not happen or was exaggerated, was acting in a manner that would advance belief in his God.

In general, it is believed the Earth and world were as rational, predictable, and subject to scientific laws and happenings 2,000, 12,000, and more years ago as they are today.

Miracles are facts before they have been explained.

DID MOSES PROPERLY SERVE HIS GOD?

The Exodus is the story of the departure of the Hebrews from Egypt, descried in the Book of Genesis. Moses reported in presenting the Ten Commandments, saying in effect, "these are rules to be followed to live a truly spiritual life under God". They seem to have been patterned after the Egyptian Book Of The Dead, in which the pharaoh declared good certain events he did and certain evens he did not do, making him eligible for entrance to the "Underworld".

There is question posed by some anthropologists, (and presented in Google) whether Moses did, in fact, go up to Mount Sinai and had conversation, by either actual voice or mental/soul communication, with God, and further whether the Exodus happening about 1200 to 1250 B.C. by Moses even existed. However, even if not fully accurate, the event would have been reported with a morally correct purpose: To draw his people to a better appreciation of God. Some present archaeologists have abandoned scholarship study of the Exodus of 1,200 to 1,250 B.C. Modern theories tend to concentrate study on an earlier Exodus prior to 1,440 B.C. (Source: Google, Exodus)

Moses, at the Exodus, portrayed the kingly ways of the Egyptian pharaohs. Moses seemed to have developed an Egyptian pharaoh-learned way in personality that affected how he was involved, both in leadership and in the dramatic.

Recall that Moses, as an infant, was found floating in a river, in a reed basket, by a daughter of a Pharaoh, who took the baby Moses to her home, where Moses was raised in a home of royalty. He surely was loved and perhaps even a bit "spoiled" as he matured to adolescence. As a young man he was assigned to assist in directing slaves in performing their work projects. The ancient Old Testament Bible relates that upon seeing his leader-supervisor mistreating the slaves, he physically struck the leader as protest against the harmful treatment of the Israelite slaves, and in fact changed allegiance from royalty, to become a leader of the Hebrew slaves.

He reportedly organized in secret an exodus of 600,000 slaves (Genesis 12.29), and their animal herds, to leave Egypt to go to the land of Canaan. This involved a few problems in leadership, one being how to secretly move out a sizable group of people in the Exodus before the Pharaoh's guards would be aware. He then had to lead the people across the Red Sea, a seemingly impossible task. Again, the story goes, he called upon God to hold back the waters of the Red Sea, allowing his followers passage through the muddy seabed to land on the opposite side. When his followers had passed through, and the Pharaoh's military pursuers came to pass through the opened waters, he had God release the waters to drown the pursuers, a rather dramatic incident in its telling.

It is numerically extraordinary that the Biblically-reported 600,000 Israelites and their animal herds, would logistically be able to complete the journey: Six hundred thousand people, traveling 10 abreast, and allowing an average of 6 feet apart in length, not counting livestock, would form a procession over 60 miles long.

Secondly, there is no evidence that Egypt suffered loss of this many people that would cause a demographic and economic catastrophe there.

Third, there is no evidence that the Sinai Desert ever hosted (or could have hosted) this extent of people and their herds, nor of a massive population increase in Canaan, estimated to have only 50,000 to 100,000 at the time. The view of mainstream, modern Biblical scholarship is that the Exodus story was written, not as history, but to demonstrate God's purpose and deeds with his chosen people to Israel. It has been described as attempted recording of history, written as religious philosophy.

The evidence of the origin of Israel is "overwhelming" and leaves no room for an Exodus from Egypt, or a 40-year pilgrimage through the Sinai wilderness. For this reason, most archaeologist and Egyptologists have abandoned the archaeological investigation of Moses and the Exodus as a "fruitless pursuit". It has become increasingly clear that the Kingdoms of Israel and Judah has its origins in Canaan, not Egypt. (Source: Google Wikipedia, Exodus)

It somehow took 40 years to reach Canaan from Egypt under leadership of Moses, his brother Aaron, and later Joshua, a key leader. It must have been a bit dramatic with a sizable group of people and flocks of sheep and other animals. They were faced with times of hardship, hunger, and thirst. These times and conditions were reportedly met with miraculous events told of discovering Manna growing out from desert rocks, and fish reportedly enough to feed his followers.

After the reported 40 years the Hebrew group, traveling in the desert, were weakening in belief of the newly presented God, and were reverting back to be followers of Baal, their former god. Moses, having been raised in the household of an Egyptian pharaoh, realized that he would need to communicate to his followers that he possessed a direct communication with God, and presented himself in a way that would meet the desired ends of influencing his Hebrew followers of fleeing from Egypt. Moses reported that he had communication with God, and in a dramatic form of God speaking from a Burning Bush, and a stone-written detail of the commands of God.

At a mountain, Mount Sinai, the story goes, Moses climbed, alone, to a spot where he met God in the form of a Burning Bush, and God created, by burning or etching into rock tablets, the Ten Commandments. The weary-from-traveling Hebrews were reverting to their former pagan ways and forsaking the monotheistic god. Enraged by the Baal-worshipers idolatry, the stone tablets, in anger, were either thrown down the mountain side by Moses to the disbelieving followers, or shown to the unruly and discouraged followers in attempt to change their Heathen-like ways.

Some sources state that Moses never reached Canaan, having died in Moab along the route. His brother, Aaron, his wife Miriam, and Joshua completed the leadership to Canaan. This was a trip filled with drama, and with miraculous events again reported, akin to the dramatic events he commanded in presence of Pharaoh Ramses to demonstrate the power of his God, and reporting that Pharaoh also told of extraordinary events.

The point of this presentation is that excessive drama in reporting seemed to be a part of the life of Moses. The drama may have been real, but in all probability, it was somewhat exaggerated, perhaps for the purpose of showing the power of his God.

There is question regarding the accuracy of this reporting by Moses considering the realism of survival and challenges during the 40 year Exodus. Those questions must be tempered with required realism. We must conclude that Moses was inclined to report events in a manner with dramatic appeal.

The meeting with God on Mount Sinai, the conversation with God, the Burning Bush, the defining and imaging of the Ten Commandments, and the fate of those stone tablets into the future, as well as the previous meeting with Pharaoh, should be questioned to ferret out the true incidents.

This requires separation of myth from fact in order to bring truth in reporting to religious history, thereby strengthening religion in the minds of current religious followers, rather than leaving obvious questions in the history of religion that turn away rational people who require realism in what they will believe regarding God.

It may be difficult for some modern-day Christians to disagree with the so-called miracles experienced by Moses as reported in the Book of Genesis. Regardless of the source, The Ten Commandments has set rules for over 3,000 years for cultures to follow as guidelines in morality, ethics, spiritualism, and adherence to a monotheistic God. It may be that we actually owe to the Egyptian " Book of the Dead" for formulation of the Ten Commandments.

In many ways Moses seemed to characterize personality traits, perhaps parallel to that of Pharaohs. He seemed to demonstrate not only leadership abilities, but also traits of being self-centered, and even what we call today "theatrical", as were some of the Pharaohs in order to impress their followers in both godliness and government. Indeed, a Pharaoh maintained his near godliness with close association and worship of the panoply of the various Egyptian gods. The Pharaoh designated his life to be the center of spiritual attention and was not beyond fabricating tales that would amplify his own greatness. This was the attitude in the household and empire-leading culture in which Moses experienced and learned how to exhibit the desired way of royal living.

In general, Books of the Old Testament were written transcriptions of the oral history perceived at that early time. This applies especially to the early chapters in the Book of Genesis, written by Moses. Dramatic characteristics are told in the Book of Genesis, whereby in a previous event, Moses posed to the Pharaoh Ramses (which of the 7 pharaohs named Ramses is not clear) magnificent and hard-to-believe occasions involving God-invoked plagues of locust, widespread drought, turning of water to blood, turning a rod into a snake, a plague of frogs, a plague of gnats, swarms of flies, death of Egyptian cattle, boils and sores on Egyptian people, hail, three days of darkness, and death of the first-born of people and animals.

An event preceding the Exodus, of the Pharaoh ordering that doors of certain violating subjects be crossed with the red blood of sheep-lambs, seems overly dramatic.

The leadership style of Moses in directing Hebrew slaves would seem somewhat dictatorial in his ruling by decree, an assumption of privilege granted by his self-appointed authority.

This leadership characteristic, or manner, was further demonstrated in his reporting of events upon Mount Sinai. In the Biblical story, Moses proceeded by himself up the mountain, so there were no witnesses to the event. Moses could report the event as he desired.

The cause of Moses's ascending Mount Sinai was to encounter God, with an event that would hopefully defeat the increasingly active return of the worship of the god, Baal, by the Hebrew followers.

Moses, in the Biblical account, met God, who was supposedly present in the form of the Burning Bush. God supposedly revealed the contents, or rules, that should be the laws of the Hebrews. The Commandments were supposedly etched, letter by letter, upon the stone tablets. (Where flat, slate-like stones were found on the mountain is not revealed in the writings.)

In the report, the two tablets were thrown down the mountainside in anger to the midst of the errant and increasingly active worshipers of Baal, broken, and supposedly other tablets were re-created by Moses.

These tablets, heavy as they would be, were supposedly placed in an Ark, to be hand-transported with the exile-driven Hebrews to Canaan, to be treasured and revered later in the Temple as a place of worship. It would seem more logical that the Ten Commandments were written on sheep skin or papyrus then on heavy stone, but also would make the documents subject to age deterioration.

It is a point of wonderment concerning 1) how the stone tablets were imaged, on both the first set of tablets, as well as a possible second set of tablets, and 2) in what language the tablets were written, supposedly in Hebrew, but God in an omnipotent scope, could have written in any of many languages present at the time, and 3) the inscribed stone tablets have so far, in over 3,000 years, never been found.

This was an event, witnessed and reported solely by Moses himself, not unlike a Pharaoh back in Egypt might stage and report a self-involving event of wonderment in order to affirm his holy image to the Pharaoh's followers.

The Egyptian "Book of the Dead" was a document in which each pharaoh would make record of what good in his living he had done, and what evil he had not done, all to be presented upon his death to the "Keeper of the Underworld" and would thus ease his entrance to the Underworld. The rules, as outlined in the Ten Commandments, are an amazing parallel to the rules of the Egyptian Book of the Dead, perhaps remembered by Moses in devising the Ten Commandments. These were rules with merit, but possibly not originally etched in stone by God.

As an alternative to the details, the event could well instead have been modified: Moses took it upon himself to climb Mount Sinai alone. He would there encounter a Burning Bush, or another object, to be an embodiment of God. In the mind of Moses, this representation of God could have been real and actual.

It is proposed that God did not in fact image the two stone tablets, but instead told Moses (by thought transmission) of the Ten Commandments, in silent communication, soul to soul, reported them to Moses, whereby Moses himself wrote the message, not on stone, but perhaps on animal skin which was a way of recording events in that day. Animal skins would have been much easier, and much lighter to transport down the mountain, and later be carried in the Ark during the balance of the passage to Canaan. The animal skins would have long ago deteriorated in 3,000 years, accounting for an absence of physical being or evidence of the event.

As alternative, Moses, like a Pharaoh in the royal home in which Moses had experience, produced the story to his followers to renew their faltering faith in God which Moses had previously revealed to them, and would override faith in Baal and other gods, which had been prevalent at that time and place.

Regardless of the source, the "Ten Commandments" have remained for over 3,000 years as a moral, philosophical, and spiritual guide for mankind to follow in religious life.

The Manna told of in the Bible (Exodus 16:31 and other), mysteriously appeared from the rocks that served to feed Moses and the Israelites fleeing from Egypt. This may have been a form of Lichens.

Lichens are ancient, abundant, and often misunderstood growths, many times seen in forests, but also in dry areas. They are not even plants, but grow or appear on the surface of trees and rocks. For hundreds of years naturalists didn't quite comprehend what they were. Originally these odd forms were thought to be part of a plant kingdom. Eventually microscopy enabled scientist to identify lichens as composites of mutually beneficial fungi and algae.

Fungi are the more dominant role and cultivate by photosynthesizing algae for food, in return providing them a shady, moist, vitamin-rich environment. Scientists have classified lichens based on their species. Historically lichens are symbiotic in cooperation with ancient algae and fungi. Lichens thrive in regions of the earth where other vegetation would die out. Lichens are found further north and south than any other plant, and at altitudes up to 20,000 feet. They grow in the desert and on the surface of various items of matter. There are 15,000 species of lichens. Some grow 9 feet tall and others are smaller than a pinhead.

Many persons have observed a simple form of lichens in woods, known as "British Soldiers", with a distinctive bright red cap atop green stalks. Another variety is "Old Man's Beard" that can run 3 feet long and hang from trees in the manner of "Spanish Moss"". A colony of lichens might look like a plant, an uncomplicated fungus, or even a patch of rust.

Lichens are by no means rare and could very well have been the Manna that served as needed food for some of the 600,000 Israelites fleeing Egypt on their way through the Arabian Desert. (Source: Discover magazine, November 2009).

The actions of Moses at Mount Sinai, and in the desert, were a way to communicate the desires of God to the Hebrews on earth. Moses should not be vilified for choosing this route, but be honored for having accomplished the perceived desires of God and disseminating that information to humanity. It is not the intent to vilify Moses, but to provide a logical alternative sequence of events that provide reasonableness, logic, and possible truth in the history of an Exodus.

The drastic actions of Moses in advocating monotheism in his unfaltering belief of one God was a turning point in world history and an event to model for people believing in one God, reflecting to Judaism, and for development later of Christianity, and later Islam, the three comprising 90 percent of the world population. In this event it can be said that, "the end justifies the means".

A CONTROVERSIAL IDEA: WAS CHRIST BRAIN DEAD AT THE CRUCIFIXION?

> *The death of Jesus Christ at the crucifixion and his return to life is a cornerstone of Christian belief. The reader, of course, is free to believe, or even consider the possibility, but it is presented here as an alternative scenario only as a medical possibility, in pursuit of Truth, and may actually strengthen the faith in Christ by disbelievers who question the event of rising from the dead.*
>
> *Whether or not Christ died and arose from the dead is not critical to belief and appreciation of Him as a miraculous and wonderful person, teacher, and embodiment of God the Father, the Son, and the Holy Ghost, as presented in the Neocene Creed of year 381 A.D. Certainly, by his actions in life, Christ was a human who possessed at least a portion of the soul and the essence of God within him.*

To understand stress occurring in Christ, we must recognize that the body was, after all, human. Today we know much more of this emotion called medical stress than was known 2,000 years ago during the traumatic event of Christ's crucifixion. We can intellectually stand back, so to speak, and take a clinical view of just what was happening with the physical, human body of Christ. Stress was not known or named 2,000 years ago as a mental/physical condition. People having symptoms of this affliction were known just as "bothered" in their thinking process.

Stress occurs within all mammals including humans, to varying extent in the body and brain when events happen which the brain cannot readily accept as deviations from its normal state. Our brain becomes overloaded in its attempted ability to assimilate and solve situation problems, with arising emotions of disappointment, fear of the unknown, anxiety, dislike, withdrawal, anticipation, depression, and other reactions in the body.

The brain and emotions arising medically cause change in the physical body. The stress response system is a natural and highly adaptive survival system that will divest energy from within the body in event of immense problems or danger situations. Chronic stress can turn the stress response from a safety mechanism into a real problem for our physical and mental well-being. When we have a thought, an emotion, a memory, things change with the body affecting hormones, pancreas, thymus, nervous system, and release of hormonal messengers in our sympathetic nervous system for emergency arousal, affecting the brain hypothalamus (the brain-stem area that regulates our body), the limbic system (which deals with emotions), and the brain cortex (that deals with abstract reasoning). The brain reacts to a real physiological event that changes the body's functioning in glands, hormones, and blood stream.

Medical Stress affects the way our brain behaves at the neurological level during periods of emotional trauma. It affects the body, in the cardiovascular system, blood pressure, heart rate, muscles, and other powerful physiological factors.

It can be questioned whether Jesus Christ was medically brain-dead upon the crucifixion. His body could well have been in state of medical shock, giving all external appearance of being lifeless. Electronic measurements of brain-dead status, as opposed to merely heart or pulse stoppage, were developed in the 20th century and are now the standard for determining true death. Modern embalming, which incidentally assures total death, initially became standard procedure in the mid-1800s, although use of instrumentation may turn out to be a more positive method of determining death status.

Two thousand years ago, medical science, elementary as it was, knew little of common death, as opposed to a modern determination of cessation of all activity in the brain, following apparent death of the body muscles, organs, breathing, heart pulsation, and other actions of body organs.

In fact, ancient Egyptian medical knowledge held the heart, not the brain, as the locus of life. During mummification the matter of the brain was meticulously extracted by tools and destroyed, or at best, placed in a canopic jar. Medical knowledge at time of Christ was generally near that stage, with no knowledge that death is not complete until all electrical waves within the brain have ceased.

Neither was state of medical shock understood at time of the crucifixion. Shock provides all outward appearances of lifelessness, but in fact, if not brain-dead, may allow the body to actually live on. As gruesome as it may seem, perhaps many warriors of the time found on a battlefield, may have appeared dead and were buried, still technically alive by not being brain-dead. Even the Biblically-stated event of a Roman soldier puncturing the body of Christ with a spear in order to quicken death would have produced additional stress in the body of Christ.

This phenomenon may also have occurred in the Biblical story of raising Lazarus from a state of apparent death. There perhaps was communication between the soul (of Jesus) to soul (of Lazarus).

Regarding the Crucifixion, true, it would have been a miraculous recovery, but not unheard of in modern medical science of recovery by a body, for Jesus Christ to regain adequate strength and organ action in the 36 to 40 hours reported between Friday afternoon, all of Saturday, and a portion of Sunday. Christ subsequently may well have exited his own tomb by his own efforts, later made himself known to certain followers as related in the Bible, and later died, unrecognized at the time by other people as the crucified Jesus Christ. His soul would then have transpired into Afterlife. There has been some archaeology reporting of the tomb of Christ's family, including the remains of a person in a tomb cavity, marked with a name of "Jesus".

According to the Dead Sea Scrolls there was a group, the Essenes, with whom Jesus lived, and which called themselves Christian even before the crucifixion of Christ. They believed and followed his teachings and preaching, including the Sermon on the Mount. Belief in the wonderment of Christ, and definition of Christ in relationship to God, therefore was core before the Crucifixion and Resurrection. Resurrection is not necessarily core to belief in Christianity, but is merely an affirmation of the Christ/God existence. It is regarded as a physical proof of the Christ/God dual existence. The core of belief in Christ should be his existence, teaching, and actions while he was on this Earth.

During a crucifixion process, while hanging from a crossarm by the hands nailed to the crossbar, the interior organs of the body slowly fall, descending and accumulating in an unnatural agglomeration within the body. This creates physical stress within the living body. Continued, the body will die by internal strangulation. If not relieved at a proper time, the body experiences physical and biological stress, with outward signs representing death. (There was an element of hurry to remove the body of Christ from the cross before the coming of sundown, the start of the Jewish Sabbath, and place him in the designated tomb.)

In terms of modern-day medical knowledge, the experience might well be described as medical or stress shock. Shock sometimes occurs when a human body undergoes, for example, sudden electrical voltage common to electrocution, or other traumatic event short of brain death. In state of shock, the person is commonly not brain-dead.

Medical Shock occurs when the body perceives a dire threat, and it begins an automatic progressive shut-down of systems in an effort to protect the brain and heart. This ultimate life-saving response is today called medical shock, and can quickly lead to death unless corrective measures are taken. The body of Christ was supposedly nailed to the crossbar, and the cross erected vertically, at about one o'clock P.M. on a Friday and removed about four o'clock P.M., being then placed in an above-ground tomb, taken as deceased. The body of Christ lay in the tomb for the balance of that day, all of Saturday, and into Sunday. When the body became missing from the tomb, supposedly a rock of some unknown dimension and weight at the entrance to the cave-like tomb was removed. It would be possible, though miraculous, that lying in the tomb, the body of Crist recovered, at least somewhat, from the symptoms of medical shock and regained ability for movement in life. Again, it would have been medically possible that Christ himself was able to partially recover, medically, and exit the tomb himself, rolling the rock, if that in fact existed, aside to allow exit from the tomb, later to meet with persons, giving them the impression of his arising from a state of death.

His body may later have truly expired. Some archaeology findings hold that the remains of Christ were later deposited in a family tomb, present with other members of his family.

This portrayal is intended not to subtract any qualities of Christ. His actions and declarations in life prior to crucifixion were truly amazing and worthy of honor with beatification, worthy of the worship by believers that followed his crucifixion. His mind and soul could well have been a portion of God, a mainstay of Christian belief.

This portrayal is intended to provide rationality, and to eliminate disbelief in minds of agnostics, and to advance religious doctrines to understandable and believable terms consistent with functioning of the human body, albeit with the essence of God in the mind and soul in Christ.

GNOSTICS

This concept is also somewhat related to the stated views of the Gnostics of the day. They believed, among other beliefs, that Christ did not arise from the dead. Gnostics had a set of religious beliefs and spiritual practices common to early Christianity, Hellenistic, Judaism, Greco-Roman mystery conditions, Zoroastrianism, and Neo-Platonism. They possibly had an advantage of contemporaneous observation and analysis of the event.

Jesus was identified by some Gnostic sects as an embodiment of the Supreme Being who became incarnate to bring gnosis (knowledge superior to and independent of faith) to Earth. Others deny that the Supreme Being came in the flesh, claiming Jesus to be merely a human who attained divinity through gnosis and taught his disciples to do the same.

Typically Gnostic systems are loosely described as being "dualistic" in nature, meaning having a view that the world consists of two fundamental entities, the relation of God and the world, and that of man and world. Gnostics proposed other strong beliefs. (Source: Google)

OTHER MIRACLES

Two biologic events in physical development of humans remain unexplained.

The first is the act of joining soul, and all of its characteristics, to the physical hominoid body. In study of the cosmos, earth, and species history, this initial incarnation probably happened in the species Homo erectus at about 600,000 BC. The second biologic event is the insemination of Mary to give birth to the baby Jesus.

These events, I believe, are attributable to manipulation of the human genome. To me, this is the only rational explanation, and even that still seems miraculous.

Incarnation by the soul to the animal body of man would conceivably be performed slowly by genetic manipulation, and occurring over a period of several generations, or even thousands of years, again starting about 600,000 years ago with the species of Homo erectus.

The second event of Mary's impregnation is contrary to general physiological understanding. However, today insemination could be performed naturally by artificial means and manipulation of the genome. Yet, in development of animals and plants, botanists and geneticists have manipulated ovaries of plants, flowers, and animals by placement of male sperm to produce new varieties of plants and animal life. Invasive plants often interact naturally with domestic plants to produce new strains of the cross breeding. Rose growers, race horse, and animal breeders continually experiment to produce new strains. Research has shown that manipulation of the plant or animal genome is in fact another way to impregnate flowers, plants, animals, and perhaps humans. The genome of Mary conceivably could have been manipulated by God, or bio-electrically by her own soul and sexual organs, in communication with her body and reproductive organs.

UNANSWERED QUESTIONS

1. After the Crucifixion did the soul of Christ, with the essence or spirit of God, transfer to another body and person (reincarnation)? Ibn Abdullah Mohammad (lived 570 to 632 A.D), was religious leader of a then-peaceful, but protective, Islamic belief. Could it be that Mohammad was reincarnation of the soul of Jesus Christ? Was the soul of Muhammad the Biblical Second Coming of Christ? Both Muhammad and Christ were mystical men, surrounded by mystical lore, religious leaders, and believers in one God (the Islamic Supreme god is Allah and is held to be the same as the Christian God. (Both God and Allah may be one and the same entity). The soul of Muhammad may be the reincarnated soul of Jesus Christ.

    The current and long-standing conflict within Islamic beliefs should not be attributed to Mohammad, but to the philosophical difference between the remaining sects for right to inherit the position of Mohammad after his death. Much dogma and practices of today were proclaimed, not by Muhammad, but by the various sects of his followers. It is noted that many conflicting religions and philosophies in Christendom existed following the life and death of Jesus Christ, resolved for accepted Christian dogma only by a series of conferences to decide recognized religious issues. Exploration and adoption of this presence would create much unity and peace in our world.
2. The second event is of Mary's impregnation with Jesus, contrary to general physiological understanding today. It can be shown that insemination could be performed naturally by artificial means and manipulation of the genome. In development of animals and plants, botanists and geneticists have manipulated ovaries of plants, flowers, and animals by placement of male sperm, to produce new varieties of plants and animal life. Invasive plants often interact naturally with domestic plants to produce new strains of cross-breeding. Rose growers, race horse and animal breeders continually

experiment to produce new strains. Research has shown that manipulation of the plant or animal genome is in fact another way to impregnate flowers, plants, and perhaps even humans. The genome of Mary conceivably could have been manipulated by God, or even by Mary's own soul, in communication with her body and reproductive organs.

The time before Jesus Christ was a recognized history of events—the cosmos, stars, galaxies, Sun, planets, Earth, animal life, and development of species leading to Modern Man. After the time of Christ, history focuses on individual men and events, their importance, and accomplishments.

IS CHRISTIANITY THE APEX OF RELIGIOUS THOUGHT?

Christianity will be defined in different ways by different people. However, one definition is belief in one God, and that Jesus Christ is the Son Of God, or even that God was personified in Jesus Christ.

A main division in religion is that of Catholicism and historic, unseemly practices by religious leaders, with promotion by Martin Luther of a new religious belief to be called Protestantism.

Within Protestantism there are many religious subdivisions, each with their own basic differences in religious philosophy. Some can be mentioned of Lutheran, Methodist, Baptist, Unitarian, Gnosticism, Episcopalian, Presbyterian, Anglican, Moravian, Amish, Mennonite, Quaker, Mormon, and many others on a world-wide scale.

Catholicism also has many sub-groups worldwide, although there is a more tight and strict central administration from the Vatican with regional control. The Papacy is charged with maintaining a unified regional ecclesiastic adherence throughout the world, perhaps more so than administration of any of the various Protestant churches.

Jewish religion and its various sects has variations in religious thought, practices, and adherence.

Even the Islam religion, represented by one-fifth of the world's religious population, maintains the basic belief that there is but one God (called Allah in the Islam religion). Islam believes in Jesus

Christ and that Muhammad is the messenger prophet of Allah. It is believed that the Christian God and the Islamic Allah may be one and the same entity.

After the death of Muhammad in 632 A.D., different sects developed. The Shiite sect generally believed that a successor Caliph to Muhammad should have been selected from the blood line of Muhammad, whereas the Sunni Sect believed choice should be made from among the best qualified and most pious, regardless of blood lines. Interpretation of correctness in adherence to prescribed doctrines stated in the authentic Qur'an (which means recitations) differs dramatically, and many tribal groups have even different blending of religious thought. In most countries of predominant Moslem religion the majority sect is Sunni.

Large worldwide groups adhere to Buddhism, Confucianism, Hindi, and localized populations that follow deviations of belief in one God, with belief that living an admirable life on Earth is the only necessity for "goodness" in secular life, with some questioning even if there is an Afterlife.

Some form of Christianity is followed by most of the Earth's religious believers. It follows a common thought that Christianity, a division of ancient Judaism and forerunner to Islam, is the apex for religious belief, incorporating various beliefs in Catholicism, Protestantism, Islam, Judaism, that is probably three-quarters of earth's population.

At first, Neolithic man, living in caves, painted images on cave walls depicting the hoped-for results in hunting of animals for food that would be directed from some unseen and unknown power. This was as early as 75,000 years ago. No name is recorded for that god and power. Perhaps many gods were worshipped, each maintaining power over certain facets of not just prowess in hunting, but of all aspects of life, including formation and birth of a man-child, the growth of that man-child from a Neolithic baby to a grown youth who would join with his elders in hunts and fishing. He would join in pre-language assemblages around community fire-groups, where each elder would portray his ideas of what the unseen god or gods did, and when that god's power appeared.

About 10,000 to 8,000 B.C. the culture, technology, and thinking processes of man on earth changed from cave dwellers to Modern Man in an agrarian, agriculture-oriented, communal, and more social existence.

Late in ancient ages, perhaps around 10,000 to 8,000 years ago or after, a central god had been clarified and was given the name of "Baal". Baal was openly recognized and worshipped, to be satisfied as developed by myth, demanded by sacrifices of a tribal animal, or even a person—sometimes a child, sacrificed in the worship of Baal. Baal was considered paramount in guiding the fortunes of pre-historic life. It was the conventional religion until about 1,000 B.C.

The folklore of Biblical accounts, reduced to writing starting about 600 years B.C., was present in communities such a Sodom and Gomorrah, and at construction of the Tower of Babel, affecting child birth, everyday living experiences, and at the essential hunting and fishing experiences.

The belief in Baal was present at organization of the Exodus from Egypt, even though some leaders and participants told of a new presence called "Yahweh", which would later be called God, with a new mythology constructed around that existence.

The existence of their god Baal was still paramount in the minds of many. This continued belief in Baal was a factor in the decision of Moses to travel up on Mount Sinai to encounter his God to receive, it is told, an engraving on stone tablets of the Ten Commandments. Moses went alone up Mount Sinai to encounter God and report the encounter.

The Commandments were in many respects parallel to the Egyptian Book of the Dead, written by the Pharaohs, which told for the benefit of the Keeper of the Underworld who would meet a deceased pharaoh and decide his fate, based on events of the pharaoh, told of deeds done as good, or bad deeds and acts not done. Moses, having been found as a baby in river reeds and adopted by a Pharaoh's daughter and raised in the household of the pharaoh, was well aware of the existence and content of an Egyptian Book of the Dead as a guide to moral and ethical living.

Strength was given by Israelites to the power of Yahweh as God, with a temple constructed to honor God and to worship his being.

A lore in religion was developed by some Hebrews for the future coming to Earth of a representative of God, to be a Messiah on Earth, which was manifested by the birth of the Christ child Jesus. From this event, with the insight and teaching of Jesus, a new concept in religion developed to be known as Christianity, later to become the leading religion on Earth, but having many local interpretations for its application on Earth.

From about 600 A.D. when the doctrine of Jesus Christ was accepted by King Charlemagne, to be followed by his subjects, Christianity gained strength in numbers, belief, and influence in the developing world, save only philosophy contrasting by earlier religions of Buddhism, Confucianism, and Hinduism, with Islam developing around the life of a prophet named Mohammad in Arabia, a different version and practice of the developing worship in Christianity and of Mohammad.

The beliefs and practices of Islam became more influential and was practiced in a growing population, culminating in a series of wars, the Crusades between 1250 and 1400 A.D., between the Holy Roman Empire and the forces of Islam, the result being inconclusive with Islam continuing in existence to today, and certain fanatical practices of Christianity to eliminate non-believers in God occurring during the Inquisition period of 1400 to 1600 A.D.

RADICAL SCIENCE

I place this chapter in the book, I suppose mostly to show that not only the various religions of the past, that even today, have followed hard-to-understand tales in their existence, but that science, too, has explored extreme possibilities. Some of the original pursuits in science, although thought radical at the time, proved to be breakthroughs in scientific thinking. Others require further proof.

Albert Einstein was certainly a great scientist, one of the greatest in a long line including Isaac Newton, Archimedes, Copernicus, and others. Nevertheless, other qualified scientists have taken exception to interpretations and proofs that Albert Einstein, and other men, have installed into accepted science, precepts that have become bedrock laws of today's science.

Some scientists believe that science of Physics appears to be leading us, not to resolution, but into an "Alice In Wonderland" world of increasingly bizarre theories, each further removed than the last, from our experience of the everyday world. Galileo Galilei was threatened by the Church of Rome in excommunication for his revolutionary ideas of the cosmos. The wife of Charles Darwin wept profusely because she believed her husband, with his outlandish presentations, could never meet her in Heaven because of his unconventional thoughts of evolution in the mid-1800 years.

In recent years some cosmologists have posited that our universe is just one among an untold number of universes that exist from quantum sources. Theoretical physicists have looked to exotic mathematics of string theory, which suggests the existence of seven other dimensions of existence beyond the four we already experience in our everyday life of height, width, depth, and time.

Experimentalists have built the $19 million Large Hadron Collider on the mountainous border of France and Switzerland, in part to understand why we observe only a portion of what our theories of matter and antimatter predict. (Source: Discover magazine, April 2010, pgs. 34-37) At present some doubters exist. They represent minority views, but perhaps it is justifiable to listen to what they have to say. Advocates of operating the Large Hadron Collider may conceivably be wrong, leading to catastrophe, but if scientists are correct, it will usher in a great advance in physics and world knowledge.

A certain scientist, now a professor of Physics at the University of California, ran into a paradox which he calls, "the clock ambiguity". In our ordinary experience, he states, we tend to take the passage of time for granted. He states that at the succession of moments following the Big Bang the universe was anything but simple. He studied a time when the entire universe was compressed into a space the size no greater than a grapefruit. He was trying to understand the process in the infant universe when it had existed in that form for only a small fraction of a second. He attempted the difficult position of isolating the dimension of time. This required to "undo" the unification of time and space, a key achievement of Einstein's Theory of General Relativity.

The quest to reconcile the quantum view of mystical existence with the real world and views of matter has occupied academic study in physics for decades. At the moment, in year 2012, string theory seems to be a most promising way forward in this vision. In the current vision, matter is made up of "strings" vibrating in the cosmos at different frequencies that exist in 11-dimensions of space-time. So far, no single version of string theory has yielded a consistent or complete answer.

Another scientist, of the Perimeter Institute for Theoretical Physics in Waterloo, Ontario, Canada, takes issue with certain tenets of present-day physics, advocating the belief that our universe is just one of many. The so-called "multiverse theory" of the universe arose, like string theory, as a way to explain some mysteries in cosmology. He holds that String Theory cannot be made to work in a world of only four dimensions. In response String Theory theorists posit the existence of at least seven additional dimensions that are presently hidden from us.

Some scientists have proposed that at inflation into space shortly after the Big Bang, a small shard of space-time underwent extremely rapid expansion, ultimately to become matter that we see around us. That theory opened up the possibility that other bits of space-time went through their own inflation, creating a tremendous number of "pocket universes", and the notion that our observable universe is just one of perhaps an infinite number of cosmic domains, each with its own version of the laws of physics. Somewhere out there, this theory goes, exists pocket universes, each having its own set of different physical laws.

A second shard, or numerous other shards of space-time, also underwent rapid expansion to become different entities of antimatter. Possibly from this antimatter soul there occurred paralleling development of life in physical matter that also occurred on Earth.

Another scientist, not a degreed physicist, advances an idea of "reductionism", the belief that a whole can be understood by the predictable behavior of the parts. He explores how novelty (or unexplained science), rather than certain timeless laws, could play an expanded role in cosmology and physics. Perhaps evolution has taken the universe from the relative simplicity of its early moments to the full glory of our modern universe. His followers believe it should be possible to look back into the early universe and measure changes in physical laws.

Another instance of radical, but conceivable, physics is the possibility of "reverse causality". This explores that science, in a distant future, can be influenced by prior acts. The end result of a certain development in an organism or mechanical entity can be defined by transitory, preliminary physics occurring at an earlier time. This would hold that the state of human beings today evolved from historic transitory phases. For example, the mind of humans today was formed in many previous generations by programmed changes.

In a materialistic example of "reverse causality", an automobile requires and undergoes progressive phases in year models associated with developments that occurred in various earlier times. It is also like a moving picture depiction, when the last moments of a story are unclear until the preceding story parts are shown, that add understanding to the climax.

Another recent theory is one of "entanglement". In entanglement, two electrons present in individual persons establish a kind of "telepathic link" that transcends space and time. The bond endures no matter how far apart they may be. You, and each person in whom you become in communication with, therefore become engaged in entanglement, releasing a quantum bond between individuals, whether family members, loved ones, enemies, or just acquaintances. (Source: Scientific American magazine, article by George Musser.) There is evidence in modern life that Doctor Edgar Cayce was able, from several miles distant, to communicate and diagnose patient maladies and needed cures.

Each of these scientists, and many more, have advanced unique and challenging views of how the universe has physical and antimatter laws, and perhaps how those laws have changed since the first Big Bang. The history of science development has consistently shown that truth in physical laws has been developed and refined by irregular and new ideas that may run counter to physical rules previously accepted.

In this book I have, or will, introduce several original thoughts or theories of evolution of the earth, species, and development of events leading to Modern Man in his body, brain, mind, genome, spirituality, soul, and culture.

Some ideas are radical or differ greatly from accepted knowledge in science, anthropology, theology, physiology, mind, brain, and soul. My concepts, I believe, are based on logical events. These represent alternative theories and supplant age-old myths and legends. I believe they are each reasonable and possible. If only one of these proposals turn out to be true, there will be tremendous change in accepted knowledge of how we became human.

Some ideas presented here are "unorthodox" and may "fly in the face" of presently understood knowledge of learned men, but each idea will stand the test of reason and logic within our ever-changing existence on Earth. Ideas presented can be considered an alternative action to published events.

The following have been proposed as original thoughts of the author:

1. An existence of cosmic energy to be gathered in a Black Hole occurred before the Big Bang. Matter in space did not exist, only cosmic energy to provide unification of elementary particles.
2. Basic, inert minerals from the Hadean Era changed to organic molecules through chemical combinations with oxygen and other chemical elements. (For example, iron and oxygen combine to form Ferrous Oxide; carbon and oxygen combine to form Carbon Dioxide. There are many evidences of carbon "sinks", such as coal or oil deposits. The lists of chemical formation of other molecules and compounds seemed endless, happening in pre-history and duplicated in modern chemistry laboratories as rust and wood rot.
3. A new species of man, an extension of Homo erectus, at about 600,000 B.C., possessing basic elements of soul, developed in the parade of species on Earth.
4. Incarnation of soul to an existing body of Homo erectus, at 600,000 B.C., provided man with early spirituality.
5. A second look is made to the acts of Moses and formation of the Ten Commandments, as reported by writers of the age, and his reported meeting with God at the Burning Bush on Mount Sinai. There is considerable parallel of the Egyptian Book of the Dead, as a guide to spiritual living, that became familiar to Moses while he was raised in the home of an Egyptian pharaoh. This parallelism was reflected in the decreed Ten Commandments. Biblical writers of the Moses epic may have ascribed to Moses unfounded powers and theatrics commonly described of Egyptian pharaohs.
6. A second look at the crucifixion of Christ, that he may not have suffered medical status of being "brain dead" at the crucifixion, but instead suffered modern-known medial shock, resembling death, a phenomenon undreamed of 2,000 years ago.

7. The revelation at 610 A.D. that the life of Mohammad, at age 40, may have been the Biblical "second coming of Christ". Mohammad created and led a religion that believed in God and Jesus Christ. The religion spread to thousands, and now millions of people on Earth. The prophet Muhammad is recognized in the Islamic following as a "Messenger", and died 22 years later. Note: It is the opposing philosophical views of theological inheritance, recorded in history, of the successors of Mohammad that created a controversial reputation in Christian countries. Some doctrines stated in the Koran, like in the Christian Bible, have been added to and interpreted differently by followers over a period of hundreds of years.
8. A portion of God exists and is within our own mind and soul.
9. Heaven, or Afterlife, is present on the face of Earth, but in another dimension of Space/Time.
10. "Ancient Homo erectus" is perhaps a different species than "Modern Homo erectus", mainly in the presence and development of soul.
11. Soul was incarnated to man about 600,000 years ago, occurring in a species that could be called. "Modern Homo erectus".
12. Soul exists. There is evidence, tangible and real, both theorized and observed. This book includes discussion of this life-changing phenomena and differences between animals of the forests and human beings. The many evidences of soul are detailed in another chapter
13. Soul of man works in conjunction with the brain, the mind, the genome, and enzymes of the body to affect physical changes, and with God, or his Heavenly "organization".
14. The "Great Wall of Galaxies" at the extreme distance of our cosmos may be another universe, having its own gravity, perhaps even in another dimension.
15. Many dimensions exists, some scientists proclaim 11 different space/time dimensions. Others proclaim an unlimited number of dimensions.

16. Afterlife exists, with absence of our physical bodies and our five senses. Communication is by thought transmission. Our soul may be reincarnated to some future life.
17. Glacial Build-up and Global Warming, is caused mainly by the location of Earth in the cosmos, as our Sun, Solar System, and arm of our own Milky Way Galaxy rotates in the universe through periodic relatively colder and warmer cosmic locations. We are currently entering a new colder glacial period, tempered by warming human actions of our culture.
18. The female of our species, with the many physical and mental differences, may actually be a sub-species of the male (or vice versa), as proven by the physical, mind, and brain differences. At one time there was no differentiation of sexes in animals, but instead were homophobic, with self-insemination, as can be observed today in certain animals, yeast, bacteria, virus, and microbes. First evidence of sexual reproduction appeared about 1.2 billion years ago. All life prior was self-reproducing.
19. In evolution, God proposes and nature disposes: Evolution is planned, but is implemented by: 1) natural needs and development, and 2) action of the genome and enzymes, directed by the soul and bio-electrically to the genome.
20. Mind and brain may be compared in terms of the modern computer: The brain is basically an electrical processor of the neuro-system, with nerve action and coordination, similar to hardware of a computer. The more complex comparison is that of the human mind with the computer software. The mind, not the electro-biologic brain, like computer software, processes relationships, judgments, and thoughtful analysis, together with appropriate emotions.
21. Man, Earth, the cosmos, and culture are all developed in an organized system. They were not left to exist in developmental chaos, as might be observed in odd sea-life of the deep, where there is complete absence of common light, of Sun's rays or energy, or even as imagined on some strange new world of the cosmos.

22. Glaciers, in their formation during a cosmic cooling phase, transform much total energy from elsewhere on Earth, much like making ice, or keeping food cold in your refrigerator or freezer requires transfer of energy. Likewise, when glaciers thaw in periods of rising Earth temperatures, latent, stored energy is released to the world. History reveals that progress in technology, innovation developments in biology, and culture, such as word forming for communication, written language, hieroglyphics, counting, and other changes in culture, happened mostly during earth temperature warming periods. We can observe the periods of 1) say 30,000 to 15,000 B.C., freezing and cool periods with only minor changes during the Pleistocene, and 2) since 15,000 to 9,000 B.C. and later (a warming period and many advancements in culture).
23. Many species have developed in pre-man, animals, and plants of the forests over the 6 billion years of life on Earth. Many species became extinct, mostly from inadequacies, and competition within nature and the environment. Some species could not endure change in atmosphere, food supply, climate and temperature, and competing species.
24. In addition to the existence of body, brain, mind, soul, genome, enzymes, bio-electricity, and nerve system, there is a magnificent communication network within the many functions. For example, a sensation causing body pain will transmit, by nerves, to the brain, which in turn communicates to the mind, that makes a learned judgment to relieve the cause of pain. Continuous causes of uncomforting or inadequacies will cause eventual change in the genome, which in turn causes bio-electricity of the body, through enzymes to change teeth, lips, bone structure, skin coloration for reflectivity, or numerous internal body tolerances. Soul provides the communication between a Superior Being (God, Allah, or other manifestation from various cultures) and the minds of human individuals, with actions and thoughts, resulting in our consciousness.
25. The realism of now-recognized Quantum Mechanics has existed from time of first life. The actions resulted, in life

and in humans 15,000 years ago for example, as it does today. The body, and events of nature were as real then as they are today. The elements were just not individually identified and described as well as they are presently.

26. Science and theology are compatible. Certainly the system of science exists and, I believe, a system of Supreme Being exists. In essence, God formulates an intelligent "plan of action", and the system of nature, involving the genome, causes the necessary changes.

27. Science today has proven the existence, not only of matter that we can see, hear, and touch, but also of antimatter, that is mystical (not understood) and beyond our general concept of items of the earth.

28. The author proposes the origin of antimatter was at moment of the Big Bang, and division of electron forces from positive-charge protons. Negative-charge electrons formed matter of the world and universe. Positive-charge protons formed antimatter, and in further dimensions from that known for billions of years as matter. The additional dimensions are being currently explored in science.

29. Presence of the Agrarian culture about 10,000 to 8,000 B.C. was a logical and explainable development, occurring from availability of new abilities, and need to live and work in cooperation with their human counterparts.

30. A warring culture and activity developed somewhat naturally due to unreconciled differences between groups having different values and ethics, agricultural and geographic needs of one group from another, and an inherent drive of a king or group to expand their own personal or kingdom realm.

31. Perhaps the most significant idea presented is that a new species may have been found in its infancy. There seemed to be in past reporting an existence of Homo erectus from 1.6 million years ago. Upon my analysis, early Homo erectus was somewhat unrefined in his soul capacity. At the time of 600,000 B.C. there was introduction of a new and different Homo erectus, most significant in possession of soul, and in

technologic and cultural advancements. Perhaps this was a moment of Creation and incarnation of soul.

32. It seems logical that Homo neanderthal, and even Homo sapiens, are really sub-species of Homo erectus, which continued as a basic species of humanity, and to our living species today as Modern Man.

33. We each are witness today in what may be the formation of an entirely new species, or a "quasi-specie". Let us call this new species "Homo roboticus". Since about the 1950's our industrial culture has developed machinery that is combination and coordination of mechanical operations, electronics, and intelligence, its activities directed by a developed program and a series of computer chips. Robots presently exist in performing routine jobs, replacing human beings who formerly worked in automobile assembly plants, doing operations of welding, drilling, painting, polishing, testing, lifting or other repetitious routines involved in manufacture of automobiles and other products.

Today robots independently and intelligently vacuum-clean carpets or mow our lawns, while man sits leisurely away in a supervisory mode. Engineers are developing automobiles that drive and park themselves, with only the presence of supervision personnel to provide mind-like guidance. Computers can perform accounting operations with only a minimum of human input. Robots are on a threshold of intelligence whereby they will in the future have the capacity to reproduce themselves in manufacturing, and in drawing blueprints showing necessary modifications. They will have ingenuity to develop improvements into successive generations of their "species", and ability to determine, transport, and supply elemental parts. Even today, complete functions and departments exist that produce goods without operating human personnel. At some future time intelligent robots conceivably could form a "civilization" of their own, perhaps even being a threat to the welfare of Modern Man.

A SCHEME OF EVOLUTION

There is no doubt that evolution, the change and development of Earth, its first animals, plants, people with body, mind, culture, and spiritual being happened. It is only a question of how it happened. Did it come about from natural, self-selecting devices, or did it happen from some overall planned wonderment?

There seems to be fact that changes resulted from intelligent direction in the origin of RNA and DNA, going back to the year one in life history on earth. Reading the genome of a particular life form, whether it be plant or animal, fish, bird, or reptile, is like reading a book of history concerning development of that life form. The genome of mankind, now over 1.5 billion genes long, was once much simpler, perhaps only 100 genes long, that provided for only the basic functions of life for intake of food for energy, digestion, and a waste expulsion system, but grew as life forms became more complex, to its present complexity of millions of genes that make up the genome, not only in mankind, but also in primates, plants, fish, reptiles, and amphibians, and development in flora of grasses, plants, bushes, and trees. Microbiologists have shown the genome of some primates are within about 99 percent the same as for modern humans.

What caused this change and similarity? It seems true that some development occurred through natural sources. Green algae, that somewhat slimy substance found today on sedentary waters, is much the same today as in our year one.

Bacteria too, although greater in variety today, were once extremely simple in biology, as were microbes and other single-cell prokaryote organisms, like sponges and yeast.

Some development occurred simply to support first activities of animal life, such as circulatory systems to distribute and use the energy evolved from food intake, and digestion in first animals on Earth, with other systems to support systems in life.

But what prompted change in the system of life biology? Our culture would still be back in the first development before physical machines if intelligence had not been applied to convert energy of steam, acting upon a piston, to push a rod, that converted lineal motion to circular motion, such as in the steam engine, allowing the

Industrial Revolution, and later the steam-powered ocean ships and railroad transportation, progressing to modern electrical, electronics, and computer chip technology of today. Such is the parallel of genome development.

Intelligence is the key. Analogy is made of today's junk yard. Items in a junk yard would remain inert but for the intelligence of some mechanical or design genius person who could combine parts to perform in some planned or derived system. This intelligence was also the moment of inertia in providing change in biology of life in humans, and continues through additional complexities in evolution in life of animals, sea life, plants, bacteria, humans, and in all forms of complex life.

The source of intelligence that instigated change in biology of life is discussed later. It seems inconceivable that change occurred strictly by accident, or by advantageous mutations of change in RNA and DNA of genomes.

Again, an analogy in industry, perhaps parallel in plant growth: A worker would not provide overall knowledge to affect change. Bodies of coordinators, planners, administrators, managers, and supervisors, as well as engineers, product imagers, and even marketers, will affect first and subsequent happenings to bring about intelligent and planned changes.

We do not positively know if this is the way events actually happened. No one really knows this because of the ancient times. Likewise, no one really knows the accuracy of interpretations, even of knowledgeable people in some given field. Differences of opinions exist even among experts in these fields. Most of the events were never factually recorded, except in Earth attributes of older rocks, mountains, identifiable climate history, species and fossil interpretation, technology development, and observed cultural changes. It is impossible to project one's self into an ancient man's brain for judgments. Conclusions following observation may differ from individual to individual.

For interpreting certain acts told in the ancient writings and recorded lore, it is sometimes well to "step back", to critically evaluate reports in respect to reasonableness, logic, and physical authenticity of so-called miracles, and to question possible myth and lore, all in balancing spiritual faith with realism.

WONDEFUL WORLDS IN REFLECTION

We are nearing the end of our journey of exploring the origin, development, and future of our complete selves in a scope spanning the history of the cosmos, from the very first incidents before the Big Bang, right up to the time of Jesus Christ. To explore the wonderful (and sometimes not so wonderful) accomplishments over the past 2,000 years is beyond the scale of this writing.

It has been a momentous trip, with some ideas original from the author, review of generally accepted principles of origins, development, and advancement of our culture. Presented here is a review, a unified history of existence of the physical entity of the cosmos or universe, the origin and "life" of stars and galaxies, the Solar System, energy and matter, Earth, moons, and planets, the presence of antimatter, and extra dimensions in space that may be "home" to Afterlife, soul, and God, all during our everyday living, as well as in an anticipated Afterlife.

In this writing we have explored many subjects, some of which are basic to various religions. Some are deemed to be myths and magic and should be replaced to reflect true dogma. We have also explored the "holy grail" that supports certain aspects of science and proven by some of the greatest minds to exist in history.

We have explored possibilities regarding the dynamics of soul within man on earth, in an afterlife, and perhaps even reincarnation of soul to a later generation of man. We have explored a time in civilization tracing a culture, and as practiced in ancient Africa, Asia, Europe, China, and lands leading to, and beyond the Malay Archipelago, and even to the Americas.

We have not solved the question of whether origin of all humanity centered in Africa, or whether humanity arose, somewhat simultaneously, in the Americas, evidenced by ancient fossils, relics, and artifacts of 12,500 B.C. found at Clovis, New Mexico, and in various parts of the world, with dating comparable to that of ancient

Egypt and Mesopotamia. Neither, have we determined when and how species of the Caucasian race evolved, likely from the Russian Georgia region near the Caucuses Mountains and between the Caspian and Black seas.

EXPLORED AND FOUND

Several different and original ideas and concepts have been proposed. Most ideas explored are based on conventional and proven scientific principles. Some unconventional ideas and original thoughts of the author, are believed to be more logical, realistic, and void of myth and magic. The author believes they have merit and should be explored further.

1. There was an existence and activity in the cosmos before the Big Bang.
2. The original Black Hole contained energy and not matter.
3. The collected energy of the original Black Hole was Vacuum energy.
4. Gases of energy expanded into the cosmos, with limits of the cosmos increased to contain the expanding galaxies. First elements from the Big Bang were hydrogen and helium.
5. Dark Energy and Dark Matter are proposed as excess vacuum energy from star formation.
6. Energy converts to matter upon change in pressure and temperature ($E=MC^2$).
7. Space was originally compact. Space expanded in the cosmos as needed.
8. Variable space has defined limits to accommodate newly created stars and galaxies. Perhaps a particle that creates space will be found from research.
9. Gravity attracted stars within galaxies and galaxies into galaxy clusters.
10. There were comparable Big Bangs in different galaxy locations. Multiple Big Bangs occurred in multiple galaxies.
11. In their lifetimes stars exploded in supernovae, were re-formed by gravity, their composition changed at extreme pressures

and temperatures, and created new chemical elements within the stars. Supernovae elements and minerals were broadcast into the star system, to be attracted to the planets.

12. Cosmic gases cooled and congealed, developed gravity, collected space particles, and grew, forming planets that developed orbits around the Sun to form the Solar System. Some asteroids and comets continued to orbit. Stars are the "mothers of the universe" in producing elements presently found and mined on Earth.

13. After about 200 to 400 million years from the Big Bang, stars began to give light in form of photons. Our Sun was formed with later Solar System of orbiting planets, asteroids, and comets.

14. For 7 billion years the substance of Earth was Hadean (physically resembling the heat of Hell) forming concentric layers of elements at the Mantle, cooling, and gradually decomposing from geologic formations, volcanic action, erosion, and atmospheric conditions. Action of water waves along shores, and severe atmospheric conditions, turned rocks into sand, and with later chemical combinations, to soil.

15. Life on Earth started from indefinite theories of conditions, but did form first single-cell organisms (prokaryote) of amoeba, yeast, bacteria, and microbes that grew in cellular complexity, to cell structures having cell walls and nuclei (Eukaryote cells).

16. First life appeared on Earth, and possibly life on other planets, probably now extinct.

17. A genome system developed of RNA, then DNA, from yet undefined means.

18. Many species of animals, pre-man (primates, hominids, hominoids) and Homo species developed, leading to modern man.

19. Any life and advancement on alien planets could be eons of time behind, or ahead, of Earth.

20. Our Moon formed from extraction from Earth, caused by cosmic collision with Earth of a Mars-size space traveler during its molten or semi-cooled stage. (It would be highly

coincidental that moons of other planets were formed in the same way, leaving in question the origin of those moons.)

21. Geologic plates formed at earth's mantle, and carried whole continents to other locations on the globe.
22. Collision of geologic plates formed mountain ranges on Earth.
23. Water, wind, chemicals and atmospheric acids carved environmental and geologic changes of rivers, lakes, and sea shores.
24. A genome system developed early on Earth forming RNA and DNA, which reacted with enzymes to form proteins, the building blocks that changed organs of animals, plants, and fish.
25. The human body, brain, and mind are testimonials to intelligent evolution. Workings of the animal body, brain, soul, and mind, with evolutionary changes, is more exacting than perhaps life forms found in deepest seas, or in the age of dinosaurs. Present life on Earth is inconceivable to have happened by a series accidental and undirected developments and lucky mutations. If evolution had followed chance mutations, life forms would be as grotesque in appearance.
26. The complexities of parts of the body and brain are miraculously designed, "engneered", and planned, including: functioning of the heart, blood circulating system, neural system, brain, digestive system, glands, muscles, skeleton, liver, kidney with purification system, lungs, with our oxygen/carbon dioxide exchange system, eye sight, hearing, taste, reproductive system, and a myriad of sub-systems of the body needed for intelligent guidance. Over time, evolution provided as much as environmental needs required, but original intelligent direction was required for organized development.
27. About 10 to 8,000 B.C. culture changed to an agrarian system with agriculture-related land use, crop planting, harvest, animal herding, and trade development.

28. Major cultural, social, technical, and intelligent changes came about following the Agrarian Revolution, existing at least through the time of Christ.
29. Many kingdoms developed from an agrarian, urban culture, with warfare to build empires, and technology to follow.
30. Moses may have been overly dramatic at Mount Sinai, a leadership technique learned as a youth while living in the royal home of a pharaoh in Egypt.
31. The Great Flood of Noah's time could have been caused by earthquake at the Straits of Hormuz, allowing waters of the Arabian Sea to flood into the Persian Gulf and river system.
32. Christ may not have been brain-dead at Crucifixion and in the tomb.
33. Elements of early soul derived from annihilation of positive-charge elementary particles at Unification at the Big Bang, creating antimatter and later, souls.
34. Early elements of soul became an integral part of Homo species about 1.7 million to 600,000 B.C., with many elementary qualities of emotions commencing at that time.
35. Soul continues existence in afterlife following human death.
36. Hell is defined as denial of existence to Afterlife. Hell is not at the center of Earth.
37. Reincarnation to future generations of humans after death is a possibility.
38. Qualities of soul can be defined and are presented in this writing.
39. Species leading to man developed over millenniums in a planned continuum through intelligence of a superior being, but features were formed by environment and genome.
40. Groups of souls, of people that excel with special innate abilities in periods of intellectual development, may be prompted by cross-stimulation of souls and persons living at or near the same time. (Many great persons lived at about the same moments in history.)

41. Prayers to God may be superfluous, with desired results arranged by communication within our own mind, genome, and external soul relationships.
42. Locus of Heaven could be on surface of Earth, but in a different dimension.
43. Evolution of soul may be attributed to a "God Gene" in the genome.
44. Soul and mind are guiding forces to ethics, morality, emotions, and beliefs of individual men and women.
45. A personal Guardian Angel may guide our decisions of mind and soul.
46. Soul has a continuing life beyond secular death on Earth. There may be future reincarnations of soul to man (as infant, or sometimes when older).
47. Soul and genome interact through the mind to affect control and changes of the brain, which are transmitted by enzymes and protein for change to body or brain activity.
48. Scope of science orientation and independent scope of religion dogma can be separated (see chapter), eliminating much conflict of interpretation between the two.
49. Greater population support for actual religious and science understanding will result in unity.
50. Conclave of all religions and science interest could be called, similar to religious conclaves of 2nd and 4th centuries A. D., to resolve a unified Christian doctrine.
51. Idea of a miraculous deity is much older than any present religion.
52. Science does not dispute some articles of conventional religious dogma.
53. The Big Bang created and left cosmic radiation background energy (COBE). This has been mapped and recorded in astronomy.
54. Functions of the mind include display of emotions, feelings, judgment, concept of a greater power, abstract reasoning, appreciation, communal and familial concerns, love, mercy, sorrow, empathy, and a host of other senses and emotions elsewhere, all as qualities of the mind and consciousness.

55. There exists extensive theories in science concerning development in Physics, Chemistry, Elementary Particles, Cosmos, Anthropology, Cultural Development and in other fields of study.
56. Several theories have been developed in religious beliefs, spirituality, and faith. Some theories have merit and others are unproven. Yet, no significant studies or revelations in the subject of soul or afterlife have been made in the field of hard science.
57. Some non-human animals and inanimate objects are thought by some Shamans of native American Indians, and other cultures, to have soul. The Hindu religion places great religious reverence in the sacred Brahman bull.
58. Afterlife exists as a normal process of the life cycle. An activity of unknown type, but perhaps including reincarnation, takes place in Afterlife. (The author is currently developing a third book in this series, about "Afterlife".)
59. Culture changed to agrarian in agriculture-oriented land use, crop planting and harvest, animal husbandry, and trade development.

AUTHOR'S REVIEW OF THE HOMO ERECTUS DIASPORA

Another close species at the time of first Homo erectus (1.8 million years ago) was Homo ergaster, who emerged about 1.7 million years ago and disappeared 100,000 years later. Some anthropologist hold that Homo ergaster was either 1) supplanted by Homo erectus because of the latter's superior qualities and abilities to continue living, or 2) that H. ergaster interbred with H, erectus, and by 1.6 million years ago was indistinguishable as a separate species. Charts show a branch of H. erectus migrated northward into the Russian Georgia area about 1.7 million years ago, as the mainstream of migration proceeded eastward. (Source: Prehistoric Life, DK Publishers, 2009) The mainstream of migration at this time were Negroid and migrated to

Northern Australia, Borneo, Papua New Guinea, and Pacific islands, perhaps even by water voyages on large rafts, outrigger canoes or crude boats from east Africa, accounting for Negroid race of ancient people in northern Australia, New Guinea, and some Pacific Islands.

Extensive commentary on this subject is recorded at Yahoo.com: Homo habilis "handy man" or "skillful person" is the oldest known species of the genus Homo at 2.3 to 1.4 million years ago. Homo ergaster fossils are dated as early as 1.8 to 1.3 million years ago. Ancient Homo erectus fossils have been dated as early as 1.8 to 1.6 million years ago. It is thought by various anthropologists that Homo ergaster either 1) became extinct through attrition, or 2) merged into the more successful Homo erectus species.

NAUTICAL ANCIENTS

Ancient man was not always bound to the land. Nearly everywhere that mankind lived there were bodies of water, of lakes in variety of sizes, rivers, seas, and oceans. Lakes existed with far shores beyond eyesight. Lakes and rivers were origins of food, of fish, clams, mussels, and a wide variety of water-living creatures.

Man learned to travel out onto lakes on rafts, canoes, outrigger canoes, and later even small ships with an upright pole for a mast and a form of sail to catch the winds, and steering mechanism to maintain a course, in addition to oars powered by several men (or women) pulling in unison.

On larger bodies of water ancient man traveled to islands, unseen from his starting shore, taking one or several days for their journey, and learning to take food provisions for the trips in addition to fish to be caught. Several such islands were in the Mediterranean Sea and also off the coast of East Africa.

Man seems to be an inquisitive explorer by nature. Eventually water vessels were used to travel to both known and unknown islands and lands. This means of water travel was employed in migration to Ceylon, various islands leading to Papua New Guinea, and beyond to northern Australia, New Zealand and Polynesian islands.

Specifics of navigation became known by observing bright stars, changing directions of winds, observation of birds associated with land, to locations previously unknown. At times both male and female ventured together as families to locate and settle in new places and for new lives.

Some of Black race sailed in this manner eastward from the shores of East Africa, touching lands of Ceylon and islands of the Indian Ocean and to the Asian Archipelago, northern Australia, New Zealand, and progressing further to unknown lands.

A similar venture would take place thousands of years later by natives extending from Java, Fiji, Tahiti, and to new islands of the South Pacific Ocean, and the author believes to the Galapagos Islands and the western shore of South America. Still later, Christopher Columbus, and other explorers, would sail and find lands unknown to the world.

H. erectus is not notable in archeologic findings in Russian Georgia (due to poor preservation conditions) for over 500,000 years, until about 600,000 years ago when there was migration through India, southeastern Asia, and Indochina. It is believed H. erectus at this time was of Caucasian race (the cause of separation of Caucasian from Negroid race is not known), and from the Caucuses Mountains area between the Black and Caspian Seas. Branches split and some migrated westward to Europe, northeastward to Russia and Siberia, and to northern China (as Denisovan Hominids), and southward, only 2 or 3 hundred miles to Mesopotamia, to become the Kurds of Kurdistan, (still a culture in northern Iraq), and rich in the growth of history of Mesopotamia and as a Caucasian race. Egypt possessed Caucasians in the Delta region of Lower Egypt, as well as Negroid race in Upper Egypt from the original eastern Africa origins. Interbreeding of these two races produced people of ruddy complexion, and other features, some in Egypt, Mesopotamia, Greece, the Balkans, Italy, and North Africa.

About 600,000 years ago, in Russian Georgia, H. erectus, the author believes, became unique in archaeological history, demonstrating advanced qualities in not only race (Caucasian), but also in mental abilities including recognition of spiritual presence of a Superior Power, reasoning, logic, love, affection, familial responsibilities, forgiveness, charity, communicating, and other emotions.

At about 350,000 years ago a sub-specie of H. erectus developed and became known as Homo Neanderthal, and later Homo Sapiens, with archaeological findings showing migration and settlement to additional locations westward through Switzerland, France, northern Italy, the Pyrenees Mountains of France and Spain, Germany, and later to the British Isles, southwest to western and south Africa, and northeast to Russia and Siberia, to be ancestors of Eskimos, and North American Indians (Denisovan Hominoids). Caucasian Homo erectus also migrated eastward through India, China, Indochina, Java, Papua New Guinea, to Polynesian islands of the Pacific Ocean, and subsequently to the west coast of South America (possible ancestors of Maya people), then northward through present Ecuador, to Central America, with fossils and artifacts (Clovis point tools and weapons of at least 12,500 years age) in the present southeastern and central west of North America, and stone cutting tools of 20 to 22,000 years ago (paralleling the tools of the Solutrean culture in southwestern Europe of that time) uncovered at a site near Chesapeake Bay at the central North America east coast.

H. erectus may have met and mingled with another migration coming from the north that originated from arctic Siberia. They had traversed the lands of the Alaskan Bering Strait, exposed during an ice age of low water level, then fanning out in their North American migration eastward as Inuit Eskimos, and southward along mountain valleys of the northern Rocky Mountains chain, plus travel by water southward along the western coast of America. Several nations of Indians developed as Blackfoot, Apache, Chinook, Sioux, Navaho, Cherokee, Iroquois, Shawnee, Huron, Chippewa, and many others. Many Indian nations warred, somewhat continuously, with other Indian nations for hunting grounds, and for retaliation in past tribal infringements against their possessive nations.

ORIGIN OF RACES

The reasons and cause for advent of the Caucasian race, and other races of the world, is not fully understood, but the following sequence in biologic history is presented as the geographic source.

In biologic inheritance and genealogy, traits displayed by genes depend on the interaction between genetics AND the environment. (Source: Wikipedia, The Free Encyclopedia/Genetics). An allele of skin coloration, and other biologic variations, can develop in genetics, but be altered by environment.

People of a more northern environment than Africa will mutate in their genes to a white variation, or allele, once exposed for several generations to a northern environment. The change is hereditary, creating a new population of Caucasian race.

A branch of Homo erectus in the eastern migration, about 1,700,000 years ago, proceeded northward, off the primary eastern migration path, into the mountainous, cooler, and environmentally different area of the Caucuses Mountains, and the Black and Caspian Seas of the Crimean Peninsula in the Ukraine, now an autonomous republic in present-day southern Russia. The people emerged, after genetic changes, several generations, and thousands of years later, as a Caucasian race.

The same process could account for sectional variations in the people of India, Indochina, Chinese, Mongolian, later Korean and Japanese, Denisovan hominoids, and "Hobbits" of Papua, New Guinea.

Other traits developed that differentiated Caucasians from Negroid race, in addition to skin pigmentation, such as Negroid propensity for Sickle Cell Anemia, and other variations specifically known to the medical profession.

The process would account for development of various races and sub-races of the world, such as occur in India, China, Korea, Japan, Scandinavia, the Balkans, North American Indians, Mayan, Polynesian, various sections of Africa, Mesopotamia, Egypt, central European, Nordic, Mongolian, Mediterranean (Spanish, Italian, Greek), all of which could be segmented in race or sub-race by visual appearance in a crowd of people.

GLOBAL WARMING

Global warming occurs, but its cause is mainly not of domestic and commercial abuse but of our changing location in the cosmos, of our Sun, and all of its planets, including Earth.

Our Solar System is located within a spinning arm of a bar spiral galaxy. That spinning arm has changed location in the cosmos constantly over billions of years from "average" temperature at one location of the cosmos, to "warmer" temperature locations (producing global warming), and colder temperatures (that produce growth in glaciers of the world and generally colder climate with glacial conditions). This alternating condition has existed for eons and has been recorded in ice borings in Antarctica and Iceland showing temperature history for the past 420,000 years. Our latest glacial period, the Wisconsin, started about 30,000 years ago, with a cooling period peaking about 20,000 to 15,000 years ago. Earth, since, is presently located in a cosmic location of a warm period, but entering again into a glacial period, probably to reach its lowest average temperature in about 20,000 years.

Warmer locations are caused by variation in areas of cosmic gases of energy (infrared and cosmic background radiation caused from the early universe). The phenomenon is likened to travelling through one, two, three, or more heavy and light rain storms in a drive across the country, some good periods and some lesser or bad.

So far, science or astronomy efforts have few corrective action plans to this cooling period, and planet Earth looks forward to increasingly low temperatures as we enter a predicted colder location in the cosmos, countering with warming temperatures produced by industrial and domestic processes.

The question arises of whether the warming trend will last for many years (like thousands) or will be of short duration. Historically, records show existence of both short and long periods of global warming, and glacial periods. Some charted past temperature variations have been over extremely short (or long) periods.

THE PAST 2,000 YEARS

Since the time of Christ there has been little change in the physical earth, anatomy or biology of man. In 2,000 years there has been change only in culture, religions, world exploration, colonization, political theories, warfare, philosophies of man, and technology development. From Hebrew and Christian beliefs there has been growth in Christianity, introduction of Islam philosophy. Confucianism, Buddhism, Hindi, and other forms of worship have developed around the world.

In 2,000 years there has been substantial change in cultures of people, and a variety of religions and spiritual thought. There has been extensive exploration of the physical world, which continues even today at Antarctica and regions of the North Pole. New species of man and civilization has been discovered in just this past century of the Hobbits (Homo florensis) of New Guinea, and the Denisovan Hominids in northern China and Java. Cross-breeding has produced unification in the myriad of species of man that have located throughout the world.

Developments on Earth have led to further technology and exploration in our universe and to the cosmos. We are exploring for the first time the "inner world" of minute elementary particles, as well as antimatter and multiple dimensions.

Academic research has produced advancements in medical sciences, which has developed tremendously over 500 years, providing life expectancy of an increasingly many to reach 100 years and more.

Science in industry has produced manufacturing robots, with accompanying systems that serve as judgmental minds. Advancements in chemistry have increased a thousand-fold since time when man tried, as alchemist, to produce gold from the elements.

TO THE FUTURE

Are we the next phase, the next generation of a dying culture and species? Every religion, every culture, has pondered where and when our cosmos began, and perhaps when our civilization will end.

Some scientists talk of cosmic catastrophe in which, at the speed of light, our physical existence would be exterminated. (Some religious sources would be quick to parallel such extinction with the Biblical "End of the World" prediction.) In theoretical analysis, souls are in antimatter and evidenced in Afterlife. After millions of future years, existing in another dimension of Space/time, souls would continue on to exist despite catastrophe in the physical world.

Are we a universe within another universe? Another universe would be in another dimension, presently unknown to mortals. Is the Great Wall of Galaxies at the edge of our universe a beginning of another universe?

Will planet Earth continue in existence until an expanding Sun will exterminate all life and after several millions of years of life, again to develop a Hadean Earth?

COLONIZATION IN THE COSMOS

Will galaxies collide in the far future, or at least come close enough to create catastrophe on Earth? Will our world supply of non-renewable energy resources and other minerals, that took billions of years to create, finally expire, and will our civilization have the wherewithal to develop substitutes for them? Our Sun has a limited life (it is expected the Sun is about half through its expected lifetime) but when will first signs of expiration begin? Will planets be identified that would serve as an alternate to Earth, and will science create "cosmic lifeboats" to transport our Earth civilization to that point?

In exploring for possible life on far-away cosmic planets of stars scientists look for 1) liquids, 2) organic molecules, and 3) energy. These factors would be the same as in looking for life, or past life, on planets of our solar and cosmic system.

New advances in science continue to be made in geometric proportions of development and would affect our movement toward what makes us human. But, like 8,000 and 2,000 years ago, we are limited today by what exists today. New efforts in science, philosophy, theology, spiritualism, extra dimensions, and other fields of study today, will lead us further through the specifics of true knowledge of reality and ability within extra-dimensions.

CERN

People of the world, and in particular men of science, are entranced concerning experiments to provide high energy physics research. Situated at the northwest suburb of Geneva, Switzerland, on the Franco-Swiss border, CERN will provide high energy physics research. CERN is the European Organization for Nuclear Research, which is the world's largest particle physics laboratory.

This is a momentous effort by the world of Science. It could produce information that may open the doors to communicate from our world of matter to the mysterious world of antimatter, and added dimensions of space/time, with unimagined pathways that could follow.

In one sense, conditions of the Big Bang may be duplicated in order to accomplish these goals. Leading scientists have assured the populace that all experiments will be controlled to prevent an incident of world catastrophe.

Negative-charge particles and positive-charge particles will be collided and antimatter produced. These particles of antimatter will produce tangible knowledge of a new world—that of antimatter. Antimatter will have many different qualities. They may be in dimensions of space beyond those four presently recognized of height, width, depth and time. Just what those additional dimensions will show, how they are comprised, and what their qualities are, plus many other characteristics, will be discovered.

Whatever qualities that antimatter are discovered to have are the same qualities that existed 14.7 billion years ago at the Big Bang. They may explain dreams, a better description of soul, forces of nature that are now mysteries, explanation of Afterlife, the genome, activities of soul in Afterlife, an existence of other dimensions, of reincarnation, and a further existence undreamed of previously. This is the scenario in existence since the advent of Modern Man and before, the previous generations of souls.

BIOLOGIC CHANGES

Throughout the biologic history of man there have been momentous and sometimes unexplainable changes. Among these are changes that caused variation from a Black race in Africa to a Caucasian race in southwest Asia, and further from Caucasian variations to Oriental, Indian, Native North American, Maya, Polynesian people, and other subtle deviations in biology on Earth.

Statistics of average height of American soldiers in World War II show men were significantly taller than those of World War I, only about 33 years later, a change probably instigated by diet improvement occurring during the interim. Change came about that provoked innovation and exploration by man, having an internal drive within him to further understand his existence in this world. Change in diet, enforced by environment, hunting, or available eating will result in biologic and physical change.

HOMO ROBOTICUS

So too, we may currently be experiencing now the start of a new phase in development of man. We might label such change even as a species, to be possibly named "Homo roboticus". Since the 1950's, man has devised mechanical tools, controlled to perform motions in operations, powered by small electric motors, controlled at first by a central remote unit to cause the desired operations, and were later automated by performing several operations in a sequence.

Later, industry was host to robotic operations that were programmed into a series of individual industrial operations of assembly, turning, rotating, spray painting, labeling, and other routine operations previously performed by workers. Robots have been applied to domestic operations of mowing a lawn, and to vacuum clean rooms in a human's house. Economics and value of leisure, not evolution, was the motivating force.

The science has evolved tremendously, to be mechanized into more delicate and precise operations, such as in medical surgery, in using sensors that guide and control robotic pressures and actions that seem ingenious in their scope. Today robots are being created that can think, act, and relate to humans. This begs the question of whether humanity is ready for such a dramatic change. It is like a new species, a new way of life, has been introduced to our ways on Earth.

It may well be that the age of "Homo roboticus" is arriving fast upon our culture, with an innovation fostering yet additional innovations, perhaps someday soon to replace at least routine operations of man, and a future that incorporates emotion and judgment to robotics, much like emotions were included to describe "what makes us human". Conceivably, a robot would be designed to produce a duplicate of itself, or even a robot with improved features. (A parallel concept, as a "transhumanist movement, holding better technology will enable us to replace more and more body parts—even your brain", is presented in Smithsonian magazine, April, 2012)

Sources:

1. The Great Courses (Several, lecture courses and professors)
2. Google Wikipedia, The Free Encyclopedia
3. Western Civilizations (College History Textbook)
4. King James Bible
5. Harper Study Bible
6. The Book Of Mormon
7. The Glorious Qur'an
8. Archaeology magazine
9. National Geographic magazine
10. Scientific American magazine
11. Discover magazine
12. Nature magazine
13. Encyclopedia Britannica
14. Physics Annual Review, The University of Michigan
15. Goode's School Atlas
16. The Last Two Million Years, Reader's Digest Association
17. Prehistoric Life, DK Publishers, 2009
18. A wide variety of general reading

AFTERLIFE—AN INTRODUCTION

This is a short portrayal of my upcoming third book in a series describing my view of the process in Afterlife.

The first book, "Do We Live in Two Worlds?" was an exploration of the concept that we each have two lives, one secular on earth living our "regular" life wholly in a world of matter. Our second life is one of soul, mind, and spiritualism, with the associated relationship of brain, enzymes, and the genome. We explore the realism for God's existence, the event of Biblical Creation and other creations, development of physical Earth, the creation of soul, and the evolution of Man. The qualities of soul and mind of man were detailed, described in opposition to when only body and brain actions of animals served merely to allow physical life and continuation of that physical life. We briefly explored the future after worldly death.

In this second book we explored the possibility of an existence of the universe prior to the Big Bang. We described a Hadean Earth and early cosmos, first life, and millions of years following in the progression of life and spiritual development on Earth, that led to having physical forms of plant, animal, primate, hominid, hominoid, and human species, leading to making us human.

In a third book, "Afterlife", we will describe a fictitious person, George Cunningham, who is killed suddenly and his soul proceeds to an Afterlife. George is evaluated by a Heavenly character, with questions posed to explore his morality and quality of judgment. Many earthly questions in George's mind are answered. Based on his response, George Cunningham was grouped with like-minded souls where he was provided a future in Afterlife.

In this present book, first and later life on earth, as defined by archaeologists and anthropologists analyzed from discovered fossils, are successively told. We explore transcendence from inorganic metals and minerals, to organic compounds for a procession of life, cells, and molecules to form living organisms. We explore formation of stars, galaxy, and our solar system, and how to make a planet.

Many theories in science have been presented over the past 3,000 years or so and some presented here. The possibility of several Big Bang events is presented, one in each of numerous galaxies. A reasonable, logical, and practical view of different creations in the history of earth, together with how both science and various religions each possessed their own fields of concern, each religion independent of the others, without communication in conception of physical or metaphysical existence. Science has seemed unsympathetic to conflicting religious points, and religions have held to mystical reasons for unexplained events.

The author proposes a clarification of reported Biblical events to divide myth from fact as might be presented in alternative possibilities.

We discuss the topic of soul, to clarify the qualities of soul, the theorized history of soul, how soul relates to the living man, and when and how souls and genes of the genome interact.

In questioning and presenting alternative possibilities, the author intends no disrespect for existing religions, practices, or dogma. The author feels that his presentation will instead strengthen a bind between the world of God and individuals in the science of mankind. Modern man has become increasingly oriented to science, with constant demand for believable and provable facts. To this demand, the author offers explanations presented in these books.

In Afterlife there is no recollection or recall of worldly secular occupation knowledge, erased from memory at moment of worldly death. There is no professional memory or recognition for engineers, scientists, mathematicians, chemists, administrators, philosophers, psychologists, or other fields of knowledge that were known on Earth. There is no professional devotion in mind except a whole devotion to the Supreme God.

This was the status of man at the Biblical Garden of Eden as God gave spiritual life to man. Prior to Eve symbolically eating an apple from the Tree of Knowledge, all knowledge and devotion was focused on God, and God alone.

Questions are raised by scientists and philosophers whether these instances of man's creation in the Garden of Eden is accurate and factual. But there is enough symbolism in the oral history of first actions of human man that exist to provide an analogy in spiritual life. The oral history of ancient life and spirituality was composed by ancient man to the best of his limited abilities. Physical creation of man coordinated with elements of developed soul and mind and had existed in increasing extent for 600,000 years. The author believes there was an actual moment when God and secular man communicated in spirituality on Earth to instill and create spiritual man, to coordinate with acts of the human genome that would change body and mind, which in turn would be finally recognized and comprehended as complete man 6,000 years later in the 21st century A.D.

Individual books may be ordered from Trafford Publications, 1-888-232-4444, by E-Mail to: customersupport@trafford.com, or directly from the author at 1-734-662-9233. Your comments are welcome.

Books and Articles by the Author

Flatlander in the North

Timelines of the Physical Earth and of Evolution

A Bit of Philosophy (Unpublished)

We Were the Youngest

Thoughts (Unpublished)

Magnificent Change (Unpublished)

Do We Live in Two Worlds?

English Kings and Queens (Unpublished)

Indians in the Americas (Unpublished)

Indians, Old West and Civil War (Unpublished)

Timeline—The French Period in North America (Unpublished)

When We Were Colonies (Unpublished)

Memories (Unpublished)

History of World Countries (Unpublished)

Notes Concerning the History of Thought Regarding the Soul Within

Man and the Spirit World (Unpublished)

Bible, Science, and History (Unpublished)

The Dansville House